W9-CBJ-599

Process Synthesis

PRENTICE-HALL INTERNATIONAL SERIES
IN THE PHYSICAL AND CHEMICAL ENGINEERING SCIENCES

AMUNDSON *Mathematical Methods in Chemical Engineering: Matrices and Their Application*
AMUNDSON AND ARIS *Mathematical Methods in Chemical Engineering: Vol. 2, First-Order Partial Differential Equations with Applications*
ARIS *Elementary Chemical Reactor Analysis*
ARIS *Introduction to the Analysis of Chemical Reactors*
ARIS *Vectors, Tensors, and the Basic Equations of Fluid Mechanics*
BALZHISER, SAMUELS, AND ELIASSEN *Chemical Engineering Thermodynamics*
BOUDART *Kinetics of Chemical Processes*
BRIAN *Staged Cascades in Chemical Processing*
CROWE ET AL. *Chemical Plant Simulation*
DOUGLAS *Process Dynamics and Control, Volume 1, Analysis of Dynamic Systems*
DOUGLAS *Process Dynamics and Control, Volume 2, Control System Synthesis*
FOGLER *The Elements of Chemical Kinetics and Reactor Calculations: A Self-Paced Approach*
FREDERICKSON *Principles and Applications of Rheology*
FRIEDLY *Dynamic Behavior of Processes*
HAPPEL AND BRENNER *Low Reynolds Number Hydrodynamics: With Special Applications to Particulate Media*
HIMMELBLAU *Basic Principles and Calculations in Chemical Engineering, 3rd edition*
HOLLAND *Multicomponent Distillation*
HOLLAND *Unsteady State Processes with Applications to Multicomponent Distillation*
KOPPEL *Introduction to Control Theory with Applications to Process Control*
LEVICH *Physiochemical Hydrodynamics*
MEISSNER *Processes and Systems in Industrial Chemistry*
NEWMAN *Electrochemical Systems*
O'HARA AND REID *Modeling Crystal Growth Rates from Solution*
PERLMUTTER *Stability of Chemical Reactors*
PETERSEN *Chemical Reaction Analysis*
PRAUSNITZ *Molecular Thermodynamics of Fluid-Phase Equilibria*
PRAUZNITZ, ECKERT, ORYE, AND O'CONNELL *Computer Calculations for Multicomponent Vapor-Liquid Equilibria*
RUDD ET AL. *Process Synthesis*
SCHULTZ *Polymer Materials Science*
SEINFELD AND LAPIDUS *Mathematical Methods in Chemical Engineering: Vol. 3; Process Modeling, Estimation, and Identification*
WHITAKER *Introduction to Fluid Mechanics*
WILDE *Optimum Seeking Methods*
WILLIAMS *Polymer Science and Engineering*

Process Synthesis

Dale F. Rudd

Department of Chemical Engineering
University of Wisconsin

Gary J. Powers

Department of Chemical Engineering
Massachusetts Institute of Technology

Jeffrey J. Siirola

Tennessee Eastman Company

PRENTICE-HALL, INC.
Englewood Cliffs, New Jersey

Library of Congress Cataloging in Publication Data
RUDD, DALE F.
Process synthesis.
(Prentice-Hall international series in the physical and chemical engineering sciences)

Includes bibliographies.
1. Chemical processes. 2. Chemical engineering.
I. Powers, Gary, joint author. II. Siirola, Jeffrey, joint author. III. Title.
TP155.7.R83 1974 660.2′8 73-3331
ISBN 0-13-723353-1

10 9 8 7 6 5 4 3 2 1

Printed in the United States of America

PRENTICE-HALL, INC.
PRENTICE-HALL INTERNATIONAL, *United Kingdom and Eire*
PRENTICE-HALL OF CANADA, LTD., *Canada*

Contents

Preface

The development of an industrial process requires skills in synthesis and analysis. In the words of Webster, synthesis is "the combining of often diverse conceptions into a coherent whole" and analysis is "an examination of a complex, its elements, and their relations." Synthesis deals with the creation of artificial things that have desired properties; analysis deals with the understanding of how things are and how they work.

Since World War II, engineering education has moved strongly toward analysis, with courses dealing with individual process operations and phenomena. Transport Phenomena, Unit Operations, Process Control, Reaction Engineering, and other engineering science courses greatly strengthened engineering education by showing how things are and how they work. Unfortunately, there was not a parallel development of courses dealing with synthesis, the combining of these diverse concepts into a coherent whole. This deficiency has been recognized for years, but the remedy awaited the development of sufficiently general principles of synthesis about which to organize educational material.

In the late 1960s and early 1970s, research in process synthesis established the broad outlines of the field. It became apparent that a careful interlacing of synthesis and analysis is the proper way to approach process development.* Each synthesis step sets up an analysis problem—the solution of which provides the data required for the further synthesis steps. Further, it became apparent that the elementary concepts of synthesis and analysis ought to be taught early in the undergraduate curriculum.

With this goal in mind, book writing becomes a pedagogical experiment to present the material in a coherent and attractive form suitable for the students' first exposure to an engineering course. This material has been class tested on freshmen and sophomores at the University of Wisconsin, and to a lesser extent at other universities, as a one-semester course meeting three days a week for sixteen weeks. Upon completion of the course the students are able to synthesize process flow sheets that are often quantitatively identical to current technology.

In particular, the students understand the history of processing, screen reaction sequences for economic feasibility, make a pretty good material balance, allocate materials to support the process chemistry, understand the beginnings of separation technology, select the separation phenomena likely to lead to economic processing, and use thermal energy balancing to achieve economy. These concepts integrate smoothly within the broader goal of teaching the development of process technology.

Chapter 1, The Engineering of Process Systems, emphasizes the history of processing and outlines a way of thinking useful in discovering the technology to solve new processing problems. The pattern of process synthesis is then developed in detail in Chapters 2 through 6.

Chapter 2, Reaction-Path Synthesis, deals with the chemical changes upon which much of processing is based. The student learns to mix and match different chemical reactions to reach more desirable sources of raw materials or to alter the by-product spectrum. Skills are developed in the early economic screening of alternate reaction paths. Though chemistry is a conspicuous part of this chapter, we do not aim to teach chemistry, only to teach the assessment of chemical change.

Chapter 3, Material Balancing and Species Allocation, is a natural outgrowth of Chapter 2. The student learns to allocate the species from raw material sources, to reaction sites, and to product destinations, and in this way outlines the overall material flow within the process. The major part of this chapter deals with the principles of material balancing, a skill we have found to be essential in process development. Toward the end of the chapter we introduce the elements of engineering judgment needed to extend material balancing to the synthesis of new technology.

When several species in a single source are to be allocated to different destinations, the species must be separated from each other. Thus, Chapter 3 introduces the need for Chapter 4, Separation Technology. This chapter largely is descriptive with the aim of introducing the students to the idea of a separator as an exploiter of differences in the properties of the species to be separated. At the end of the chapter, the student tries his hand in the invention of simple separators.

Chapter 5, Separation Task Selection, focuses on rules for selecting the separation

*For a review of research in process synthesis, see J. Hendry, J. D. Seader and D. F. Rudd, *AICh E Journal,* Vol. 19, No. 1, January, 1973.

phenomena. Why is it logical to save the difficult separations for last, to separate the most plentiful components early, to separate, by distillation, species one-by-one as overhead products? Heuristic rules of task selection are developed and then applied to the synthesis of the main features of modern technology.

Chapter 6, Task Integration, deals with methods for increasing the economy of the skeletal processing plan formed in Chapter 5. We show how economy is reached by the design of equipment which reuses material and energy. The simplest principles of energy balancing are introduced, along with industrial procedures for energy management. At this point we are beginning to go beyond the limits of a first course, but a short excursion beyond is not altogether bad.

Chapters 7 and 8 introduce no new concepts, for they deal solely with application to real problems. In Chapter 7, Fresh Water by Freezing, we show how the Vacuum-Freezing, Vapor-Compression technology for desalination can be arrived at using these elementary principles of process synthesis and analysis. In Chapter 8, Detergents from Petroleum, these principles are applied to a completely different processing problem, again to show the student how the broad outlines of modern technology are accessible to the initiate. To broaden this text, we made a conscious effort to include problems in food processing, waste treatment, minerals processing, and a variety of other fields.

The first three chapters involving historical perspective, reaction path synthesis, material balancing, and species allocation have been used as a short orientation course for freshmen. The full course can be offered anywhere early in the curriculum, for we think of this as a replacement for the traditional course in *material and energy balancing* with a shifting to thermodynamics courses and unit operations of some of the material now taught there but not included in *Process Synthesis.* Finally, Chapters 6 and 7, dealing with energy integration and the application to the synthesis of processes to obtain fresh water by freezing, can be omitted to offer a course of intermediate difficulty. One would then go from Chapter 5 directly to Chapter 8.

This course began to evolve several years ago when two of the authors were still graduate students at the University of Wisconsin. The John Simon Guggenheim Memorial Foundation and the University of Wisconsin supported the senior author during the middle year of writing, and the final details were completed in the classroom with the help of our beginning students.

DALE F. RUDD
GARY J. POWERS
JEFFREY J. SIIROLA

Process Synthesis

1

The Engineering of Process Systems

Historically and traditionally, it has been the task of the science disciplines to teach about natural things: how they are and how they work. It has been the task of the engineering schools to teach about artificial things: how to make artifacts that have the desired properties and how to design. . . . Every one designs who devises courses of action aimed at changing existing situations into preferred ones.

Herbert A. Simon,
The Sciences of the Artificial

1.1 INTRODUCTION

Civilized man relies ever increasingly on the *artificial*. Structures, clothing, tools, foods, medicines, methods of transportation, and other artifacts are of his own creation. Very often the materials used in the construction of his artifacts or the forms of energy that power them are themselves artificial. Natural, or *raw*, materials intentionally are transformed to possess more useful or desirable properties to form the variety of things seen about us; to name a few, petroleum products, paper, rubber, synthetic fibers, pharmaceuticals, glass, metals, paints, plastics, electricity, and even pure water.

But just how does one go about generating an economic way of converting low-grade vegetable protein into a meat substitute for human consumption? Or of transforming the nitrogen in the atmosphere into fertilizer, of recovering copper from a low-grade ore, of processing the sewage and solid wastes of a city, or of cleaning metabolic wastes from the blood of a victim of kidney failure? These are the kinds of materials-processing problems with which this text deals, and for which the engineering profession has significant responsibility. We examine here the general principles of synthesis applicable to the development of processes in a wide variety of industries.

1.2 LARGE-SCALE, LOW-COST PROCESSING

We begin by examining the end product of engineering design, three operating systems for the large-scale low-cost processing of materials. These examples have been selected to illustrate the breadth of application of process engineering, and to illustrate all the features with which we shall be concerned.

While studying these examples, note especially the chemical reactions that are exploited, the routing of the different chemical species through the processes, the manner in which materials are separated from each other, and how economy is improved by the reuse of material and energy. These are important details in process engineering.

Nitrogen Fixation

Of the three major elements required for plant growth (nitrogen, phosphorus, and potassium), nitrogen is the most important on a weight basis. Nitrogen is assimilated by most plants in the form of inorganic ammonium ions or nitrate ions. Prior to this century, the major source of this nutrient was through the biological decomposition of plant and animal wastes. Direct utilization of the more plentiful atmospheric nitrogen, called *nitrogen fixation,* was limited to certain plant families (legumes). Agricultural development was thus severely restricted.

In 1895 Ragleigh discovered that nitric oxide, NO, could be formed by passing air through an electric arc. Nitric oxide oxidizes reversibly to nitrogen dioxide, NO_2, with the evolution of heat. In an analogy to the nitrate fertilization that occurs naturally during an electric storm, Norwegians Birkeland and Eyde in 1900 developed a process based on these reactions. Air was passed through an electric arc in an electromagnetic field at 3000°C, the product NO was quenched rapidly to favor the production of NO_2, and this material was absorbed in a calcium hydroxide solution to produce a calcium nitrate fertilizer, known as *norge saltpetre.* By 1918, this reaction path, illustrated in Figure 1.2–1, was the base for twelve synthetic fertilizer processes. This provided a major improvement in agricultural production.

$$N_2 + O_2 \xrightarrow[3{,}000^\circ\ F]{\text{electric arc}} 2NO$$

$$2NO + O_2 \rightleftharpoons 2NO_2 + \text{Heat}$$

Favored by Low Temperatures

$$Ca(OH)_2 + 2NO_2 \longrightarrow Ca(NO_3)_2 + H_2O$$

In Aqueous Solution — Norge Saltpetre

Figure 1.2–1. Birkeland–Eyde Reaction Path for Fixing Atmospheric Nitrogen

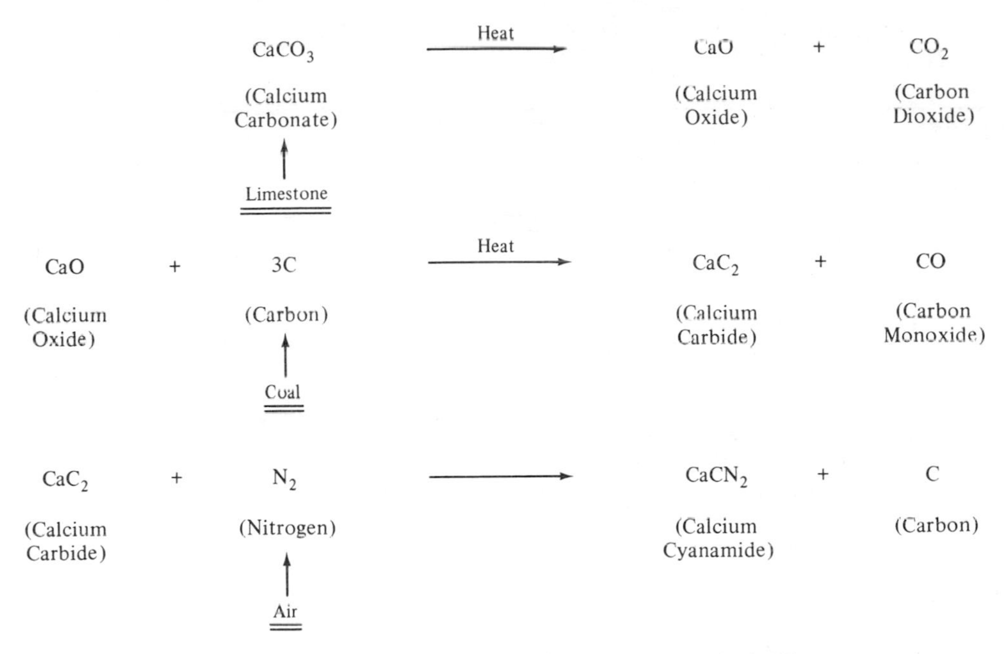

Figure 1.2–2. Cyanamide Reaction Path for Fixing Atmospheric Nitrogen

A second historically important scheme for fixing atmospheric nitrogen involves the reaction of alkaline earth carbides with nitrogen to form cyanamides. This reaction path was commercialized shortly after the Birkeland–Eyde process, and now there are a score or more processes in commercial operation. Although an expensive source of nitrogen, the cyanamides also serve as weed killers. Figure 1.2–2 illustrates the reactions involved in producing this nitrogenous fertilizer from limestone, coal, and air.

In 1918 Fritz Haber received the Nobel prize in chemistry for developing a process for the production of ammonia from nitrogen and hydrogen. The reaction path based on this synthesis is shown in Figure 1.2–3 and is the basis for most of the world's nitrogen fixation. Included in Figure 1.2–3 are several paths for the preparation of the pure nitrogen and hydrogen needed for ammonia synthesis, and several paths to a variety of fertilizers.

Figure 1.2–4 contains a schematic diagram of the equipment used by the M. W. Kellogg Company in the large-scale implementation of the reaction path from air and a light petroleum product to ammonia. Although the Haber process reaction path is half a century old, engineering improvements occur continuously. In 1967, Kellogg received the Kirkpatrick Award for the most significant chemical engineering achievement for its process, which reduced the cost of ammonia by one half.

Here we observe the exploitation of reaction paths to perform a useful and essential service, and we observe how new processes appear as new and more efficient

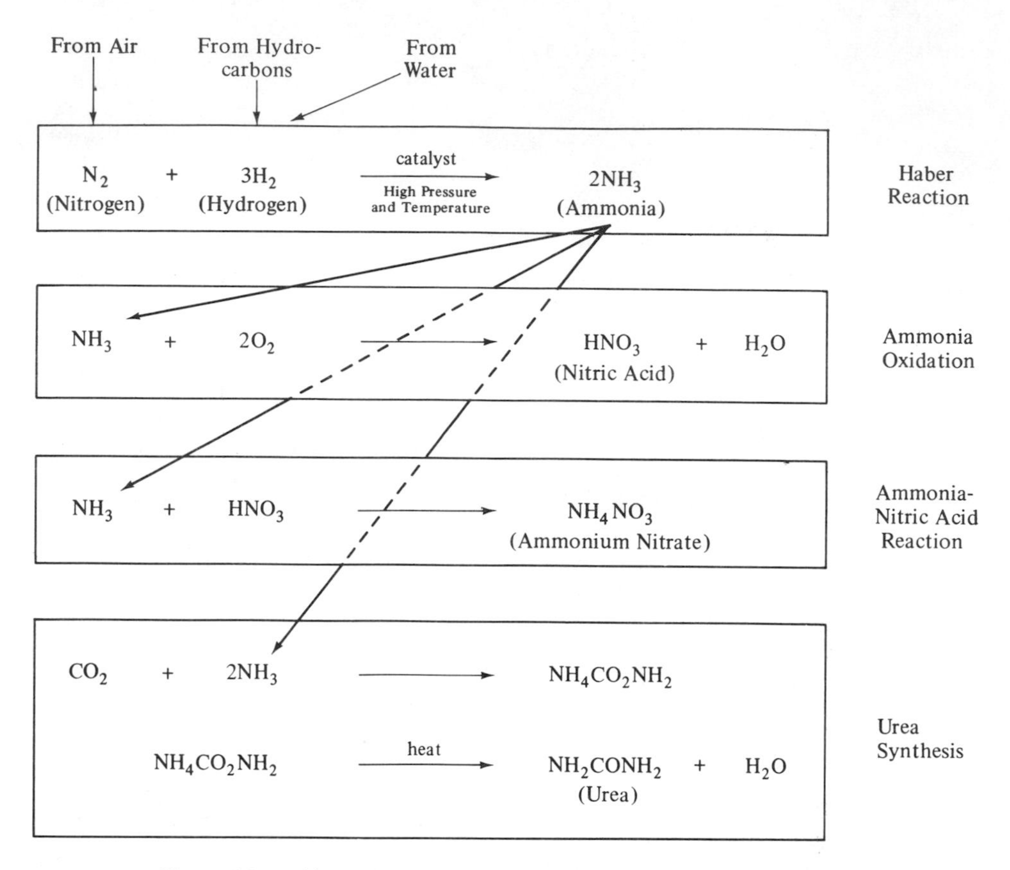

Figure 1.2–3. Alternative Reaction Paths Through the Haber Process for Fixing Atmospheric Nitrogen

ways can be discovered to engineer old reaction paths.* Next we look at a second example, to get some perspective in this field.

Protein Engineering

Increased efficiency in protein production could be accomplished if the currently popular process, the food animal, could be replaced as the converter of vegetable protein to animal protein. For example, oil-free soybean meal is half good-quality

* E. E. Van Tamelen recently found a reaction path between atmospheric nitrogen and ammonia that occurs at room temperature and pressure (*Chemical and Engineering News*, March 24, 1969, p. 48). This may hold the promise of still more effective fixing of nitrogen.

protein; when fed to cattle it produces protein in the form of steak at only 7 per cent efficiency. Soybean proteins are not extensively utilized directly by man because of unpleasant taste and the presence of indigestible carbohydrates. The direct use by man of protein in soy, cottonseed, peanut, sunflower, and other oil-seed meals would do much to provide for the world's protein needs. Table 1.2–1 shows the economic incentive of utilizing these sources.

In 1969 General Mills, Inc., received the Kirkpatrick Award for the chemical engineering advances employed in the development of edible protein foods based on

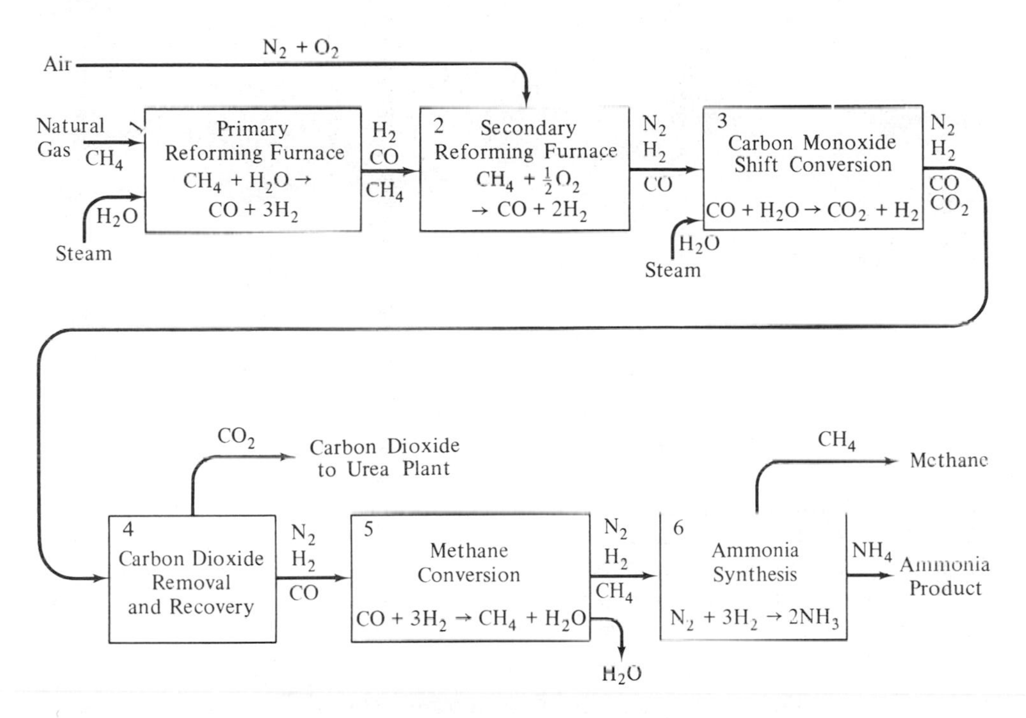

Industrial Ammonia Process is based on the high-pressure catalytic fixation method invented in 1914 by Fritz Haber and Karl Bosch, which supplied Germany with nitrates for explosives in World War I. This flow diagram is based on the process developed by the M. W. Kellogg Company. As in most modern plants, the hydrogen for the basic reaction is obtained from methane, the chief constituent of natural gas, but any hydrocarbon source will do. In Step 1 methane and steam react to produce a gas rich in hydrogen. In Step 2 atmospheric nitrogen is introduced; the oxygen accompanying it is converted to carbon monoxide by partial combustion with methane. The carbon monoxide reacts with steam in step 3. The carbon dioxide is removed in Step 4 and can be used elsewhere to convert some of the ammonia to urea, which has the formula $CO(NH_2)_2$. The last traces of carbon monoxide are converted to methane in Step 5. In Step 6, nitrogen and hydrogen combine at elevated temperature and pressure, in the presence of a catalyst, to form ammonia. A portion of the ammonia product can readily be converted to nitric acid by reacting it with oxygen. Nitric acid and ammonia can then be combined to produce ammonium nitrate, which, like urea, is another widely used fertilizer.

Figure 1.2–4a. From Air and Hydrocarbons to Ammonia, the Current Base for Nitrogenous Fertilizer Production. (Courtesy of *Chemical Engineering*, November 20, 1967, p. 113, McGraw-Hill.)

Figure 1.2–4b. Ammonia Synthesis Facility at Billingham works of Imperial Chemical Industries, Ltd., England has three 970 ton/day units. (Courtesy of The M.W. Kellogg Company.)

spun soy protein monofilaments. Figure 1.2–5 is the schematic diagram, or *flow sheet*, for the process to convert soy flour into meat-analog proteins, using basic technology developed for the production of rayon and other synthetic fibers. The conversion of soy protein in this process occurs at 70 per cent efficiency, rather than at 7 per cent efficiency, as in cattle.

TABLE 1.2–1 Estimated Cost of Various Protein Sources

	% Protein	$/lb Protein
Soybean meal	44	0.09
Defatted soy flour	50	0.14
Fish protein concentrate	80	0.23
Soy protein concentrate	70	0.26
Torula yeast	48	0.36
Peanut flour	60	0.42
Skim-milk powder	37	0.54
Fresh milk	3.5	1.60
Meatlike spun soy protein	2.3	1.70
Beef	16	4.00

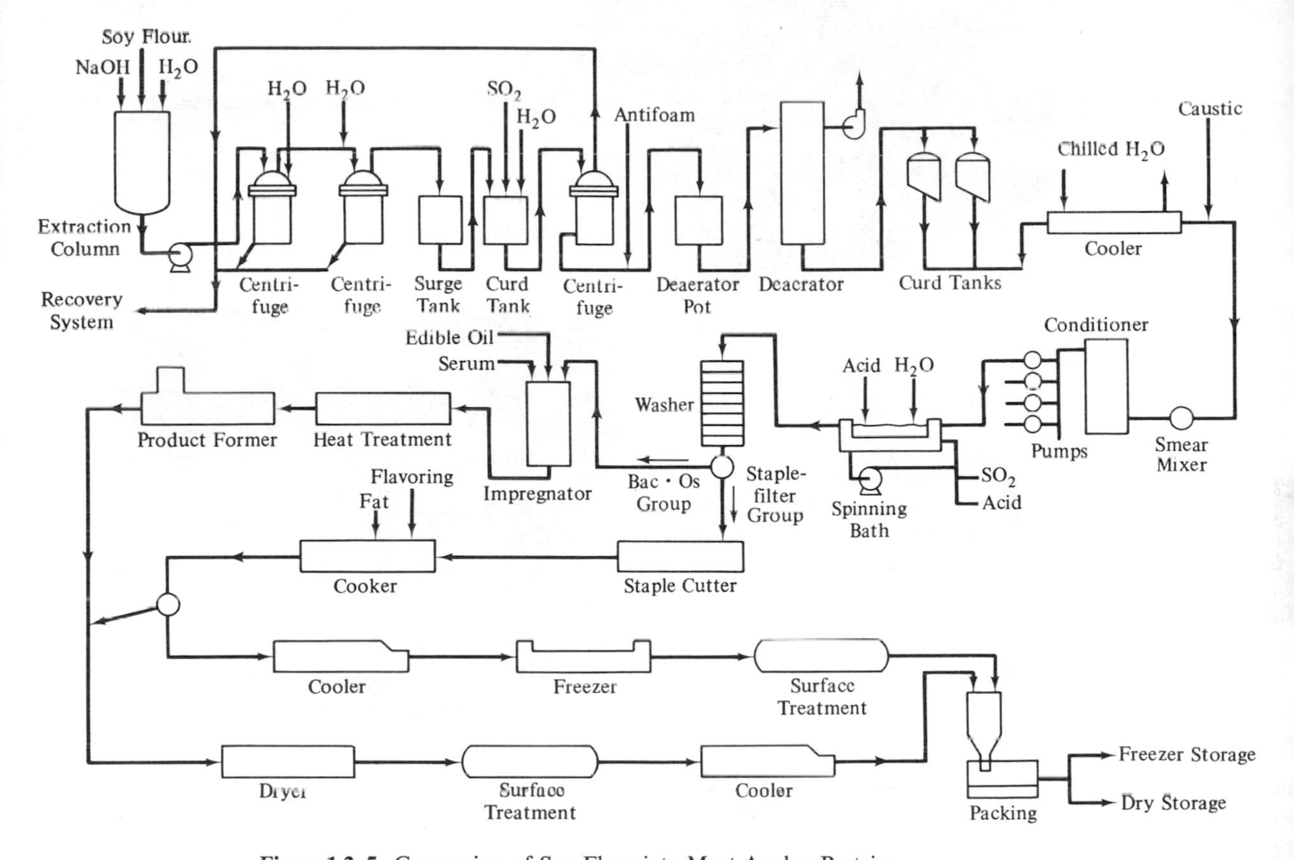

Figure 1.2–5. Conversion of Soy Flour into Meat-Analog Proteins

Solubilized protein material is extracted from defatted soybean meal under alkaline conditions and the insoluble material is removed by centrifuging. Sulfur dioxide addition lowers the pH, causing protein to precipitate. Additional centrifuging, washing, dilution, and other steps yield a purified slurry that is 95 per cent protein. This is then redissolved at pH 12 and spun through spinnerets into a bath of pH 4, solidifying into continuous filaments. After stretching and heat treating, a ribbon of monofilaments is washed and squeezed dry. To form a bacon analog, the protein ribbon is impregnated with flavoring, coloring, and other additives, and is slit, crosscut, dried, and coated with vegetable oil. To form analogs of beef, ham, chicken, and seafood, the protein fibers are cut short and extruded with a serum containing other additives through a cooker–extruder from which the product emerges as a continuous slab.

In 1969 General Mills had semiworks facility in operation producing 2 million lb/yr of dry food stuffs and 1 million lb/yr of frozen foods. The textured proteins

—Granules with a flavor like beef served in loaf, pattie, and ball forms present man-made protein in familiar dishes.

—Ham-flavor dice made into hot dishes, casseroles, and salad sandwiches show the versatility of man-made proteins.

Figure 1.2–6. Animal Protein-Analog Food Products from Soy Flour (Courtesy of General Mills)

were found in dietary studies to be equivalent to meat and almost equivalent to milk in nutritional merit. Some of the products are shown in Figure 1.2–6.

Waste Treatment

The sewage treatment plant at South Lake Tahoe, California, is one of the most modern in the world, transforming 7 million gal/day of sewage into water pure enough to drink. Figure 1.2–7 shows the processing necessary to do this job. The primary treatment involves the removal of large items, such as sticks, rags, and trash, and the settling of about 30 per cent of the suspended solids. This material is incinerated to produce a sterile ash for use as a landfill, and to produce carbon dioxide needed later in the processing. After this primary treatment, air is bubbled through the sewage to encourage bacterial growth, and the bacteria consume the organic wastes, producing a sludge that settles out. This secondary treatment is called an activated-sludge process.

About 30 per cent of the nation's sewage-treatment plants use only primary treatment. Another 60 per cent use both primary and secondary treatment, a combination that removes over 90 per cent of the pollutants in sewage. After both primary and secondary treatment, ammonia, phosphates, and some dissolved organic matter remain in the sewage.

In the South Lake Tahoe plant tertiary treatment begins with the addition of lime, which causes the phosphates to settle. The sewage is then aerated to remove ammonia and neutralized by the addition of the carbon dioxide recovered from sludge incineration. Detergent and pesticide molecules are removed by activated carbon treatment, and chlorine is added to kill any remaining bacteria. The sewage leaves the plant crystal clear and clean enough to drink. The cost of the entire operation is less than $0.03/day/residence, although the initial investment in the process system is measured in millions of dollars.

In the next section we outline the ways of thinking that are useful during the discovery of such processing plans.

Figure 1.2–7. Journey to Clean Water at South Lake Tahoe: Three-Stage Treatment of Sewage. (Courtesy of *Carnell, Howland, Hayes and Merryfield, Corvallis, Oregon.*)

One of the most advanced systems in the world, this plant can daily transform over 7 million gallons of sewage into water pure enough to drink. Valuable processing chemicals—lime and activated carbon—are reclaimed and solid wastes are incinerated, all without damage to the environment. Since state law requires the removal of waste derivatives—even purified water—from the Tahoe Basin, the plant's product is pumped 27 miles to Indian Creek Reservoir for water sports, trout fishing, and irrigation.

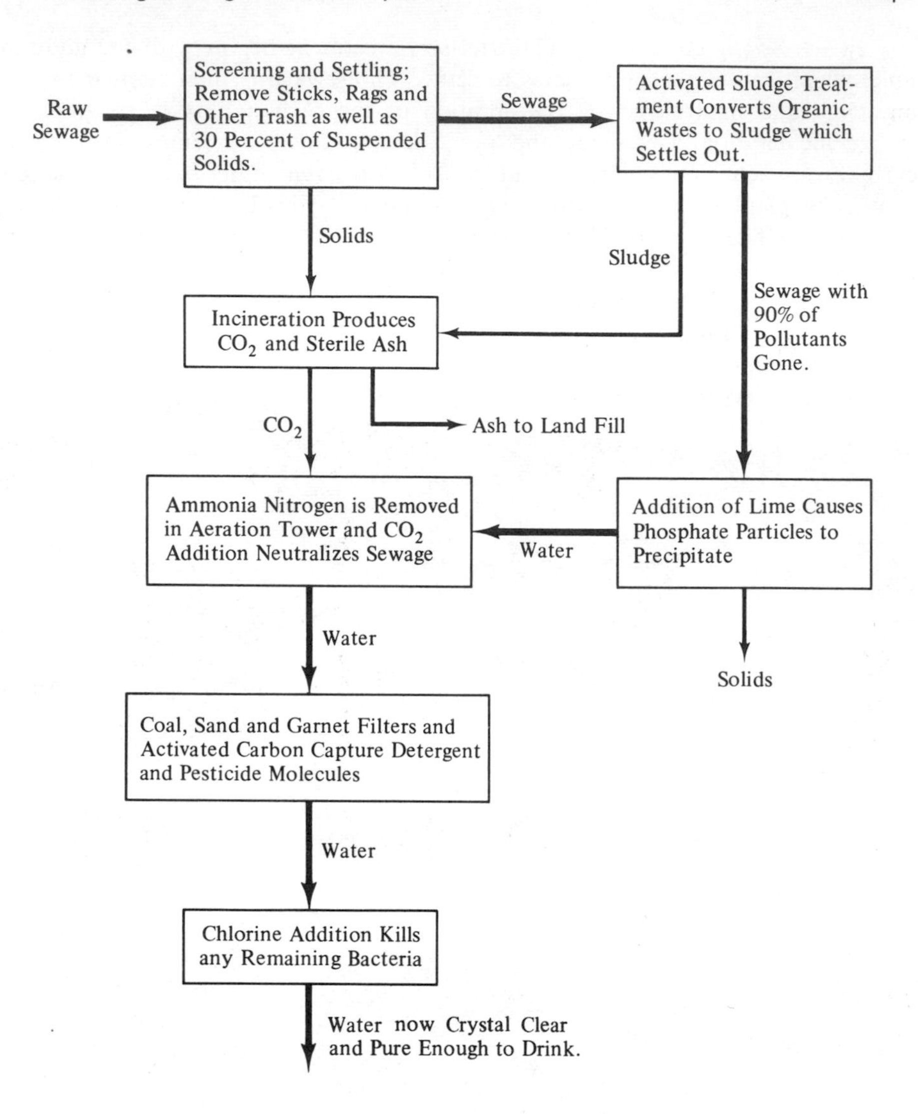

Figure 1.2–7 (*continued*). Processing Sequence for the South Lake Tahoe Treatment Plant

1.3 Pattern of Discovery

An often practiced and very effective way of solving a difficult problem is to decompose the problem into several easier problems, the solutions of which can be assembled into a solution to the original problem. The subheadings in this section are one set of subproblems into which the large and difficult problem of process synthesis may be decomposed.

Reaction Path Synthesis

Many commercial processes rely on the chemical transformation of materials, and for this reason an examination of the chemistry of processing is the starting point for process discovery. As new reactions are discovered by chemists, new products are found and better routes may open to known products.

Most products are several reaction steps away from readily available raw materials, and the sequence of reactions needed to bridge the gap is called a *reaction path*. Invariably, there are several possible reaction paths to a product. Often the path most easily performed in the laboratory is not the path most economically implemented on a large scale.

The only completely accurate way to assess the commercial attractiveness of a reaction path is to synthesize and analyze, in one's mind, the full-scale process of which it will be the basis. For this reason, the research chemist and the process engineer cooperate during the discovery and initial screening of competing reaction paths.

Example 1.3–1 Reaction Paths to Vinyl Chloride. *Vinyl chloride [the building block of poly vinyl chloride (PVC), one of the most common plastics] can be produced in the laboratory by these alternative reaction paths, among others. Can you rearrange these reactions to form still other reaction paths to vinyl chloride?*

Path 1:

$$\underset{\text{(ethylene)}}{C_2H_4} + \underset{\text{(chlorine)}}{Cl_2} \longrightarrow \underset{\text{(dichloroethane)}}{C_2H_4Cl_2}$$

$$\underset{\text{(dichloroethane)}}{C_2H_4Cl_2} \xrightarrow{\text{heat}} \underset{\text{(vinyl chloride)}}{C_2H_3Cl} + \underset{\text{(hydrogen chloride)}}{HCl}$$

Path 2:

$$\underset{\text{(acetylene)}}{C_2H_2} + \underset{\text{(hydrogen chloride)}}{HCl} \longrightarrow \underset{\text{(vinyl chloride)}}{C_2H_3Cl}$$

Path 3:

$$\underset{\text{(hydrogen chloride)}}{2\,HCl} + \underset{\text{(oxygen)}}{\tfrac{1}{2}O_2} \quad \underset{\text{(ethylene)}}{C_2H_4} \longrightarrow \underset{\text{(dichloroethane)}}{C_2H_4Cl_2} + \underset{\text{(water)}}{H_2O}$$

$$\underset{\text{(dichloroethane)}}{C_2H_4Cl_2} \xrightarrow{\text{heat}} \underset{\text{(vinyl chloride)}}{C_2H_3Cl} + \underset{\text{(hydrogen) chloride)}}{HCl}$$

Species Allocation

The several chemical reactions that comprise each reaction path under examination may be thought of initially as isolated operations which must be connected to bridge the gap between raw materials and the desired products. The conditions under which these reactions must be executed are generally quite specific. The reactors require feeds of determined purity, temperature, and pressure. Furthermore, the reactions are rarely completely specific, and a variety of products and reactants may leave a reactor. During species allocation the routing of each species from its sources to its destinations is determined with the aid of a few simple rules. Like reaction-path synthesis, species allocation does much to determine the form of the final process.

Example 1.3–2 Species Allocation for SO_2 Synthesis. *Sulfur dioxide is manufactured commercially by the combustion of sulfur.*

$$\underset{\text{(liquid)}}{S} + \underset{\text{(gas)}}{O_2} \rightarrow \underset{\text{(gas)}}{SO_2} + \text{heat}$$

The flame temperature must be lowered to protect the burner by the introduction of a cold inert gas with the oxygen.

Two alternative allocations are shown in Figure 1.3–1. The first allocation generates the problem of separating sulfur dioxide from nitrogen, and the second produces the problem of removing oxygen from nitrogen. Which might be easier?

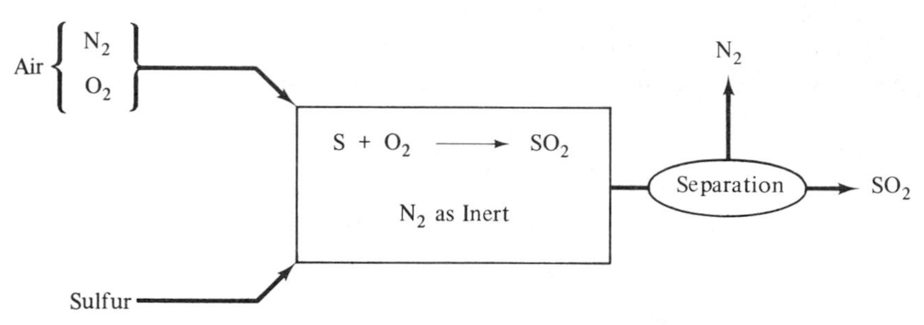

Allocation 1: Nitrogen as the Inert Gas

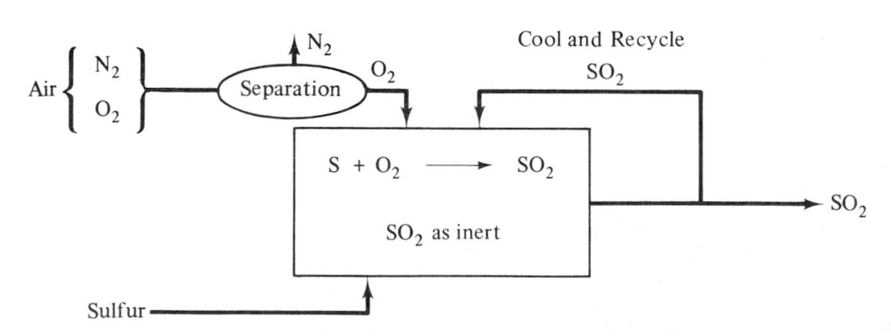

Allocation 2: Sulfur Dioxide as the Inert Gas

Figure 1.3–1. Alternative Species Allocations for SO_2 Synthesis

Separation Technology

As a result of species allocation, different materials in the same source may be allocated to different destinations. Such an allocation implies a material separation.

Separations are accomplished by exploiting differences in properties that cause the materials to behave differently in some environment. For example, fresh water can be removed from seawater by taking advantage of volatility differences of the salt and water, differences in the mobility of the salt and water molecule through a membrane, the formation of salt-free ice, differences in the chemical reactivity of salt and water, or other property differences.

Example 1.3–3 Right- and Left-Handed Shells. *On ocean beaches are often found piles of shells predominantly of right-handed or of left-handed geometry. The interaction of the waves, shells, and ocean floor causes the shells to drift to the right or left, forming these piles. How could this phenomenon be used to separate crystals of different geometry? Only certain crystal forms of some drugs are active biologically, and separation is necessary.*

Separation Task Selection

Materials can be separated from each other in a variety of ways by exploiting different combinations of property differences. Now we determine the separation method that fits best into our particular processing problem. The selection of separation tasks is accomplished by heuristic rules, which identify the separation schemes best suited to the problem at hand.

Example 1.3–4 Bromine Purification. *To serve as a raw material for fine chemicals manufacture, impure bromine containing 2 per cent chlorine and 300 parts per million (ppm) chloroform must be purified. The task sequence shown in Figure 1.3–2 is*

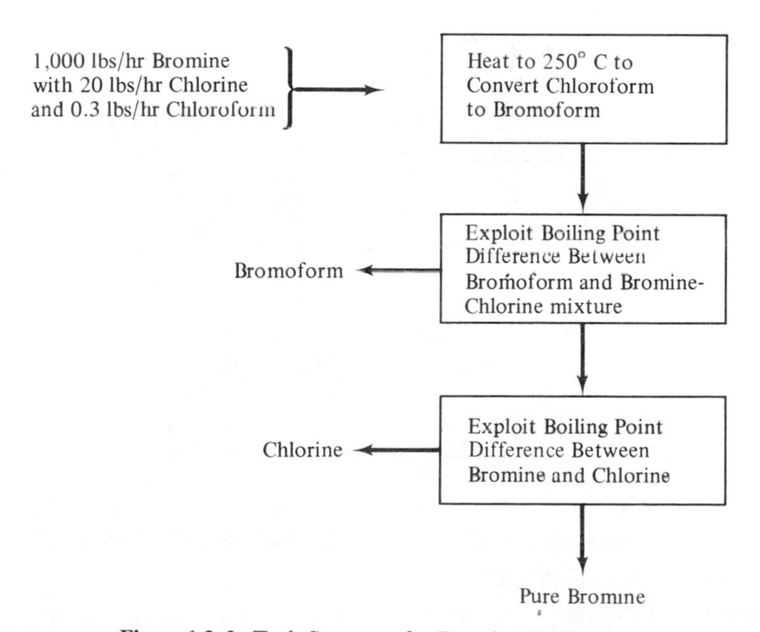

Figure 1.3–2. Task Sequence for Bromine Purification

likely to lead to efficient processing. What rules guide the synthesis of this sequence of operations?

Task Integration

During the task selections, a sequence of operations is selected for each of the several streams being processed, among raw materials, chemical reactions, separations, and desired products. However, they may not bridge the gaps economically. The cost of processing often can be reduced greatly if the need for outside sources of heat, refrigeration, electricity, water, and raw materials can be eliminated by locating sources of these within the process itself and pairing complementary tasks. Finally, all the identified tasks must be integrated into actual pieces of processing equipment.

Example 1.3–5 An Economizer in Seawater Distillation. *The task network shown in Figure 1.3–3 describes the formation of fresh water from the sea by heating seawater partially to vaporize it, collecting the vapors by a gas–liquid separation, and cooling the vapor until condensation occurs, forming fresh water. The amount of steam needed to heat the seawater can be reduced if the energy removed from the hot vapors is used to heat the incoming cold seawater. This task integration is accomplished in a heat exchanger, which, appropriately, is called an economizer. To be competitive this process must produce fresh water for less than \$0.50/1000 gallons; an economic and engineering analysis would indicate that the process in Figure 1.3–3 is still far too expensive, and process development must be carried further.*

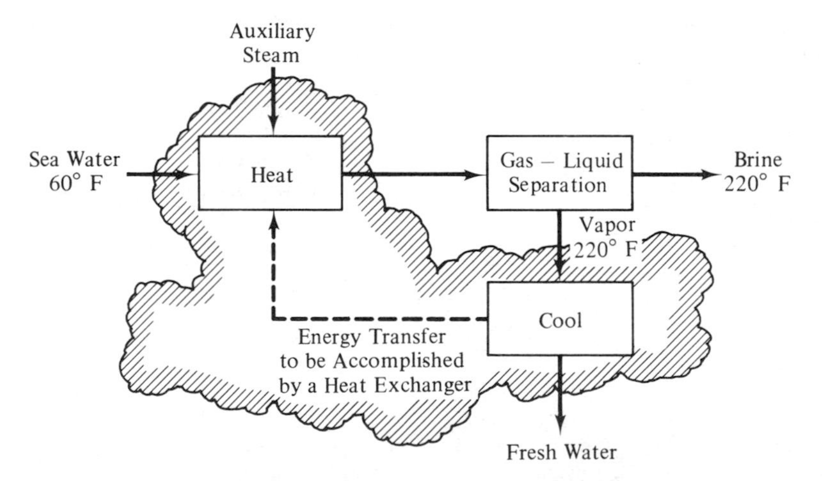

Figure 1.3–3. Integration of Energy Transfer Tasks During the Production of Fresh Water from the Sea

Some observations are in order on this approach to process synthesis:

1. Reaction-path synthesis.
2. Species allocation.

3. Separation technology.
4. Separation task selection.
5. Task integration.

Any decomposition of the problem of process discovery into smaller areas of study is necessarily an approximation, since a strong interdependence exists among the area of study. For example, the selection of the reaction path and species allocation is determined in part by ease of separation of the reaction products; the selection of the separation phenomena is influenced by the ease by which task integration might be accomplished later; and so forth. These steps in process synthesis fold back on each other and interact strongly. However, this particular ordering focuses attention on the most critical facets of the problem first, and in this way sets an effective strategy for process discovery. We have here only ordered and arranged the things to which we should pay attention. There is much to be learned on how one best pays attention to these problems.

1.4 CONCLUSION

Two broad kinds of mental activity, synthesis and analysis, occupy the engineer during the development of a new and useful process. Synthesis is the combining of diverse conceptions into a coherent whole, and analysis is the examination of a complex situation, its elements, and their relations. In the words of Herbert A. Simon, which introduce this chapter, synthesis deals with *how to make artifacts that have the desired properties* and analysis deals with *how things are and how they work.*

In this text we are concerned with problems that require for their solution a large dose of synthesis. However, after each synthesis step, which is really a gamble on a course of action leading toward a process, an analysis must be performed to find out how things are and how they work. Thus the synthesis and analysis, the engineering and the sciences, work hand in hand toward the solution of society's technical problems.

Although we limit attention in this text to the development of process technology, it is important for the student to understand the diverse role that the engineering professions play in society. Gordon S. Brown, Dean Emeritus of the Massachusetts Institute of Technology, has suggested four broad categories to describe this work function spectrum. These are described by Brown as follows:

> First is the engineer who has great talent for abstract thinking. He has the innate aptitude to detect quickly the deficiencies in the capability of existing devices or systems, and his impulse is to conceive something better and new. He can relate quite abstract and seemingly unrelated concepts in novel, purposeful ways. This engineer may be called the *composer*. His work is predominantly intellectual. He may work as a scientist, and often contributes to science. But as he seeks the useful rather than the unknown, the science with which he works is often of little concern to a pure scientist.

Second is the engineer whose great urge is to devise and build systems of engineering work by a creative, ingenious arrangement of the knowledge of his day. He is the *arranger*, *inventor*, or *innovator*, rather than the composer. His work is also highly intellectual and creative, but his major function is to devise, design, and build things that work. His concern is with engineering feasibility and economics. He works closely with the composer of the first category. He requires great competence in the sciences and the engineering sciences as he works in development laboratories, in design establishment, or in manufacturing.

Third is the engineer expert in assembling, operating, and maintaining complicated machines and engineering systems. It is this engineer who keeps our complicated system running. He can be called the *custodian* of modern technology. He is included in the largest group. He too needs competence in modern mathematics, science, and technology but not to the depth of understanding of the composer or arranger–innovator.

Fourth is the engineer engaged in activities which draw constantly on a technological background, but who *interprets* and applies technology in interface areas where the physical sciences interact with such fields as industrial management, political science, medicine, or the life sciences. This engineer does not need working familiarity with science and technology to quite the depth required of an engineer in categories One, Two, or Three, but he does need a good understanding of the field with which he is to interact.

In this book we are concerned with the development of skills in the first and second categories, while the breadth of the engineering curriculum serves to prepare the student for any area of this diverse profession.

REFERENCES

Valuable compilations of processing technology are available in such books as:

R. N. Schreve, *Chemical Process Industries*, McGraw-Hill Book Company, New York, 1967.

R. E. Kirk and D. F. Othmer, eds., *Encyclopedia of Chemical Technology*, John Wiley & Sons, Inc. (Interscience Division), New York, 1963.

This is an interesting analysis of the depletion of natural resources:

C. F. Park, *Affluence in Jeopardy*, Freeman, Cooper & Company, San Francisco, 1968.

The history of technology is an interesting area of study. Several excellent sources to examine are:

A. Ihde, *The Development of Modern Chemistry*, Harper & Row, Inc., New York, 1964.

Charles Singer et al., *A History of Technology*, Oxford University Press, Inc., New York, 1954. (A monumental history of technology. Five volumes covering from ancient times to the 1900s.)

L. T. White, *Medieval Technology and Social Change*, Oxford University Press, Inc., New York, 1962, paperback. (A well-documented and readable history, which illustrates the important role of technology in the development of the eastern hemisphere.)

F. S. TAYLOR, *A History of Industrial Chemistry*, William Heinemann Ltd., London, 1957.

WILLIAM HAYNES, *American Chemical Industry, A History*, six vols., Van Nostrand-Rheinhold Company, New York, 1945–1954.

These popular works by William Haynes,

This Chemical Age, The Miracle of Man-Made Materials, Alfred A. Knopf, Inc., New York, 1942, and

The Chemical Front, Alfred A. Knopf, Inc., New York, 1943,

are interesting, although somewhat dated, sources for the history of the twentieth-century chemical industry.

The impact of technology on society is documented in professional journals. See, for example, "Historical Relations of Science and Technology," *Technology and Culture*, **6**, 547ff., 1965.

PROBLEMS

The first problem is designed to test the reader's inherent skill in process synthesis and to give some personal meaning to the techniques developed in the text. The remaining problems require the facilities of a modern library, and involve the study of processes of historical significance.

1. **DRY ICE.** As a member of a large food-processing organization you are responsible for the cooling of perishable foods during rail transportation. Solid carbon dioxide (Dry Ice) would be a suitable cooling source, if it could be manufactured locally at a low enough price.

   While contemplating this possibility during a walk through the food-processing plant, the gentle bubbling of the fermenters in the brewery division gives you an idea, the white plume of vapor rising from the chimney of the power plant attracts your attention, and the rotary kiln producing lime from oyster shells raises possibilities.

   By the end of the day, three processes are in mind for the large-scale continuous production of Dry Ice. What are the flow sheets?

2. **PENICILLIN PROJECT.** The crash program for the production of penicillin during World War II is one of the most interesting development programs in history. It was recognized that one life could be saved per gram of penicillin produced, but it was not known if any could be produced in significant amounts.

   In 1928 Alexander Fleming noticed that a stray mold, *Penicillium notatum*, produced a substance that killed a culture of *staphylococcus* that he had grown. In its impure form this substance, one of the many forms of penicillin, was unstable and not very attractive for clinical use, and the interest in Fleming's discovery waned. However, the German bombing of England caused renewed interest, since it was found that penicillin had remarkable properties in treating certain kinds of infection. American industrial laboratories, government laboratories, and universities were brought into a massive program to develop sufficient penicillin to accompany the troops during Eisenhower's landing in Europe.

The penicillin family of compounds has the structure

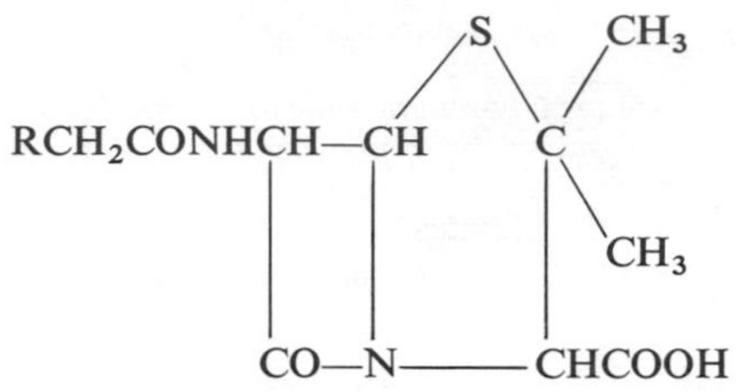

```
                         S     CH3
                        / \   /
RCH2CONHCH—CH          C
           |    |          |  \
           |    |          |   CH3
           |    |          |
          CO—N———CHCOOH
```

where the R group varies with penicillin type. The particular compound formed depends on the nature of the mold and the environment in which it is grown. For example, the mold *Penicillium notatum* growing in corn-steep liquor produces benzyl penicillin in which the R group is C_6H_5.

An outline of the penicillin project is to be prepared using the library facilities available. Much has been written on the subject. These topics should be examined:

(a) A worldwide search was conducted by the Air Force to locate molds or fungus that were good penicillin producers. What role did a moldy cantaloupe found in a fruit market in Peoria, Illinois, play, and why was this so ironic?

(b) Corn-steep liquor is a by-product of wet corn milling. In the late 1930s it was a surplus commodity, a use for which was being sought by the Northern Regional Research Laboratory of the Department of Agriculture. Why is this significant? What is deep tank fermentation?

(c) The effluent from deep tank fermentors contains only trace amounts of penicillin, only 20 to 60 ppm. This amount is on the order of magnitude of the impurities found in commercial distilled water. Furthermore, the total dissolved solids in the broth is 3 per cent, which is 1000 times the penicillin concentration. What unique property of penicillin led to the development of a separation method? What does the word podbielniak mean?

(d) Over 15,000 different kinds of penicillin have been synthesized in the last several years. What is meant by natural penicillin, semisynthetic penicillin, and synthetic penicillin? Why must penicillin facilities be kept separate from the facilities for the production of other drugs? (See A. L. Elder, *The History of Penicillin*, American Institute of Chemical Engineers, Symposium Series, **66,** 100, 1970, New York; D. Perlman, "Antibiotics—A Status Report," *Chemical Technology*, Sept. 1971.)

3. **IT FUDDIED THE MURRINERS.** In England the Towns Improvement Clauses Act of 1847 required local authorities to discharge sewage and other wastes into the rivers to keep the cities from putrifying. The potency of the river water is attested to by this recipe for beer taken from the 1702 book *Hints to Brewers*.

> Thames water, taken up about Greenwich at low water when it is free of all brackishness of the Sea and has in it all the fat and sullage from this great city of London, makes a very strong drink. It will of itself ferment wonderfully, and after its due pergation and three times stinking, it will be so strong that several sea commanders have told me that it has often fuddied their Murriners.

The cholera and typhoid epidemics were associated with the wrath of God, and not with pollution.

In the 1860s the theory was developed that disease was caused by microorganisms that can live for some time in water. In 1880 a connection was made between typhoid fever and a bacillus, and in 1883 Robert Koch isolated the Asiatic cholera bacillus. It became evident that some means of sewage treatment must be developed.

Ardern and Lockett of the Manchester Sewage Works reported the results of experiments on sewage treatment using aerated flocculant biological solids. The paper was a milestone. (See E. Ardern and W. T. Lockett, "Experiments on the Oxidation of Sewage Without the Aid of Filters," *Journal of the Society of Chemical Industries*, **33**, 523, 1914.) The news of their discovery spread, and by 1917 the Manchester Corporation had a 250,000 gal/day plant in operation in Worthington, and in the same year the 10,000,000 gal/day Northside Plant in Houston, Texas, was in operation.

This was the start of the *activated-sludge process*, which for half a century has been the workhorse of municipal waste treatment. Prepare a brief outline of the chemical reactions and separations that occur in the treatment of sewage by this process. (See, for example, C. N. Sawyer, "Milestones in the Development of the Activated Sludge Process," *Journal of the Water Pollution Control Federation*, Feb. 1965.)

4. **BAYER'S PROCESS.** Bauxite, named after the village of Les Baux in France where it was first discovered, designates various kinds of aluminum ores consisting mainly of aluminum hydroxide in amounts of about 50 per cent. These ores are found in Jamaica, U.S.S.R., Surinam, Guyana, France, the United States, and Hungary. Aluminum metal is prepared from molten and pure Al_2O_3 by electrolysis, the Hall process, due to Charles Hall, 1886.

   In 1888 Karl Josef Bayer in Russia devised a means of producing 99.5 per cent pure Al_2O_3 from crude bauxite. In the United States, bauxite, sodium carbonate, and lime water are the raw materials from which pure Al_2O_3 is formed using Bayer's process. What chemical reactions and separations occur? To what use could be put the red mud waste? (See *Chemical Engineering*, Sept. 20, 1971, p. 88.)

5. **THE KIRKPATRICK AWARD.** Biennially the Kirkpatrick Award is given to the chemical engineering achievement thought most significant by a panel of engineers. The results are announced in the publication *Chemical Engineering*. Prepare a list of the winners and the runners-up for the entire history of that award.

6. **SYNTHESIS OF GASOLINE AND FOOD.** During World War II allied bombers payed particular attention to synthetic gasoline processes in Germany. There, coal was being converted into motor fuels to overcome a severe shortage of petroleum. The reaction paths upon which these processes were based were discovered by Friedrich Bergius (1884–1949), Franz Fischer (1877–1947), and Hans Tropsch (1889–1935).

   Describe the two alternative reaction paths used to produce hydrocarbons from coal. The Germans also produced fatty acids from these hydrocarbons to overcome shortages of food fats. How was this done?

7. **SYNTHETIC FIBERS.** The first commercially important synthetic fiber was nylon 66, marketed in 1940. This was an outgrowth of the research performed by W. H. Carothers of Du Pont on the chemistry of large molecules. Nylon was introduced on the market

as a fiber made from coal, air, and water. What processing steps are needed to produce nylon from these elementary raw materials? Is this the way it is done commercially?

8. **BROMINE FROM THE SEA.** Kure Beach, North Carolina, in 1938 was the site of one of the first processes to obtain bromine from the sea. The increased demand for bromine to be used with antiknock compounds in motor fuels required the exploitation of this new source. How was the separation accomplished? What geographic features led to the selection of this site? Why was ethylene dibromide shipped from the plant rather than bromine? Is this plant still operating?

9. **COPPER PRODUCTION.** What is the technology behind these paragraphs from *The Rockies* by David Lavender, Harper & Row, Inc., 1968, pp. 296–298?

> . . . Among the mining men who traveled the Rockies looking for properties to develop was Joseph De Lamar, a onetime Dutch ship captain. For a brief time De Lamar tried running a gold mill at Cripple Creek, but was defeated by the complexity of the ores. Shifting to Utah, he bought in 1899 a substantial interest in 200 acres of copper claims owned by Enos Wall in Bingham Canyon south of Great Salt Lake. The ore, dusted in specks throughout an enormous mass of porphyry rock, was a low-grade sulphide—less than $2\frac{1}{2}$ per cent copper. The price of copper had recently climbed back to 18 cents a pound, however, and De Lamar began wondering whether, under the circumstances, a volume operation might turn a profit.
>
> To help him determine he employed as his consultant a husky, red-faced, hard-drinking thirty-year-old mining engineer named Daniel C. Jackling. A graduate of the Missouri School of Mines, Jackling had come to Cripple Creek in 1894 so poor that he could not afford stage fare and had walked the last 18 miles. In Cripple Creek he and another young engineer, Charles MacNeill, took over the mill that had defeated De Lamar and made a success of it. To De Lamar that was recommendation enough.
>
> Assisted by Robert Gemmell, state engineer of Utah, Jackling ran exhaustive tests on the copper property in Bingham Canyon. The report they came up with was far more sweeping than De Lamar had bargained for when mentioning "volume." They suggested not the tunneling and stoping operations normal to hard-rock mining but the stripping away of the over-burden and the removal of the ore with newfangled machines called steam shovels. The shovels would dump the ore straight into gondola cars traveling on tracks that followed the operation from spot to spot. Each day several trains would haul the rock over a specially built railroad to a mill and smelter site near the shore of Great Salt Lake. There it would be run through a larger mill than De Lamar had ever seen.
>
> The very thought of the expenses involved appalled him and he dropped the project, thinking probably that he had hired a madman. Finding himself abruptly unemployed, Jackling built a mill in Washington state and then returned to Cripple Creek. There he approached several prosperous friends—millman Charles MacNeill, mineowner Charles Tutt, Spencer Penrose, and a few others. After talking himself hoarse he persuaded them, in May, 1903, to go to Utah with him and ride up Bingham Canyon in a buggy. By dark MacNeill was convinced, and that swung the others. They formed Utah Copper Company

and bought 80 per cent of Enos Wall's original holdings; Wall retained a 20 per cent interest; De Lamar dropped out entirely.

To raise money the new owners carried their figures to John Hays Hammond, chief engineer of the Guggenheim Exploration Company in Denver.* After seventeen engineers of Hammond's staff had spent $150,000 testing the property and analyzing Jackling's wild ideas, the Guggenheims agreed to go ahead. They underwrote a $3 million bond issue and bought 232,000 shares of Utah Copper Company stock at $20 a share. This gave them a quarter interest in the enterprise and the right to smelt all the copper it produced.

With the money thus raised, the company purchased more ground, brought in steam shovels (the original one took its first bite in 1907), purchased seventeen locomotives for their railroad, and built a mill with the then-unheard of capacity of 6,000 tons per day.

It was the beginning of the Southwest's most notable heavy industry, the open-pit mining of low-grade porphyry copper ores. During Utah's first thirty years of operation, it produced more than 3 billion pounds of copper. Today its mills (now owned by Kennecott) can handle 100,000 tons of rock every twenty-four hours. The batteries of trains that crawl around on the gorgeously hued terraces of the demolished mountains look, on the sides of the gigantic pit, as tiny as caterpillars.

Today more than 85 per cent of all the world's copper is produced from porphyry ores by the open-pit methods envisioned by Jackling and Gemmell.

What technology is used to recover copper from this low-grade ore?

10. **A WHITE POWDER.** During the summer of 1938 a young research chemist, Roy J. Plunkett, was working in the Jackson Laboratory, E. I. du Pont de Nemours & Company, Penns Grove, New Jersey. He was interested in producing several hundred pounds of tetrafluoroethylene ($F_2C{=}CF_2$), and devised a pilot plant to produce the material from dichlorotetrafluoroethane. (What was the reaction?) The product was placed in metal cylinders and stored at $-50°C$. One day while Plunkett and his helper were vaporizing the material out of a cylinder, the flow stopped. A check of the weight of the cylinder showed that it still contained several pounds of material. Plunkett ran a wire into the cylinder, yet no gas escaped. He sawed the cylinder open with a hacksaw and found several pounds of white powder.

    What was this powder? What is the current annual production of this material?

11. **THE LEBLANC PROCESS.** The early history of the chemical-processing industries is dominated by the story of the Leblanc process. Nearly all major developments and many minor ones were related to developments in alkali trade (Na_2CO_3).

    Prior to the Leblanc process, few chemicals were produced in large amounts in an organized manner. Most chemicals were derived from local natural sources only. Potassium carbonate was obtained from wood ashes by leaching with water. Salt was produced by evaporating seawater or mining rock-salt deposits. Saltpeter for gunpowder

* The Guggenheim fortune much later led to the formation of the John Simon Guggenheim Memorial Foundation. The senior author was a Guggenheim Fellow during the middle year of book writing.

was prepared from nitrates leached from manure piles or scraped from the walls of stables or burial crypts. Sulfur came from the volcanic deposits in Sicily.

The Leblanc process marked one of the first applications of efficient large-scale chemical processing, which gave rise to modern chemical plants.

What are the chemical reactions involved in the Leblanc process?

What natural sources of Na_2CO_3 were used prior to the Leblanc process?

What motivated Nicolas Leblanc to discover this process?

Draw a simple flow sheet for the process.

12. FISHING FOR NEW COMPOUNDS. A group of amateur chemists was engaged in attempts to produce metallic calcium by reducing the oxides with carbon. A mixture of calcium carbonate and carbon was prepared and placed in an electric arc furnace and heated to 2000°C. After several hours a small quantity of the product was removed from the furnace and quenched in a pail of water. It at once gave off a large quantity of gas. Now calcium when contacted with water will give off hydrogen. Another sample was removed, and after it had cooled it was immersed in water. The gas was lighted by attaching a burning rag to the end of a long fishing pole and waving the burning rag through the gas. The gas burned with a luminous flame and gave off clouds of carbon. It is evident the gas was not hydrogen, which burns with a colorless flame. The substance was not metallic calcium. What was it? What are the chemical reactions involved? What does a modern plant to produce this compound look like?

2

Reaction-Path Synthesis

Successful solutions to industrial problems depend upon engineering judgement and experiment with the unknown and undocumented science as well as with the principles that have already been well established. This is the principal distinction between the scientist and the engineer.

Olaf A. Hougen

2.1 Introduction

Innovations in chemistry have a great effect on processing, and many important processing revolutions center about the discovery of new reaction paths. However, not every newly discovered sequence of reactions has practical significance, since the environment in which the chemist discovers reactions is significantly different than the environment in which the engineer puts these reactions to use. In this chapter we focus attention on the interface between science and engineering to identify the interaction that must occur during the synthesis of the reaction paths which link readily available raw materials with valuable products, or which link environmentally dangerous wastes with easily disposed of materials.

Although chemistry plays an important role in process development, the sophisticated chemical techniques of reaction-path synthesis will not be discussed here. However, it is necessary for the engineer to develop an appreciation of chemical syntheses, and to be able to assess the engineering significance of alternative reaction paths.

The following sections outline the creative chemistry needed to invent reaction paths, show how information is to be transferred from the chemist to the engineer, and illustrate the screening needed to weed out reaction paths that are not commercially attractive. In summary, to be of use commercially a reaction path must be technically and economically feasible. Technically the reactions must go under reasonable conditions and lead to a product of desired quality. Economically the products must be worth more than the total cost of manufacture, which includes such expensive items as raw materials, solvents, and pollution control.

2.2 STRATEGY OF MOLECULE SYNTHESIS

The chemical bonds that hold the atoms of a molecule together are the source of many of the chemical and physical properties of a molecular species. The specific behavior of these bonds determines those reactions which will occur and those which will not, and for this reason the building of a molecule is not a mechanical construction job. The atoms cannot be placed adjacent to each other and then bonded together. Rather, the chemist must learn the behavior of molecules to bring about those special conditions under which the molecules *want* to form the bonds desired. The discovery of those special conditions is a central task of research chemistry.

Often a molecule can be divided into certain units of reactivity, called *functional groups*. For each kind of functional group a wide variety of reactions is known to take place under special conditions. The arsenal of thousands of unit reactions must be brought to bear effectively in reaction-path synthesis. The information-handling problems can be enormous, for a considerable literature exists in each area of technology, ranging from pure chemistry, to food processing, to waste disposal.

The synthesis of a certain molecule usually requires a sequence of carefully chosen chemical reactions performed in a fairly rigid order. In general, larger and more complex molecules are farther removed from cheap and readily available materials, and longer reaction-step sequences are needed for synthesis. Thirty or more separate reaction steps are required in the synthesis of some chemicals.

Chemical synthesis begins with the examination of the desired or *target* molecule and with the perception of features that are of synthetic significance. Next, those single molecular changes that result in the formation of the target molecule from a *precursor* molecule are examined. (Precursors are molecules from which the target molecule might be formed and are usually chemically simpler.)

Each of the precursors formed then becomes a target molecule to be synthesized from still simpler molecules, and so forth, until readily available raw materials are identified. This results in a *synthesis tree*, any sequence of reaction steps from an identified raw material through the tree being a candidate reaction path from which a process could be developed. These ideas are illustrated by several examples.

Example 2.2–1 Synthesis of 2-Chloro-2,4,4-trimethylpentane. *The molecule in question has the molecular structure*

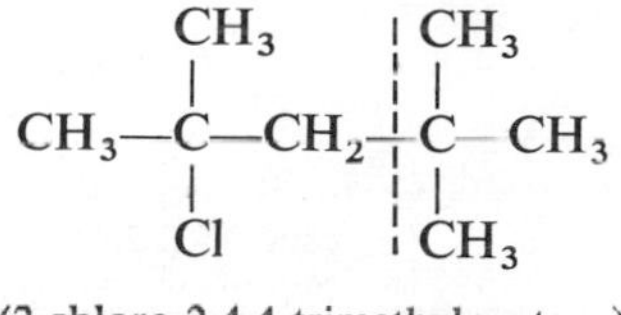

$$\begin{array}{c} CH_3 \quad\;\; CH_3 \\ | \qquad\quad | \\ CH_3-C-CH_2-C-CH_3 \\ | \qquad\quad | \\ Cl \qquad\; CH_3 \end{array}$$

(2-chloro-2,4,4-trimethylpentane)

with a carbon chain skeleton divisible into two isobutane units:

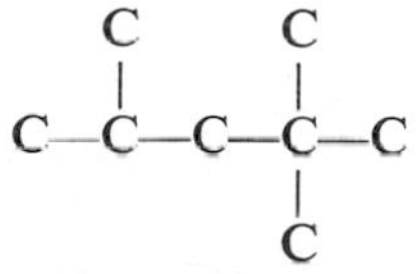

$$\begin{array}{c} C \quad\;\; C \\ | \qquad | \\ C-C-C-C-C \\ \qquad\quad | \\ \qquad\quad C \end{array}$$

This skeleton could be put together in two ways. Isobutylene with sulfuric acid dimerizes to an 80:20 mixture of 2,2,4-trimethyl-1-pentene and 2,4,4-trimethyl-2-pentene, whereas isobutylene and isobutane react with hydrogen fluoride as a catalyst to yield 2,2,4-trimethylpentane.

Reaction 1:

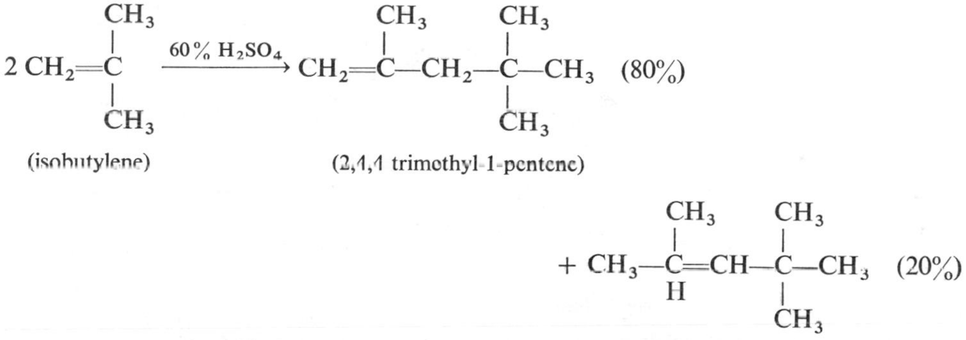

$$2\,CH_2{=}C(CH_3)_2 \xrightarrow{60\%\ H_2SO_4} CH_2{=}C(CH_3)-CH_2-C(CH_3)_2-CH_3 \quad (80\%)$$

(isobutylene) (2,4,4 trimethyl-1-pentene)

$$+\ CH_3-C(CH_3){=}CH-C(CH_3)_2-CH_3 \quad (20\%)$$

(2,4,4-trimethyl-2-pentene)

Reaction 2:

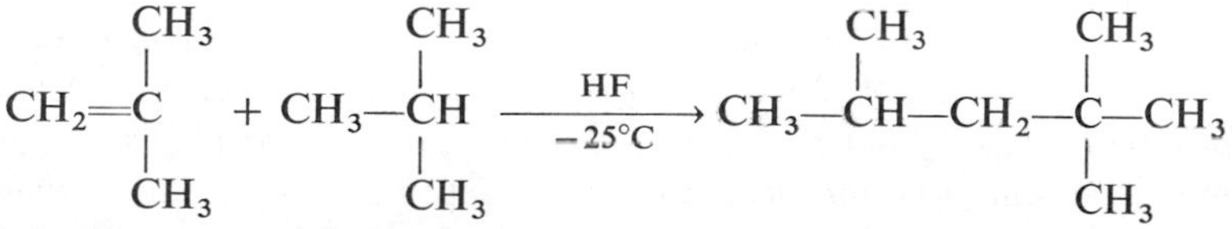

$$CH_2{=}C(CH_3)_2 + CH_3-CH(CH_3)-CH_3 \xrightarrow[-25^\circ C]{HF} CH_3-CH(CH_3)-CH_2-C(CH_3)_2-CH_3$$

(isobutylene) (isobutane) (2,2,4-trimethylpentane)

The addition of chlorine to the proper location in the octane skeleton could be accomplished by the addition of hydrogen chloride to the trimethylpentane isomers or *the free radical chlorination of 2,2,4-trimethylpentane either photochemically or thermally with chlorine, or else with sulfuryl chloride.*

Reaction 3:

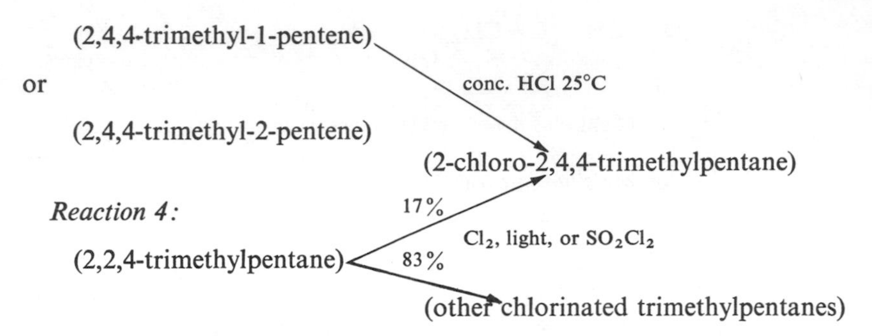

These separate reaction paths can be represented in the tree shown in Figure 2.2–1.

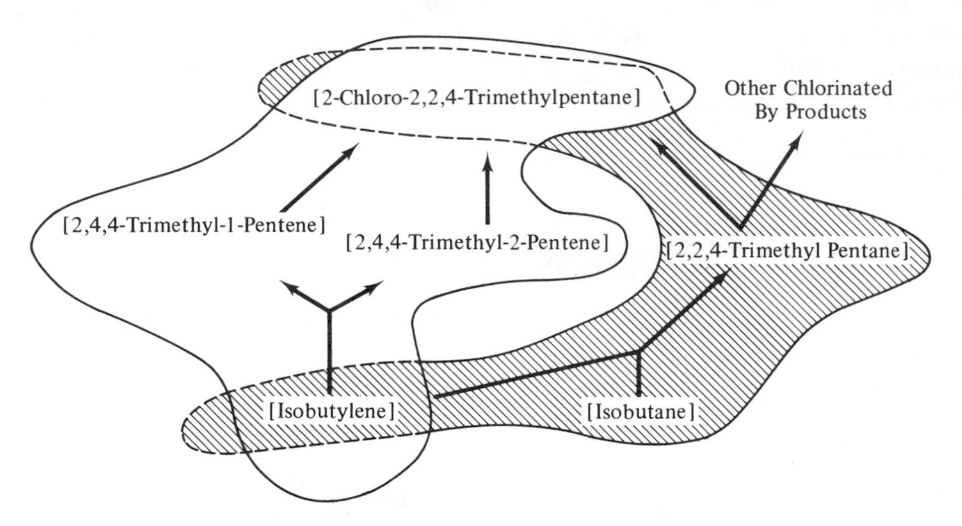

Figure 2.2–1. Alternative Pathways to a Chlorinated Octane from Isobutane Skeleton Molecules

The leftmost path beginning with isobutylene first produces the 80:20 mixture of trimethylpentene isomers, which might seem to lead to separation problems because their properties are so similar. However, the second reaction step fortunately gives the same chlorinated product from both isomers. The rightmost reaction path generates a single trimethylpentane from the isobutane–isobutylene mixture; however, the chlorination reaction is not specific. The chlorine atom attaches itself in other places as well as at the desired location in the carbon skeleton; only 17 per cent formation of the desired isomer occurs. The separation of the desired isomer might be very difficult, and one might gamble that the leftmost path would be a better candidate for commercialization. However, at this early stage of process synthesis insufficient information is available accurately to assess the reaction paths.

Example 2.2–2 Phenol Synthesis. *To illustrate further the variety of reaction paths that lead from available raw materials to valuable products, we examine five pathways to phenol. Notice the variety of opposing conditions in each reaction path and the separation problems that arise. How does one detect the better paths? Notice also that these reaction paths can be mixed and matched to form other paths to phenol (see Fig. 2.2–2).*

1. *Sulfonation process: Benzene is treated with sulfuric acid to form benzene sulfonic acid. Under the drastic reaction conditions of fused sodium hydroxide (above 300°C) the benzene sulfonic acid converts into sodium phenoxide. This fusion mass is run into water and the sodium sulfite filtered off. Under acid conditions the sodium phenoxide is converted into phenol. The generation of salt solutions to be disposed of may make this route to phenol unattractive on the large scale.*

2. *Chlorobenzene process: In a solution of sodium hydroxide at 5000 pounds per square inch (psi) and 350°C, chlorobenzene converts to sodium phenoxide. This can be acidified to form phenol. This is currently a commercial path to phenol.*

3. *Catalytic process: In the vapor phase, benzene can be catalytically chlorinated and converted to phenol. This pathway suggests a continuous high-pressure vapor-phase processing scheme.*

4. *Toluene oxidation process: Toluene can be partially oxidized using cobalt salt catalysts to an intermediate product, which in the presence of copper and magnesium*

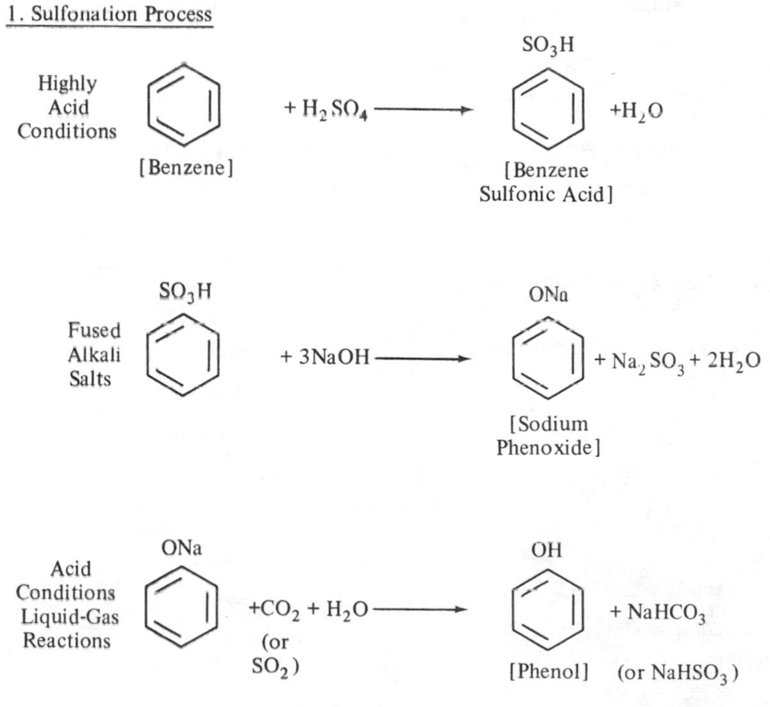

1. Sulfonation Process

Highly Acid Conditions: [Benzene] $+ H_2SO_4 \longrightarrow$ [Benzene Sulfonic Acid] (SO_3H) $+ H_2O$

Fused Alkali Salts: (SO_3H) $+ 3NaOH \longrightarrow$ [Sodium Phenoxide] (ONa) $+ Na_2SO_3 + 2H_2O$

Acid Conditions Liquid-Gas Reactions: (ONa) $+CO_2 + H_2O$ (or SO_2) $\longrightarrow$ [Phenol] (OH) $+ NaHCO_3$ (or $NaHSO_3$)

Figure 2.2–2. Pathways to Phenol

2. Chlorobenzene Process

Alkali Conditions at 5000 psi and 350° C

$C_6H_5Cl + 2NaOH \longrightarrow C_6H_5ONa + NaCl + H_2O$

[Sodium Phenoxide]

Highly Acid Conditions

$C_6H_5ONa + HCl \longrightarrow C_6H_5OH + NaCl$

[Phenol]

3. Catalytic Process

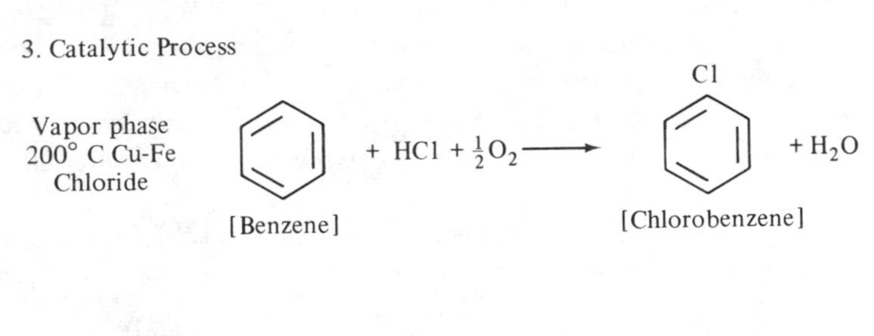

Vapor Phase 500° C SiO_2 Catalyst

$C_6H_5Cl + H_2O \longrightarrow C_6H_5OH + HCl$

[Phenol]

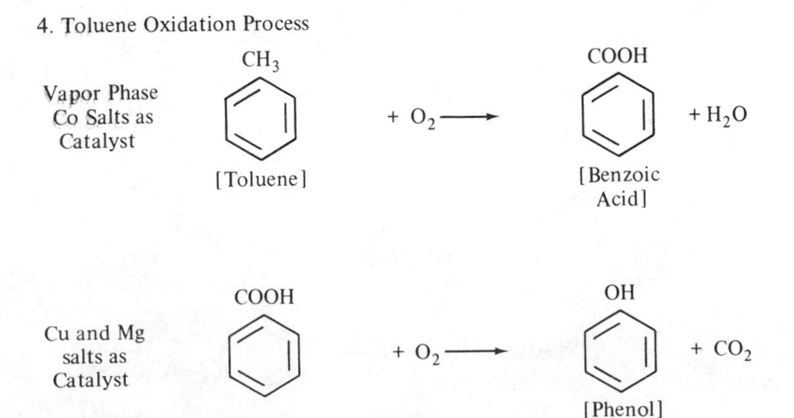

Figure 2.2–2 (*continued*). Pathways to Phenol

5. Cumene Hydroperoxide Process

Vapor Phase Oxidation

[Cumene] + O_2 → [Cumene Hydroperoxide]

Acid Conditions

[Cumene Hydroperoxide] → [Phenol] + $CH_3-C(=O)-CH_3$ [Acetone]

Figure 2.2–2 (*continued*). Pathways to Phenol

salt catalysts can be oxidized to phenol and carbon dioxide. This vapor-phase processing scheme is employed commercially rather than the benzene chlorination process previously mentioned. What makes this path easier to implement?

5. Cumene hydroperoxide process: This begins with the vapor-phase oxidation of cumene, followed by acidification to form phenol and acetone. A cumene by-product from another process and an additional market for acetone might make this route commercially attractive.

2.3 ENGINEERING DATA ON REACTION PATHS

Regardless of why a compound is to be synthesized, it is desirable to make it from readily available materials as efficiently and economically as possible. The rigid order and conditions under which the reactions must be executed in a reaction path may create difficult engineering problems that render a chemically direct path economically undesirable. That which is efficient in the laboratory may be very inefficient on the large scale.

For example, in the laboratory the sulfuric acid used as a catalyst in the isomerization of isobutylene (Example 2.2–1) might be easily disposed of, whereas on the large scale the disposal of millions of tons per year of impure acid might present some extremely difficult engineering problems. The cooling, heating, and venting accomplished so easily in the laboratory can be extremely expensive if the large-scale process is merely a scale-up of laboratory procedures. Unusual hazards may arise as a laboratory reaction path is scaled up by factors of thousands or millions to reach commercial production.

Furthermore, each step in a reaction sequence may reduce the amount of product that can be produced. For example, if in a 15-step reaction path each reaction yields only 20 per cent of the desired intermediate chemicals, the overall yield is limited to $(0.20)^{15}$ or 1 billionth of 1 per cent. If 90 per cent yields could be accomplished by improving reaction conditions or by more refined separations of intermediate products from the reaction-step products, an overall yield of $(0.90)^{15}$ or about 20 per cent might be accomplished. The manufacture of antibiotics often involves even longer reaction paths.

The final decision on the commercial value of a reaction path cannot be made in the laboratory. It is only after a full-scale process has been invented and analyzed that such decisions can be made. However, the chemist must supply certain essential data to the engineer. The data concern the common and uncommon characteristics of each reaction step.

The following questions are useful to get the interchange of information started. To bring this interchange to the level of practice, we illustrate with the discovery in the laboratory of a method for synthesizing dichloroneopentyl glycol, a compound useful as a fire-retardant coating.

1. What are the main reactions and major side reactions?

Chemist: We have been able to synthesize dichloroneopental glycol in the laboratory by the chlorination of neopentyl glycol. If we use the following shorthand notation for these glycols

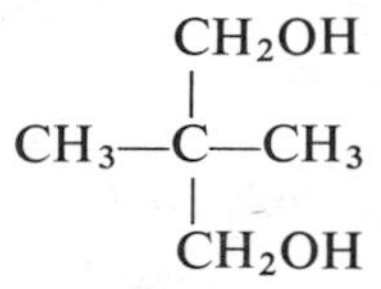

$$\begin{array}{c} CH_2OH \\ | \\ CH_3—C—CH_3 \\ | \\ CH_2OH \end{array}$$

(neopentyl glycol) NPG

$$\begin{array}{c} CH_2OH \\ | \\ CH_3—C—CH_2Cl \\ | \\ CH_2OH \end{array}$$

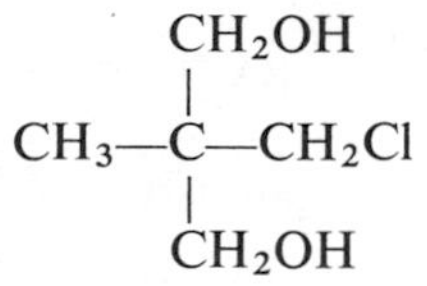

(monochloroneopentyl glycol) MCNPG

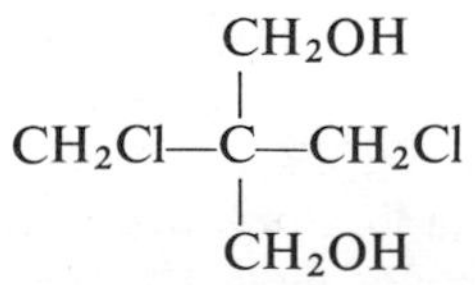

$$\begin{array}{c} CH_2OH \\ | \\ CH_2Cl—C—CH_2Cl \\ | \\ CH_2OH \end{array}$$

(dichloroneopentyl glycol) DCNPG

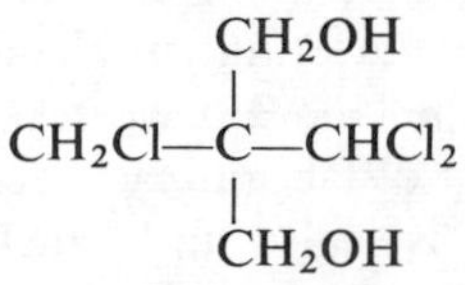

$$\begin{array}{c} CH_2OH \\ | \\ CH_2Cl—C—CHCl_2 \\ | \\ CH_2OH \end{array}$$

(trichloroneopentyl glycol) TCNPG

the main reactions and major side reactions seem to be

$$NPG + Cl_2 \rightarrow MCNPG + HCl$$

$$MCNPG + Cl_2 \rightarrow DCNPG + HCl$$

$$DCNPG + Cl_2 \rightarrow TCNPG + HCl$$

The product we want is in the middle of a reaction chain! We must control the reaction conditions carefully to maximize the yield of dichloroneopentyl glycol.

2. In what phase do the reactions take place, and what are the phases of the pure reactants and pure products?
3. At what temperature and pressure do the reactions take place?
4. Is a catalyst required? What are its properties?
5. Was a solvent used? If so, what was its purpose? Dilution, separation, heat adsorption, reaction kinetics, or safety?

Chemist: In the pure state all these glycols are solids; NPG melts at 60°C and DCNPG at 75°C. Both are soluble in acetic acid, so I ran the reaction in an acetic acid–water solution at ambient pressure at 80°C. This was most convenient in the laboratory and the reaction went all right.

Engineer: Why was the acetic acid–water solution used? Is it essential for the reaction?

Chemist: I needed to dissolve the solids in something, since it is inconvenient to blow chlorine through a beaker of solids. I had acetic acid on the lab bench, and it was used as a solvent by other researchers working with similar compounds.

Engineer: Can the reaction be run without a solvent?

Chemist: I think we need an acid solvent. In fact, the solution must be dilute. In one run using concentrated acid, solids precipitated and fouled up the reactor.

Engineer: What would happen if I melted NPG and bubbled Cl_2 through the liquid?

Chemist: It might work, but I prefer the acetic acid.

Engineer: What would happen if I ran my reaction under pressure?

Chemist: I don't know because I ran my reactions either under vacuum or at atmospheric pressure.

6. What was the quality of the reactants used? Has a run been tried with industrial-quality feed stock? What reactions would occur with common impurities? Were there any residues or was the product unusually colored?

Chemist: There was a tarlike residue occasionally. But, I was careful to select high-purity reagent-grade chemicals to avoid this.

Engineer: Here is a list of materials that might appear in the reactants under process conditions. What would happen to these? Air? Iron? Closely related chemicals found in impure feedstock?

Chemist: Oxygen would have no effect if present at, say, less than 1 per cent. Nitrogen would have no effect at all. Iron would cause a very bad color and should be kept out.

The only close relative to the main reactant that might be in cheaper feed stock is

$$\begin{array}{c} CH_2OH \\ | \\ CH_3CH_2—CH_2OH \end{array}$$

and it would be chlorinated just like NPG.

7. What yields were obtained of the product and main by-products? How many compounds have been identified in the product, by what methods, and at what concentrations? How long was the reaction run? What if the reaction was run for a longer or shorter period?

Chemist: I got a 60 per cent yield of DCNPG and the remainder was mainly MCNPG. From past experience I know that 4 to 6 hours is the best run time. I ran one sample through a gas–liquid chromatograph and found NPG, acetic acid, H_2O, Cl_2, HCl, MCNPG, DCNPG, TCNPG, four unknowns, plus some tars. The lowest concentration detectable was 500 ppm.

Engineer: Could you detect lower concentrations, say, 1 ppm or less, by mass spectroscopy, infrared analysis, ultraviolet analysis, neutron activation, or nuclear-magnetic-resonance methods? What about materials not detectable by gas–liquid chromatography, such as Fe^{2+}, O_2, or very high molecular weight residue? If I shift to a continuous recycle operation, trace materials could build up to high concentrations. What would happen if I took some of your reaction products and recycled them into the feed of a new run?

Chemist: The MCNPG would react to the desired product, and not much would happen to TCNPG, except it might form some residue. I will try a batch and see what happens.

8. What are the physical, chemical, safety, and toxic properties of all the materials involved, including intermediate species that may occur during reaction?

Chemist: HCl and Cl_2 are dangerous gases, and H_2O–HCl mixtures can be quite corrosive. NPG and DCNPG in dust or powder form would irritate the lungs. I don't know how toxic the other compounds are. None are explosive or burn readily.

Engineer: We will need to know the melting points, boiling points, solubilities in a variety of solvents, densities, vapor pressure, and so forth of all compounds, including trace materials. This is critical for engineering design. What color is the desired product?

Chemist: It should be white. The least bit of iron colors it brown, and if the reactors runs too long the product has little black particles in it.

9. Describe in some detail the laboratory procedures used during reaction and product purification.

Chemist: The acetic acid and water are placed in the reactor (shown in Figure 2.3–1) and the mixture is heated to about 80°C. The NPG is dumped in and mixed until dissolved. Chlorine gas is bubbled in just under the stirrer. A reflux condenser returns the water boiled off to the reactor and lets the hydrogen chloride and excess chlorine pass out. After about 4 hours I stop the reaction by cooling the flask in an ice bath.

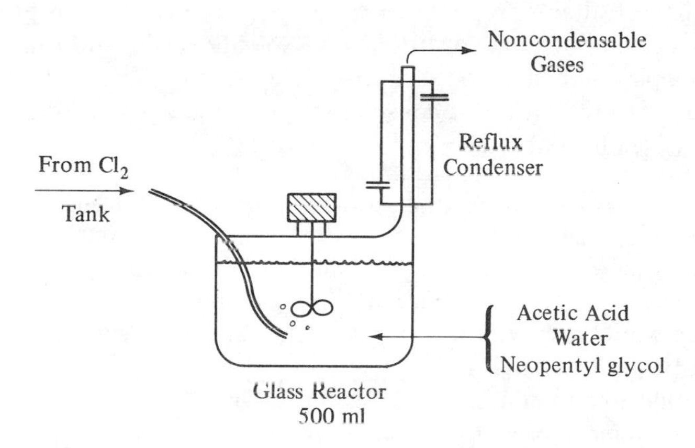

Figure 2.3–1. Laboratory Reactor Setup

The organic phase separates out and solidifies. The aqueous phase is decanted off into the sewer, and the organic phase is washed with water. Then the organic phase is dissolved in hot acetic acid and water, and is recrystallized out by cooling. The solid is washed in ethanol, dried, and ground.

Engineer: Why recrystallize in acetic acid and why use the ethanol wash? How much heat is given off? Do you need to use the steam jacket all the time? Would it be dangerous to bubble in the chlorine gas too fast? What would happen if liquid chlorine were dumped in with the reactants?

Chemist: The acetic acid recrystallization works, and the ethanol was handy in the lab. Keep the chlorine concentrations low. Also, don't keep DCNPG near water too long since it hydrolyzes.

Engineer: Could you easily make a run without the acetic acid?

Chemist: To run without the acetic acid solvent would probably ruin my reactor. I have grown to like the acetic acid, and anyway it possibly might help reduce tar formation. If you don't allow acetic acid, how can I recrystallize the product? What are you going to do, distill the product?

Sublime it? Crystallize it from its melt? These are all difficult to do in my lab!

10. Do you know of any other reaction paths to the products? Can the over-reacted material be reacted back?

Engineer: I would really like to see if there is some useful reaction in which I can use the HCl molecules formed. If not, we have as much waste HCl to dispose of as we have useful product formed.

Debriefing Session

After the questions are answered, a debriefing session is essential in which the engineer outlines the line of attack proposed to commercialize the reaction path. This often reveals misinterpretations and nearly always solicits further information from the chemist on topics not mentioned during the question-and-answer session. The chemist should be brought in for consultation during all phases of process development. A stray idea here and there can do wonders.

Engineer: I have looked over the questionnaire and am now considering several alternatives. If I duplicate your lab procedure at 10,000 pounds per day production, I would have a mess handling all the dirty acetic acid, hydrogen chloride, excess chlorine fumes, and dirty ethanol. Quite a bit of heating and cooling would also be necessary. And handling the large batches of materials might be pretty costly.

I would like to eliminate unessential solvents, to deal with pure materials. And, I prefer processing using continuous flow, separating, and recycling incompletely reacted materials. For this reason, it is essential that I have reaction data for chlorine with both liquid and powder NPG, MCNPG, DCNPG, and TCNPG. Let's suppose that these reactions go all right. If they don't, we can come back to the solvent later.

Now when materials come out of the reactor, they will be NPG, MCNPG DCNPG, and TCNPG, and perhaps I will run the reactor at conditions that give only a small yield of DCNPG and therefore even less TCNPG and higher residues. I would like to split off the NPG and MCNPG so that they can be recycled to the reactor, and the TCNPG and higher residues so that they can be disposed of. Somehow the Cl_2 and HCl have to be separated to recycle the Cl_2. What can be done with the HCl? Could we use that anywhere else in the company?

Can you think of five or six different ways that the NPG–MCNPG mixture could be removed from the NPG–MCNPG–DCNPG–TCNPG–residue mixture, and then the DCNPG from the remaining DCNPG–TCNPG–residue mixture? Maybe the TCNPG–residue mixture should be removed first?

Should the pure chlorine reactant be too harsh, perhaps nitrogen could be used as a diluent, since it should be inert. Maybe the HCl by-product could be the diluent? It would be nice if the heat of reaction could be used to melt the solids during recrystallization. Can't use ordinary equipment since iron is not allowed anywhere. Maybe I can run the reaction at such high pressures that the chlorine is a liquid and the HCl is still a gas; then I would let only the gas out of the reactor, thereby eliminating the HCl–Cl_2 separation problem at the source. . . .

We have seen how the chemist and engineer exchange information as the chemistry is to shift from the research laboratory to full-scale processing. In the next example we show the enormous differences that can exist between the test tube and the process.

Example 2.3–1 Copper Leaching in the Western United States.* *Copper in the Western United States is produced from low-grade ore bodies found predominately in igneous rocks, such as quartz monzonite, quartz porphyry, granite porphyry, and quartz diorite. The principal copper minerals found in these host rocks are chalcopyrite and chalcocite with other forms present. The total copper content often amounts to less than 0.5 per cent copper.*

Metallic copper can be produced directly from copper sulfate solutions and metallic iron by the following reaction:

$$\underset{\text{(copper sulfate)}}{CuSO_4} + \underset{\text{(iron)}}{Fe} \rightarrow \underset{\text{(copper)}}{Cu} + \underset{\text{(iron sulfate)}}{FeSO_4}$$

The copper forms from solution as a growth which can be washed off the iron surface. The copper formed in this way is called cement copper.

However, to use this reaction to reach the target molecule, copper, it is necessary to find reactions that transform the copper compounds in the ore into copper sulfate. Shown are the varieties of such reactions, all of which occur in the presence of some sort of leach solution under acidic conditions. The leaching solution is usually sulfuric acid or sulfuric acid plus ferric iron.

Azurite:

$$Cu_3(OH)_2 \cdot (CO_3)_2 + 3\,H_2SO_4 \rightleftharpoons 3\,CuSO_4 + 2\,CO_2 + 4\,H_2O$$

Malachite:

$$Cu_2(OH)_2 \cdot CO_3 + 2\,H_2SO_4 \rightleftharpoons 2\,CuSO_4 + CO_2 + 3\,H_2O$$

Chrysocolla:

$$CuSiO_3 \cdot 2\,H_2O + H_2SO_4 \rightleftharpoons CuSO_4 + SiO_2 + 3\,H_2O$$

* "Copper Leaching Practices in the Western United States," Bureau of Mines Information Circular 8341, U.S. Department of the Interior, 1968.

Cuprite:

$$Cu_2O + H_2SO_4 \rightleftharpoons CuSO_4 + Cu + H_2O$$

$$Cu_2O + H_2SO_4 + Fe_2(SO_4)_3 \rightleftharpoons 2\ CuSO_4 + H_2O + 2\ FeSO_4$$

Native Copper:

$$Cu + Fe_2(SO_4)_3 \rightleftharpoons CuSO_4 + 2\ FeSO_4$$

Tenorite:

$$CuO + H_2SO_4 \rightleftharpoons CuSO_4 + H_2O$$

$$3\ CuO + Fe_2(SO_4)_3 + 3\ H_2O \rightleftharpoons 3\ CuSO_4 + 2\ Fe(OH)_3$$

$$4\ CuO + 4\ FeSO_4 + 6\ H_2O + O_2 \rightleftharpoons 4\ CuSO_4 + 4\ Fe(OH)_3$$

Chalcocite:

$$Cu_2S + Fe_2(SO_4)_3 \rightleftharpoons CuS + CuSO_4 + 2\ FeSO_4$$

$$Cu_2S + 2\ Fe_2(SO_4)_3 \rightleftharpoons 2\ CuSO_4 + 4\ FeSO_4 + S$$

Covellite:

$$CuS + Fe_2(SO_4)_3 \rightleftharpoons CuSO_4 + 2\ FeSO_4 + S$$

Chalcopyrite will slowly dissolve in acid ferric sulfate solutions and also will oxidize according to the following reactions:

$$CuFeS_2 + 2\ O_2 \rightleftharpoons CuS + FeSO_4$$

$$CuS + 2\ O_2 \rightleftharpoons CuSO_4$$

The copper precipitation reaction indicates that a molecule of iron must be consumed for each molecule of copper produced. The weight of iron consumed per pound of copper manufactured can be computed from the molecular weights of iron and copper, 56 and 64.

$$\left(\frac{1 \text{ mole iron consumed}}{1 \text{ mole copper produced}}\right)\left(\frac{56 \text{ lb iron}}{1 \text{ mole}}\right)\left(\frac{1 \text{ mole}}{64 \text{ lb copper}}\right) = \frac{0.88 \text{ lb iron consumed}}{1 \text{ lb copper produced}}$$

This factor is of critical importance, since the iron must be purchased as a raw material to enter into the copper precipitation reaction. Since in mass production of copper, shredded tin cans are the major source of the iron, this factor is called the "can factor." The can factor is always greater than the theoretical 0.88, and in large-scale operations it ranges from 1.3 to 2.5. The lost iron reacts with acid in the leachate and appears as impurities in the cement copper.

Figure 2.3–2 shows the typical design of a gravity launder–precipitation reactor. The copper sulfate solution flows through a bed of shredded and detinned cans onto which the copper precipitates. The spent solution flows out of the reactor to be used again as a leaching solution. To recover the cement copper, the solution flow is stopped and the cans are washed with water, freeing the copper; the copper flows through the perforated screen and out into the decant basin. Later the copper is dried on the drying

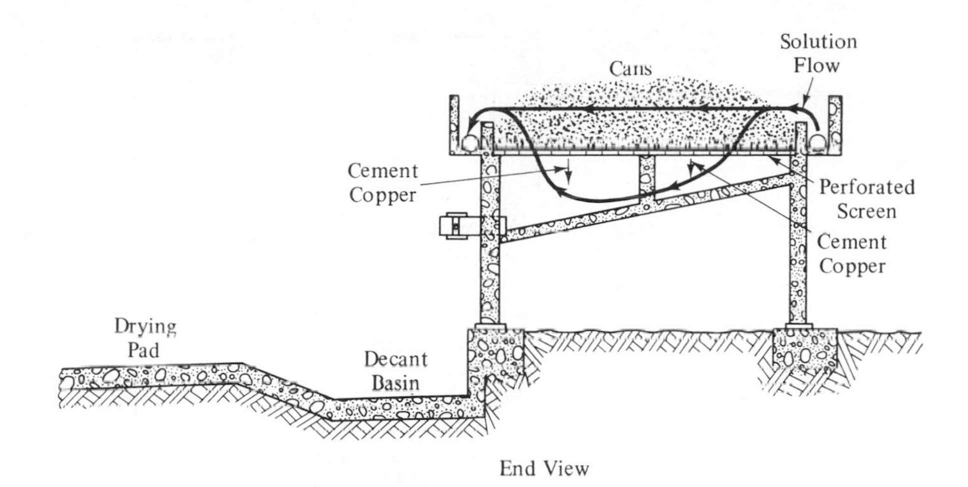

Figure 2.3–2. Typical Design of Gravity Launder–Precipitation Plant

pad. Up to 50 of these reactors might be used, each reactor being 50 ft long, 4 ft wide and 4 ft high.

The reactors in which the leaching operations take place can be quite large, for often the ore is placed in dumps over which the leach solution is sprayed. The Bingham Canyon dump of the Kennecott Copper Corporation in Utah contains 4,000,000,000 tons of ore. The dump is 1200 ft deep and has a surface area of 31,000,000 ft^2*. These operations are far from the beaker and test tube of the laboratory.*

Figure 2.3–3 is the flow sheet for the commercial use of these reaction paths, which free copper from its low-grade ore and produce the pure metal. The chemistry in the field is identical to the chemistry in the laboratory, but the materials-handling and engineering details are markedly different. *Figure 2.3–3 appears on page 38.*

2.4 EARLY SCREENING OF REACTION PATHS

Economics plays a strong role in process synthesis, since the dollar is a convenient common denominator against which various factors may be evaluated. In profit-making activities the economic incentives are obvious; they are equally important in not-for-profit activities, such as pollution abatement. Resources are not to be squandered on processing that does not upgrade the resources, or squandered on processing that could be performed more economically some other way. The score is kept in dollars.

Unfortunately, processing costs include factors that are not known until the process is nearly completely synthesized. How much steam, electricity, cooling water, refrigeration, and other utilities will be consumed by the full-scale process? How much capital must be invested in the process equipment, and how much labor will

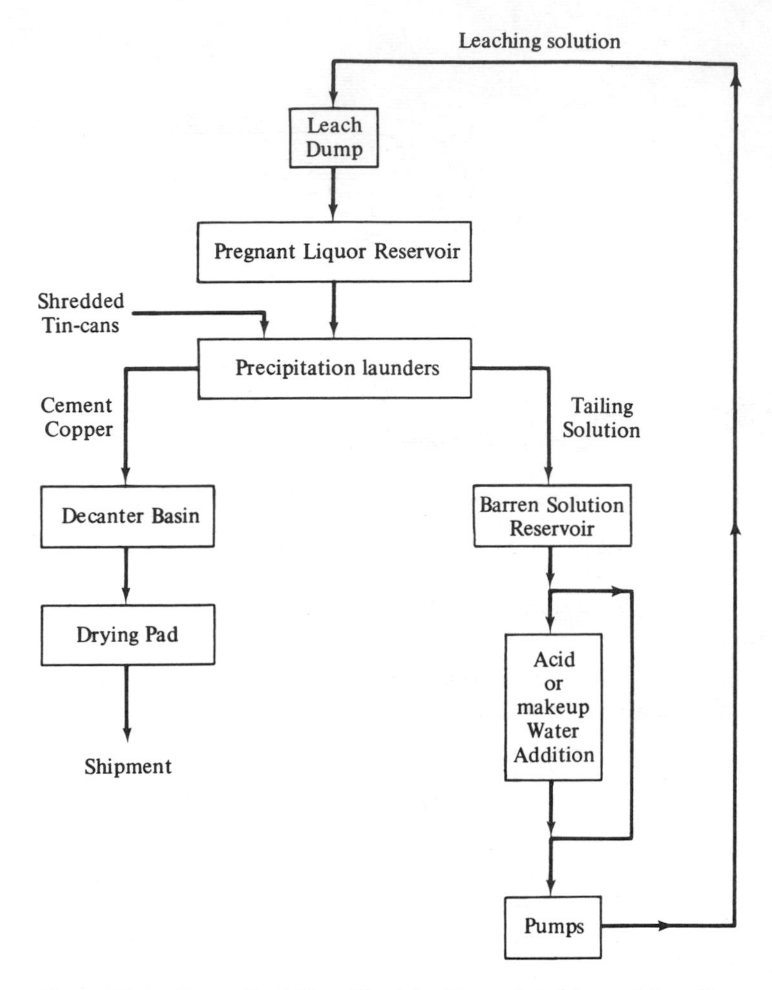

Figure 2.3–3. Generalized Flow Sheet for Dump Leaching and Iron Precipitation

be required to operate it? Are storage tanks, loading docks, and a railroad spur going to be required to supply the process? Is fire protection an important factor? What about the other cost factors that appear in Table 2.4–1?

The engineer seeks to discover the process which has a large venture profit, that is, a process which is efficient, as measured by Equation 2.4–1.

$$\begin{aligned}\begin{pmatrix}\text{annual}\\\text{venture}\\\text{profit}\end{pmatrix} &= \begin{pmatrix}\text{annual value}\\\text{of products}\\\text{manufactured}\end{pmatrix} - \begin{pmatrix}\text{annual costs}\\\text{of}\\\text{raw materials}\end{pmatrix}\\ &\quad - \begin{pmatrix}\text{annual costs of}\\\text{equipment amortization, maintenance, and}\\\text{repair; steam, cooling water, and other}\\\text{utilities; labor, supervision, and other}\\\text{factors}\end{pmatrix}\end{aligned} \qquad 2.4\text{–}1$$

TABLE 2.4–1 Costs Involved in Manufacturing Operations

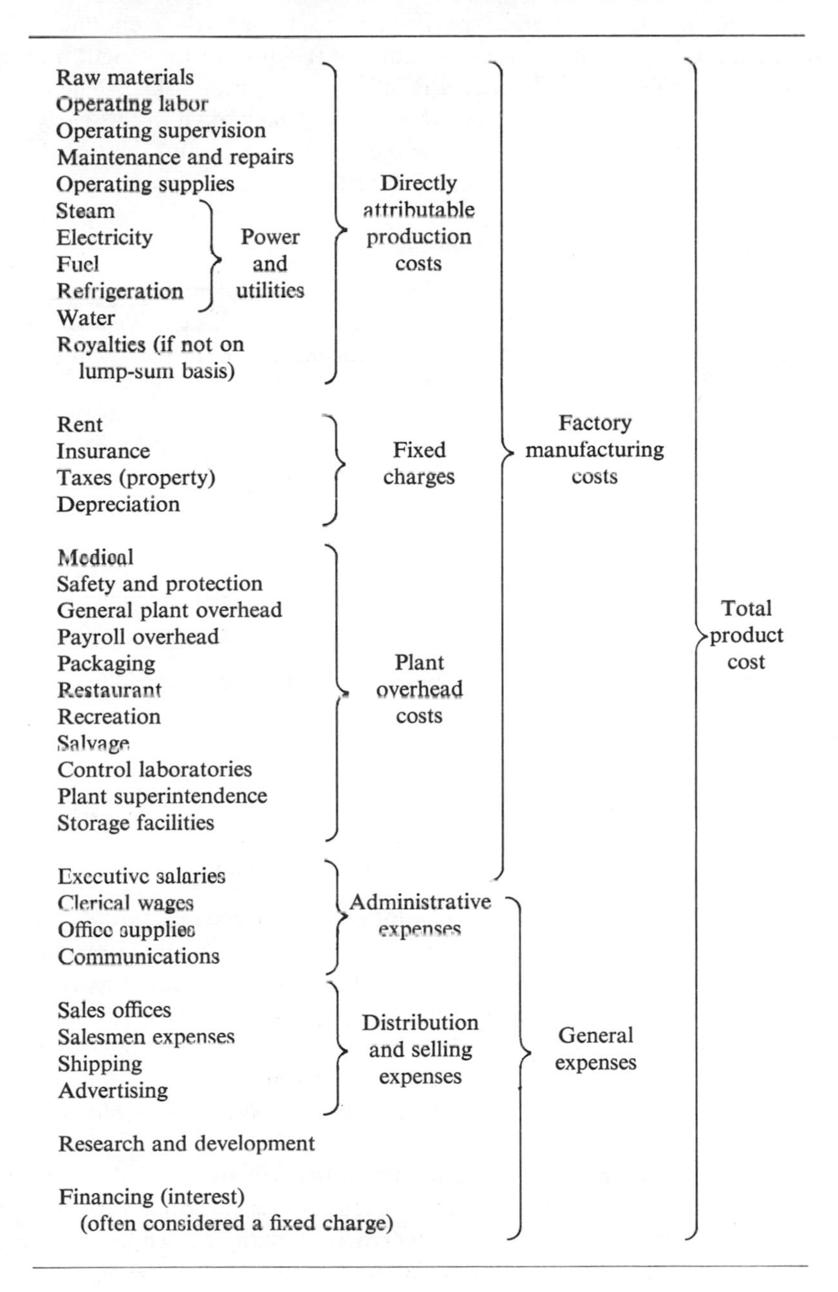

Cost item	Subgroup	Category	Group	Total
Raw materials		Directly attributable production costs	Factory manufacturing costs	Total product cost
Operating labor				
Operating supervision				
Maintenance and repairs				
Operating supplies				
Steam	Power and utilities			
Electricity				
Fuel				
Refrigeration				
Water				
Royalties (if not on lump-sum basis)				
Rent		Fixed charges		
Insurance				
Taxes (property)				
Depreciation				
Medical		Plant overhead costs		
Safety and protection				
General plant overhead				
Payroll overhead				
Packaging				
Restaurant				
Recreation				
Salvage				
Control laboratories				
Plant superintendence				
Storage facilities				
Executive salaries		Administrative expenses	General expenses	
Clerical wages				
Office supplies				
Communications				
Sales offices		Distribution and selling expenses		
Salesmen expenses				
Shipping				
Advertising				
Research and development				
Financing (interest) (often considered a fixed charge)				

The first two terms, the difference between the value of the products and the cost of the raw materials, is a *gross profit*, particularly useful in early economic screening. As defined here, this gross profit is an immutable function of the reaction path, and can be estimated quite accurately once a reaction path is established. The remaining costs depend on the engineering detail, which cannot be known in this early stage of development. At this stage we can only use the first two terms in Equation 2.4–1, and this requires an accurate estimate of material production and consumption.

Production and Consumption Analysis

We now present an orderly procedure for determining the stoichiometry, or production and consumption of materials in the chemistry of processing. We start with the example of producing vinyl chloride from chlorine and ethylene by the reactions shown.

Reaction 1:

$$\underset{\text{(ethylene)}}{C_2H_4} + \underset{\text{(chlorine)}}{Cl_2} \rightarrow \underset{\text{(dichloroethane)}}{C_2H_4Cl_2}$$

Reaction 2:

$$\underset{\text{(dichloroethane)}}{C_2H_4Cl_2} \rightarrow \underset{\text{(vinyl chloride)}}{C_2H_3Cl} + \underset{\text{(hydrogen chloride)}}{HCl}$$

The use of reactions 1 and 2 generates by-product hydrogen chloride as well as the desired vinyl chloride. If there is no convenient outlet for hydrogen chloride, the following reaction is useful.

Reaction 3:

$$\underset{\text{(hydrogen chloride)}}{2\,HCl} + \underset{\text{(oxygen)}}{\tfrac{1}{2}\,O_2} \rightarrow \underset{\text{(chlorine)}}{Cl_2} + \underset{\text{(water)}}{H_2O}$$

To analyze the production and consumption of materials by this chemistry, it is convenient to form Table 2.4–2. Each line corresponds to a single reaction of the proposed scheme. The entry appearing in the table is proportional to the stoichiometric coefficient in the balanced reaction, being positive for products and negative for reactants. All the entries in any line in the table can therefore be multiplied by a constant and still represent the reaction stoichiometry. Notice that if reactions 1 and 2 are both doubled, there would be no net production of dichloroethane and hydrogen chloride.

By adjusting each reaction to minimize intermediates and by algebraically adding the production and consumption of each species, we obtain on the last line of the table the net stoichiometry of the reaction path, shown in Figure 2.4–1. Coming into the process are two molecular units of ethylene, one of chlorine, and one half of oxygen. Leaving are two units of vinyl chloride and one of water. Here we have the data needed to assess the reaction path, using the first two terms of Equation 2.4–1.

TABLE 2.4–2 Production–Consumption Analysis

Reaction	Ethylene	Chlorine	Dichloro-ethane	Vinyl chloride	Hydrogen chloride	Oxygen	Water
1	−1(2)	−1(2)	+1(2)				
2			−1(2)	+1(2)	+1(2)		
3		+1			−2	$-\frac{1}{2}$	+1
Net	−2	−1	0	+2	0	$-\frac{1}{2}$	+1

\+ Production.
− Consumption.
( ) Adjustment of reaction amounts by this multiplying factor.

Typical of the several sources of cost data is the *Chemical Marketing Newspaper* (formerly called the *Oil, Paint and Drug Reporter*). In the July 26, 1971, issue we find these data:

Species	$/lb
Ethylene	0.03
Chlorine	0.04
Vinyl chloride	0.05
Oxygen	Use air

The reactions occur according to molecular ratios rather than weight ratios, and for this reason it is convenient to express production and consumption in terms of

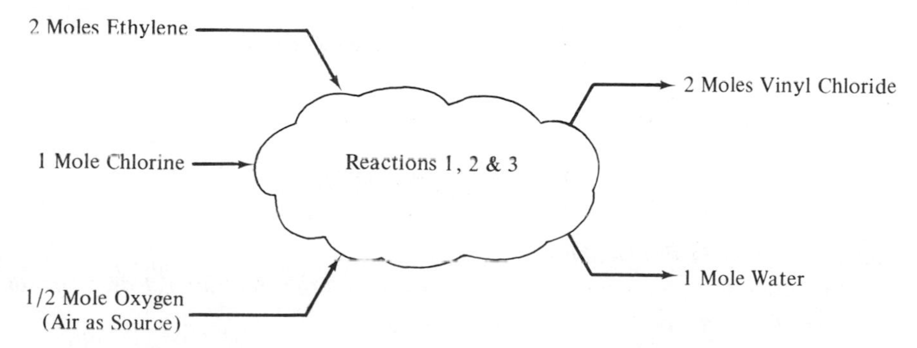

Figure 2.4–1. Net Production and Consumption of Materials for One Path to Vinyl Chloride

pound-moles, the weight of a species in pounds numerically equal to its molecular weight.

SPECIES	MOLECULAR WEIGHT	×	$/LB	=	$/LB-MOLE
Ethylene	28		0.03		0.84
Chlorine	70		0.04		2.80
Vinyl chloride	62		0.05		3.10

Now, using Figure 2.4–1, we see that for every mole of vinyl chloride produced, 1 mole of ethylene and $\frac{1}{2}$ mole of chlorine are consumed. The gross profit for this reaction path is

$$3.10 - 0.84 - \frac{2.80}{2} = \$0.86/\text{lb-mole vinyl chloride}$$

We conclude that this reaction path has the *potential* for commercialization, since the products are more valuable than the raw materials.

Example 2.4–1 The Solvay Process. *It is unfortunate that the following reaction does not occur directly, for enormous amounts of soda ash are required in the manufacture of glass, paper, and other products.*

$$\underset{\text{(limestone)}}{CaCO_3} + \underset{\text{(salt)}}{2\,NaCl} \not\rightleftarrows \underset{\text{(soda ash)}}{Na_2CO_3} + \underset{\text{(calcium chloride)}}{CaCl_2}$$

Ernest Solvay devised the following circuitous reaction path to drive this otherwise unwilling reaction.

Reaction 1. Limestone Calcination

At 1000°C limestone decomposes into lime and carbon dioxide:

$$\underset{\text{(limestone)}}{CaCO_3} \rightarrow \underset{\text{(lime)}}{CaO} + \underset{\text{(carbon dioxide)}}{CO_2}$$

Reaction 2. Lime Slaking

The addition of water to lime produces a solution called milk of lime:

$$\underset{\text{(lime)}}{CaO} + H_2O \rightarrow \underset{\text{(milk of lime)}}{Ca(OH)_2}$$

Reaction 3. Ammonia Production

The addition of milk of lime to an ammonium chloride solution releases ammonia and produces calcium chloride:

$$Ca(OH)_2 + \underset{\text{(ammonium chloride)}}{2\,NH_4Cl} \rightarrow \underset{\text{(ammonia)}}{2\,NH_3} + \underset{\text{(calcium chloride)}}{CaCl_2} + 2\,H_2O$$

Reaction 4. Ammonia Absorption

The absorption of ammonia in water generates an ammonium hydroxide solution:

$$NH_3 + H_2O \rightarrow \underset{\text{(ammonium hydroxide)}}{NH_4OH}$$

Reaction 5. Bicarbonate Production

Ammonium hydroxide, salt, and carbon dioxide react in solution to produce ammonium chloride and sodium bicarbonate:

$$NH_4OH + CO_2 + NaCl \rightarrow NH_4Cl + \underset{\text{(sodium bicarbonate)}}{NaHCO_3}$$

Reaction 6. Bicarbonate Decomposition

Calcination of sodium bicarbonate produces soda ash and carbon dioxide:

$$2\,NaHCO_3 \rightarrow \underset{\text{(soda ash)}}{Na_2CO_3} + CO_2 + H_2O$$

Table 2.4–3 outlines the adjustment of the reaction amounts, showing that the Solvay process, when in reaction balance, consumes 1 mole of limestone and 2 moles of salt to produce 1 mole of soda ash and 1 mole of calcium chloride. The other materials used for processing merely serve to lubricate the chemical reaction paths, causing the reactants to slide easily into the products. The Solvay process was a major discovery in the history of processing.

TABLE 2.4–3 Production–Consumption Analysis for Solvay Reaction Path

Reaction	Species											
	$CaCO_3$	CaO	CO_2	H_2O	$Ca(OH)_2$	NH_4Cl	NH_3	$CaCl_2$	NH_4OH	$NaCl$	Na_2CO_3	$NaHCO_3$
1	−1	+1	+1									
2		−1		−1	+1							
3				+2	−1	−2	+2	+1				
Double 4				−2			−2		+2			
Double 5			−2			+2			−2	−2		+2
6			+1	+1							+1	−2
Net	−1	0	0	0	0	0	0	+1	0	−2	+1	0

Net reaction executed by the process of Figure 2.4–2:

$$CaCo_3 + 2\,NaCl \rightsquigarrow Na_2CO_3 + CaCl_2$$

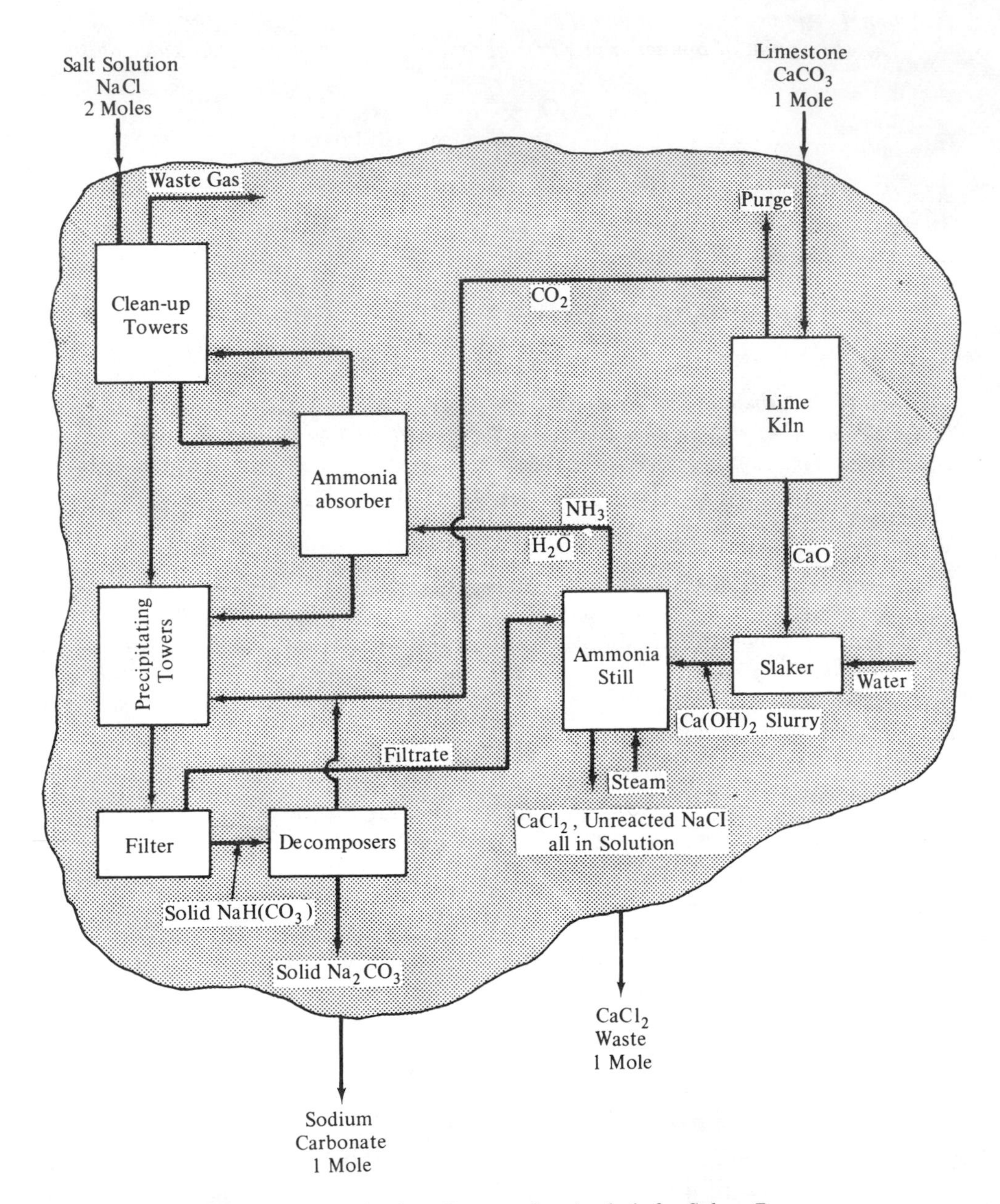

Figure 2.4–2. Production–Consumption Analysis for Solvay Process Internal detail dependent on creativity of engineer. Material crossing boundary is an immutable function of the reaction path.

We now look at the economics by assuming that the market value of sodium carbonate is $40/ton, the cost of locally available limestone is $15/ton, and of salt $20/ton. Since the production of 1 ton-mole of sodium carbonate requires the consumption of 1 ton-mole of limestone and 2 of salt, an economic consumption–production analysis yields the following. We assume the wastes are of no value.

Production of 1 Ton-Mole of Soda Ash

SPECIES	TON-MOLES		MOLECULAR WEIGHT		TONS	$/TON	VALUE ($)
Na_2CO_3	1	×	106	=	106	40	4240
$CaCO_3$ as limestone	1	×	100	=	100	15	1500
NaCl as rock salt	2	×	58	=	116	20	2320

Gross profit/ton of sodium carbonate produced

$$= \frac{(4240 - 1500 - 2320)}{106} = \$4/\text{ton}$$

We conclude that the Solvay process has the chance of life in this economic situation as long as the other production costs are sufficiently less than the gross profit of $4/ton. A market for the by-product calcium chloride would make the economics even more attractive.

2.5 REACTION PATHS WITH RECYCLE

Many reactions do not proceed to completion. In such cases, the reactants will also appear among the products. These species will have headings in two columns of the product–consumption analysis table.

Sometimes the unconverted reactants can be simply disposed of as waste. Greater economies may be achieved, however, if part or all of the unconverted reactants are recovered from the reactor effluent and returned, or *recycled*, to the reactor feed for another try at conversion.

The effect of proposed recycle should be included in the production–consumption analysis. Recycling of unconverted reactants results in a higher net process product-to-reactant ratio, or *yield*. In other words, for the same amount of desired product, less original feed is required if unconverted reactants are recycled than if they are not. In the production–consumption table, therefore, the entry for unconverted reactant in the reactor effluent should be reduced by the amount that will be recycled. Similarly, the stoichiometric coefficient of the original feed of the unconverted reactant should also be reduced by the amount that will be recycled.

Example 2.5–1 Isopropyl Acetate Production. *Isopropyl acetate, a common industrial solvent and extraction agent, can be produced by an esterification reaction between isopropyl alcohol and acetic acid. This, like most esterification reactions, is reversible and proceeds to only 60 per cent conversion:*

$$\underset{\text{(isopropyl alcohol)}}{C_3H_7OH} + \underset{\text{(acetic acid)}}{CH_3COOH} \rightleftharpoons \underset{\text{(isopropyl acetate)}}{C_3H_7OOCCH_3} + \underset{\text{(water)}}{H_2O}$$

Can isopropyl acetate be made profitably by this process if isopropyl alcohol costs \$0.07/lb, acetic acid costs \$0.09/lb, and isopropyl acetate can be sold for \$0.12/lb?

isopropyl alcohol	\$0.07/lb =	\$4.20/lb-mole
acetic acid	\$0.09/lb =	\$5.40/lb-mole
isopropyl acetate	\$0.12/lb =	\$12.25/lb-mole

Table 2.5–1 shows the production–consumption analysis for this process as written.

TABLE 2.5–1 Production–Consumption Analysis for Isopropyl Acetate Path

	Species					
	Alcohol	Acid	Acetate	H_2O*	Alcohol*	Acid*
Reaction	−1	−1	0.60	0.60	0.40	0.40
Net	−1	−1	0.60	0.60	0.40	0.40

*Waste

Because of the incomplete conversion, the reaction could have been written

$$\text{alcohol} + \text{acid} \rightarrow 0.6 \text{ acetate} + 0.6 \text{ water} + 0.4 \text{ alcohol} + 0.4 \text{ acid}$$

Therefore, the alcohol and acid appear twice in the table as both feeds and products of the same reaction. If the H_2O, alcohol, and acid in the product are worth nothing, the gross profit for this reaction path is

$$0.6(12.25) - 1.0(4.20) - 1.0(5.40) = -\$1.25/\text{mole isopropyl acetate}$$

We conclude that this scheme is not economically feasible.

However, assume that 80 per cent of the alcohol and acid in the reactor effluent can be recovered and recycled. Table 2.5–2 shows this new production–consumption analysis for this process with recycle. The amount of alcohol and acid in both the product and feed have been reduced by the amount recycled, 80 per cent of 0.40 or 0.38 lb-mole. Again, assume that the unrecovered unconverted reactants have no value; then disregarding the expense of recovering the recycled material, the gross profit for the reaction path with recycle is now

$$0.6(12.25) - 0.68(4.20) - 0.68(5.40) = \$0.63/\text{mole isopropyl acetate}$$

Therefore, the reaction path with recycle has potential for commercialization.

TABLE 2.5-2 Production–Consumption Analysis for Isopropyl Acetate Process with Recycle

	Species					
	Alcohol	Acid	Acetate	Water	Alcohol	Acid
Reaction	−1.00	−1.00	0.60	0.60	0.40	0.40
Recycled	0.32	0.32	0	0	−0.32	−0.32
Net	−0.68	−0.68	0.60	0.60	0.08	0.08

2.6 CONCLUSIONS

All other things being equal, the reaction path that has the largest difference between the value of the products and the cost of the raw materials is the most promising. Unfortunately, all things are rarely equal. The screening criterion is useful to eliminate uneconomic processes, but passing this test is not sufficient to ensure economic viability. Furthermore, many profitable processes exist that cannot be justified on published cost data, because of, for example, internal cheap sources of otherwise expensive raw materials. For these reasons, early economic analyses can provide only an approximate picture of the real process economics.

More questions have been raised in this chapter than conclusions reached, and this is natural when dealing with the interface between two strong fields, the sciences and engineering. Here we have been fact gathering and hunting clues to provide for a smooth transition of responsibility from the chemist to the process engineer. The differences between the laboratory and the commercial scale are not only those of volume, but often of processing concept. Procedures efficient and useful in the laboratory can be disastrous in the large scale, both physically and economically.

A major difference between the chemist and the engineer is their attitude toward solvents. The chemist likes solvents and uses them frequently, and the engineer finds solvents distasteful. Difficulties in handling and reacting materials are suppressed in the laboratory by working with dilute solutions. The solvent can be easily boiled off when the pure material is required, if only a quart or so of the solution is being handled. On the other hand, on the large scale it is very desirable to deal with pure materials, and to carry along thousands of tons of solvents merely to duplicate laboratory procedures can be economically disastrous.

A second difference is that in the laboratory nearly all procedures are batch, the material being kept together in a vessel until the processing is complete and then moved as a batch to the next processing step. On the commercial scale there are often advantages to operating continuously, with the materials passing as a continuous stream in pipes from one operation to the next. Often the processing is not completed during one pass through the operation, and the recycle of unreacted materials is necessary. Recycle can cause the buildup of trace materials, and for this reason the engineer is much more concerned than the chemist with the effects of small amounts of impurities on the reactions. Effects that are unimportant in batch operations can overwhelm continuous recycle operations.

The chemist likes to operate near ambient conditions, with only minor excursions in temperature and pressure. But for the engineer a wide range of temperatures and pressures is easily obtained. In many large-scale processes, processing occurs within thick steel vessels at temperatures and pressures well beyond those found in the laboratory.

The chemist generally disregards the cost of utilities, since he uses them in such small amounts. However, the economics of processing dictate that the engineer must minimize electricity, cooling water, steam, gas, solvents, and so forth.

The chemist can suppress hazards, when handling ounces of material, by dilution in solvents and by venting in hoods materials that could lead to disaster. The engineer must face these hazards undiluted when handling tons of material. This causes a decided change in procedures.

The questions raised in this chapter serve the important role of easing the transition from the laboratory to the commercial process. Often the chemist does not recognize his own reaction path after the engineer has altered it to be useful on the full scale.

REFERENCES

Little has been written on the art of discovering useful reaction paths. Only recently has the strategy of molecule synthesis received the attention needed, and these studies are now opening an important research area.

On the art of molecule synthesis:

N. Anand, J. S. Bindra, and S. Ranganathan, *The Art in Organic Synthesis*, Holden-Day, Inc., San Francisco, 1970.

On computer-aided molecule synthesis:

E. J. Corey and D. J. Wipke, "Computer-Aided Design of Complex Organic Synthesis," *Science*, **166**, Oct. 10, 1969.

E. J. Corey, "Computer-Assisted Analysis of Complex Synthetic Problems," *Quarterly Reviews*, **25**, 1971.

An enormous amount of literature is available on individual reactions. See, for example:

C. A. Jacobson, ed., *Encyclopedia of Chemical Reactions*, Vols. I–VIII, Van Nostrand Reinhold Publishing Company, New York, 1946.

Information on industrial chemistry is found in many sources, including:

R. N. Schreve, *Chemical Process Industries*, McGraw-Hill Book Company, New York, 1967.

H. P. Meissner, *Processes and Systems in Industrial Chemistry*, Prentice-Hall, Inc., Englewood Cliffs, N.J., 1971.

See also:

H. Kölbel and J. Schulze, "Influence of Chemical Innovations on Future Developments," *Technological Forecasting and Social Change*, **2**, 195–200, 1970.

PROBLEMS

1. **WASTE RECYCLE FROM PICKLING LIQUOR.** Iron oxide scale, which forms on the surface of steel during manufacture, is removed by pickling. The steel surface is washed with sulfuric acid to dissolve the scale, following the reaction

$$FeO + H_2SO_4 \rightarrow FeSO_4 + H_2O$$

(scale) (sulfuric acid) (ferrous sulfate) (water)

For most small steel plants the recovery of by-products from the waste pickling liquor is not economic, and therefore they neutralize the liquor with lime and dispose of it.

However, the Blaw–Knox–Ruther process for the recovery of sulfuric acid involves the concentration of the waste liquor by evaporation before discharging it into a reactor, where anhydrous hydrogen chloride is bubbled through it, reacting with the ferrous sulfate to produce H_2SO_4 and $FeCl_2$.

$$FeSO_4 + 2\,HCl \rightarrow FeCl_2 + H_2SO_4$$

(ferrous sulfate) (hydrogen chloride) (ferrous chloride) (sulfuric acid)

The ferrous chloride is separated from the sulfuric acid (which is returned to the steel pickling) and is converted to iron oxide in a direct fired roaster. This liberates HCl, which is recovered by scrubbing and stripping and returned to the reactor.

$$FeCl_2 + H_2O \rightarrow 2\,HCl + FeO$$

(ferrous chloride) (hydrogen chloride) (iron oxide)

Determine the kind and amount of raw materials and by-products the steel manufacturer must contend with if this process is used to handle the waste pickling liquor.

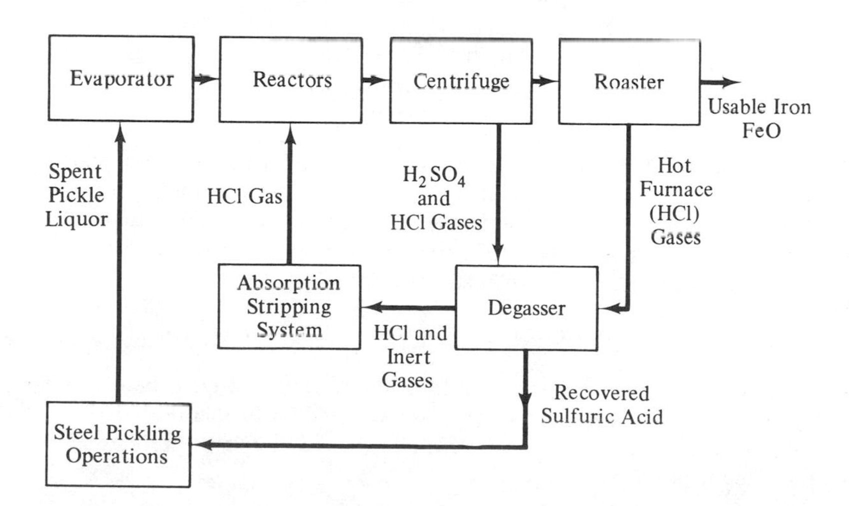

2. **ACRYLONITRILE FROM PROPYLENE.** In March 1957, James D. Idol fed propylene, ammonia, and air into a miniature fluidized bed reactor 1 ft high and 2 ft in diameter, filled with particles of a bismuth phosphomolybdate catalyst. Acrylonitrile appeared in the effluent:

$$\underset{\text{(propylene)}}{C_3H_6} + \underset{\text{(ammonia)}}{NH_3} + \underset{\text{(oxygen)}}{\tfrac{3}{2}\,O_2} \rightarrow \underset{\text{(acrylonitrile)}}{C_3H_3N} + \underset{\text{(water)}}{3\,H_2O}$$

Acrylonitrile is the basis of orlon fiber, synthetic rubbers, and a variety of other polymeric materials, and in 1957 the U.S. consumption was some 160 million lb/year (in 1971 4 billion lb/year). The alternative commercial reaction paths were, at that time,

$$\underset{\text{(acetylene)}}{C_2H_2} + \underset{\text{(hydrogen cyanide)}}{HCN} \rightarrow \underset{\text{(acrylonitrile)}}{C_3H_3N}$$

and

$$\underset{\text{(ethylene oxide)}}{C_2H_4O} + \underset{\text{(hydrogen cyanide)}}{HCN} \rightarrow \underset{\text{(acrylonitrile)}}{C_3H_3N} + \underset{\text{(water)}}{H_2O}$$

Chemical and Engineering News, July 5, 1971, cited Idol with the Chemical Innovator Award, stating that he set the world of nitrile chemistry on fire. Given these costs, indicate the effect Idol's reaction-path discovery has had on the economics of acrylonitrile manufacture.

MATERIAL	1971 VALUE ($/LB)
Acrylonitrile	0.14
Propylene	0.02
Ammonia	0.03
Acetylene	0.20
Ethylene oxide	0.07
Hydrogen cyanide	0.11

3. **PURE TITANIUM BY CHLORINATION.** The chlorides of some metals are volatile. This fact forms the basis for a number of processes to recover metals from their ores.

In the process shown below, rutile, a titanium ore, is bricketted with coke and directly chlorinated in a fluidized bed reactor. In the reactor titanium and other metals are converted to their chlorides, which are volatile.

$$TiO_2 + 2\,Cl_2 + C \rightarrow TiCl_4 + CO_2$$

The volatile material is removed from the reactor and quenched with cold, liquid $TiCl_4$. A slurry of solid $FeCl_3$ and $TiCl_4$ is recovered for further purification. Determine the tons of coke and of chlorine needed to process 1 ton of rutile, and the tons of titanium tetrachloride produced.

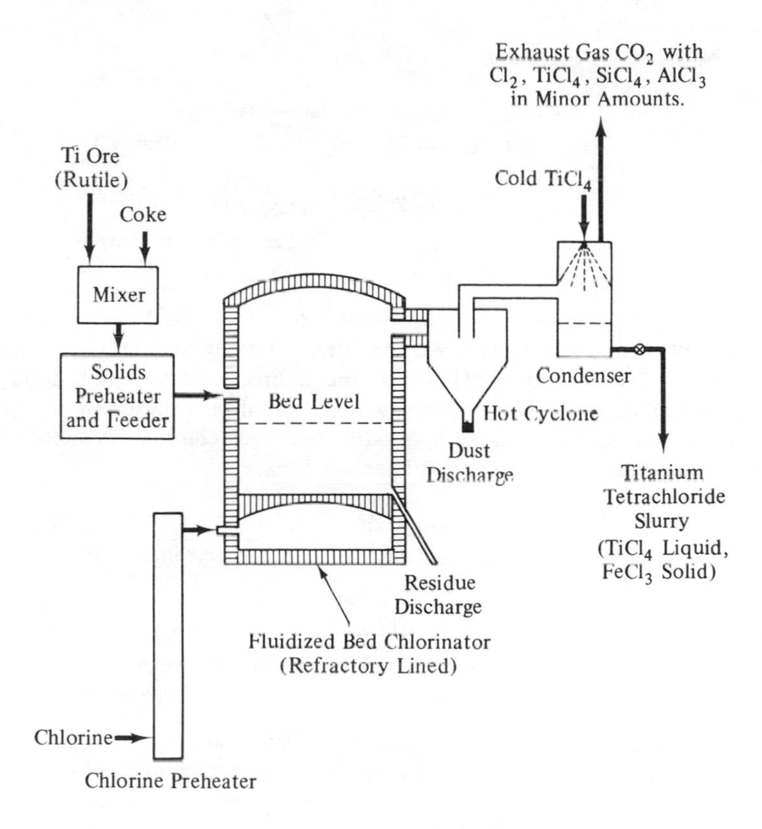

Rutile Composition

SPECIES	MASS FRACTION
TiO_2	0.90
Other metals	0.02
Residue	0.08

4. **VINYL CHLORIDE SYNTHESIS.** Poly vinyl chloride (PVC) is used extensively in the manufacture of plastic films, pipes, insulation, and so forth. This polymer is formed by the catalytic polymerization of vinyl chloride, which in turn can be produced from ethylene, acetylene, chlorine, and/or hydrogen chloride by a variety of reactions. We examine the means for producing the vinyl chloride by the *direct reaction path*,

$$\underset{\text{(acetylene)}}{C_2H_2} + \underset{\text{(hydrogen chloride)}}{HCl} \xrightarrow{\text{catalyst}} \underset{\text{(vinyl chloride)}}{C_2H_3Cl}$$

and by the *indirect reaction path*,

$$\underset{\text{(ethylene)}}{C_2H_4} + \underset{\text{(chlorine)}}{Cl_2} \longrightarrow \underset{\text{(dichloroethane)}}{C_2H_4Cl_2}$$

$$\underset{\text{(dichloroethane)}}{C_2H_4Cl_2} \xrightarrow{\text{catalyst}} \underset{\text{(vinyl chloride)}}{C_2H_3Cl} + \underset{\text{(hydrogen chloride)}}{HCl}$$

The direct reaction path from acetylene is less desirable than the indirect reaction path from ethylene, involving two reaction steps, primarily since ethylene is cheaper than acetylene and less hazardous. However, the indirect reaction path produces a by-product of hydrogen chloride, which must be utilized or disposed of in some manner. In the following reaction path the by-product hydrogen chloride is reacted with oxygen and ethylene to form more dichloroethane.

$$\underset{\text{(hydrogen chloride)}}{2\,HCl} + \underset{\text{(oxygen)}}{\tfrac{1}{2}\,O_2} + \underset{\text{(ethylene)}}{C_2H_4} \rightarrow \underset{\text{(dichloroethane)}}{C_2H_4Cl_2} + \underset{\text{(water)}}{H_2O}$$

$$\underset{\text{(dichloroethane)}}{C_2H_4Cl_2} \rightarrow \underset{\text{(vinyl chloride)}}{C_2H_3Cl} + \underset{\text{(hydrogen chloride)}}{HCl}$$

The only by-product is water. However, hydrogen chloride must be purchased, since an insufficient amount is available from the dichloroethane decomposition.

In this problem statement there are four basic chemical reactions that can be mixed and matched to form reaction paths.

(a) Determine a reaction path for the production of two molecules of vinyl chloride that requires two molecules of ethylene, one molecule of Cl_2, and oxygen. Also, this reaction path must not produce any by-products other than water.

(b) Determine a reaction path for the production of two molecules of vinyl chloride that requires one molecule of ethylene, one molecule of acetylene, and one molecule of Cl_2. No by-products are allowed in this reaction path.

(c) Screen these new reaction paths and the original ones to see if the economics make sense.

SPECIES	MARKET VALUE ($/LB)
Ethylene	0.03
Acetylene	0.20
Chlorine	0.04
Hydrogen chloride	0.40*
Vinyl chloride	0.05

* Presume this to be the purchase price and that it has no sales value.

5. **RESHAPING THE CHLORINE INDUSTRY.** In the early 1970s nearly 10 million tons/year of chlorine were produced, and over 90 per cent of the production was via the chlorine–caustic process. In this process salt is converted to chlorine and caustic in electrolytic cells by the following net reaction:

$$\underset{\text{(sodium chloride)}}{2\,NaCl} + \underset{\text{(water)}}{2\,H_2O} \xrightarrow{\text{electrolysis}} \underset{\text{(chlorine)}}{Cl_2} + \underset{\text{(hydrogen)}}{H_2} + \underset{\text{(sodium hydroxide)}}{2\,NaOH}$$

For each molecule of chlorine produced, two molecules of caustic or sodium hydroxide are produced. This close tie between the chlorine and the caustic industry reached crisis proportions as the demand for chlorine greatly outstripped the caustic market. It is becoming more and more difficult to dispose of the by-product caustic.

In the May 6, 1969, *Chemical and Engineering News* the Kel-Chlor process was announced, which is an improvement of the older Deacon process based on the following reaction path:

$$\underset{\text{(hydrogen chloride)}}{2\,HCl} + \underset{\text{(oxygen)}}{\tfrac{1}{2}O_2} \rightleftharpoons \underset{\text{(chlorine)}}{Cl_2} + \underset{\text{(water)}}{H_2O}$$

The use of this path would free the chlorine industry of the glut and nuisance levels of by-product caustic, and would also serve as an outlet for by-product HCl formed by many chlorination processes.

Determine the reaction paths and by-products associated with the following sources of hydrogen chloride.

(a) Magnesium chloride, which can be hydrolyzed by very hot steam to give magnesium oxide and hydrochloric acid.

(b) Sodium chloride, which can be, and has for years been, converted to hydrogen chloride by the action of sulfuric acid.

(c) Potassium chloride, which, as a fertilizer, can be converted to more useful potassium phosphate or nitrate to give by-product hydrogen chloride.

6. **WASTE RECYCLE IN SOLVAY PROCESS.** What net production and consumption of materials would occur if the ammonia were recovered in the Solvay process of Example 2.4–1 by drying the waste solution of reaction 3 to get NH_4Cl, heating to initiate this reaction,

$$\underset{\text{(ammonium chloride)}}{NH_4Cl} \xrightarrow{\text{heat}} \underset{\text{(ammonia)}}{NH_3} + \underset{\text{(hydrogen chloride)}}{HCl}$$

and by applying the Kel–Chlor process of Problem 5 to hydrogen chloride? Chlorine is worth \$80/ton. How does this influence the economics of soda ash production? What other by-products could be sold?

7. **SODIUM BICARBONATE BY SOLVAY CHEMISTRY.** It is proposed to produce sodium bicarbonate from limestone and salt using the technology of the Solvay process of Example 2.4–1. This involves the removal of the bicarbonate kiln. How would this modification influence the materials production and consumption? In practice, the

calcining of limestone requires the burning of 1 ton of coal per 10 tons of limestone to provide heat energy. How might this fact influence the bicarbonate process?

8. **SUPERPHOSPHATE FERTILIZER.** Phosphate rock, even when finely ground, has a limited use as a fertilizer, since it is essentially fluorapatite, $(CaF)Ca_4(PO_4)_3$, which is extremely insoluble. Superphosphate fertilizer is formed by acidifying phosphate rock.

$$\underset{\text{(phosphate rock)}}{2[(CaF)Ca_4(PO_4)_3]} + \underset{\text{(sulfuric acid)}}{7\,H_2SO_4} + \underset{\text{(water)}}{3\,H_2O} \rightarrow \underset{\text{(monocalcium phosphate)}}{3\,CaH_4(PO_4)_2\,H_2O} + \underset{\text{(calcium sulfate)}}{7\,CaSO_4} + \underset{\text{(hydrogen fluoride)}}{2\,HF}$$

The soluble monocalcium phosphate is the chief ingredient in superphosphate fertilizer, and the calcium sulfate is a diluent.

Florida, Idaho, and Tennessee are the main producers of phosphate rock, the United States having over half the world's estimated supply of 30 billion tons. Much of the production goes into phosphate fertilizers. What is the gross profit per ton of superphosphate manufactured?

SPECIES	VALUE ($/TON)
Upgraded phosphate rock (largely fluorapatite)	16
Sulfuric acid	30
Superphosphate (run of the pile) (monocalcium phosphate and calcium sulfate)	40
Hydrogen fluoride (unrecovered)	

Answer: $29/ton.

9. **LEACHING VARIOUS COPPER ORES.** How many tons of shredded cans and of sulfuric acid are required to recover 1 ton of copper by the leaching process of Example 2.3–1 from ore deposits which are predominately azurite? malachite? crysocolla? Screen these ore deposits for economic feasibility.

SPECIES	MARKET PRICE ($/TON)
Cement copper (60% Cu)	100
Shredded and cleaned tin cans	20
Sulfuric acid	30

10. **CAUSTIC MANUFACTURE.** In Example 3.3–6 sodium carbonate and slaked lime are reacted to produce caustic and calcium carbonate. Are the reported feed and product flow rates consistent with the reaction stoichiometry? What is the gross profit from this process?

SPECIES	MARKET PRICE ($/TON)
Sodium carbonate	40
Slaked lime	18
Caustic	20
Calcium carbonate	18

Why would a process be operated that shows a negative gross profit based on costs reported in the literature?

11. LEBLANC PROCESS. The present process for the manufacture of soda ash is the Solvay process, Example 2.4–1. Before this method was developed, the Leblanc process was devised by Nicholas Leblanc in about 1773. The Leblanc process was the first large-scale industrial process.

The reactions involved are

$$\underset{\text{(salt)}}{2\,NaCl} + \underset{\text{(sulfuric acid)}}{H_2SO_4} \rightarrow \underset{\text{(sodium sulfate)}}{Na_2SO_4} + \underset{\text{(hydrogen chloride)}}{2\,HCl}$$

$$\underset{\text{(sodium sulfate)}}{Na_2SO_4} + \underset{\text{(carbon)}}{2\,C} \rightarrow \underset{\text{(sodium sulfide)}}{Na_2S} + \underset{\text{(carbon dioxide)}}{2\,CO_2}$$

$$\underset{\text{(sodium sulfide)}}{Na_2S} + \underset{\text{(calcium carbonate)}}{CaCO_3} \rightarrow \underset{\text{(soda ash)}}{Na_2CO_3} + \underset{\text{(calcium sulfide)}}{CaS}$$

The Solvay process had completely displaced the Leblanc process by 1915. Given the economic situation of these cost data, would you consider reviving some sort of Leblanc process?

SPECIES	LOCAL VALUE ($/TON)
Calcium sulfide	0
Salt	14
Sulfuric acid	30
Sodium sulfate	24
Hydrogen chloride	110
Carbon	10
Carbon dioxide	0
Calcium carbonate	12
Soda ash	16

12. LOWER BOUND ON SOAP COST. Soaps are the sodium or potassium salts of various fatty acids, chiefly oleic, stearic, palmitic, lauric, and myristic acids. Tallow is the principal fatty raw material of soap making, with greases, coconut oil, and palm oil other important raw materials.

In the hydrolyzer the raw fat or oil, here glyceryl stearate, is hydrolyzed and the fatty acid and glycerine by-product are separated:

$$\underset{\text{(glyceryl stearate)}}{(C_{17}H_{35}COO)_3C_3H_5} + 3\ H_2O \rightarrow \underset{\text{(stearic acid)}}{3\ C_{17}H_{35}COOH} + \underset{\text{(glycerine)}}{C_3H_5(OH)_3}$$

In the mixer–neutralizer the soap is salted out:

$$\underset{\text{(stearic acid)}}{C_{17}H_{35}COOH} + \underset{\text{(caustic soda)}}{NaOH} \rightarrow \underset{\text{(sodium stearate soap)}}{C_{17}H_{35}COONa} + H_2O$$

If glyceryl stearate costs \$0.20/lb and caustic soda \$0.10/lb, and if the by-product glycerine can be sold for \$0.22/lb, what is the lower bound on the sales price of sodium stearate soap? The lower bound is that price which gives a zero gross profit. The actual price must be significantly higher than the lower bound to cover expenses other than raw material costs and to account for a profit.

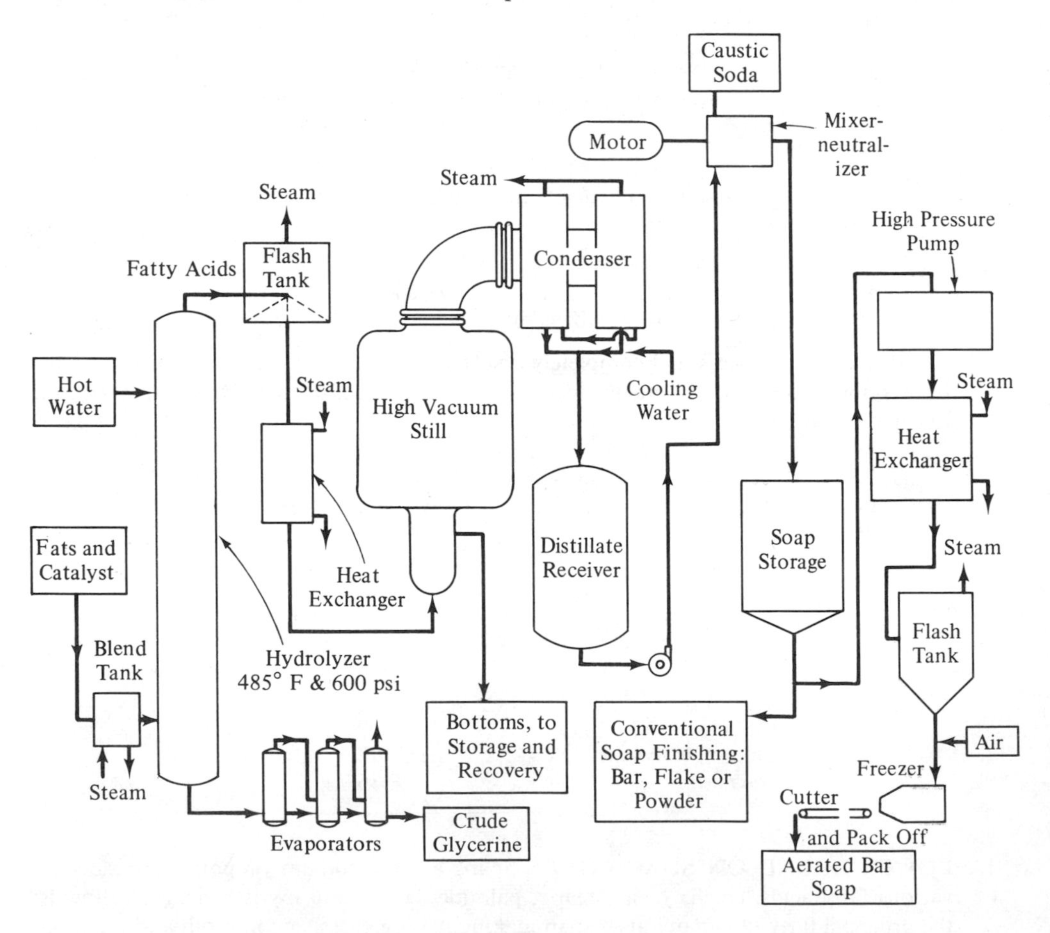

Continuous process for Fatty Acids and Soap

13. **RAW MATERIAL NEEDS FOR NITROGEN FIXATION.** The M. W. Kellogg scheme for fixing atmospheric nitrogen, shown in Figure 1.2–4, uses natural gas, air, and water. Determine the complete consumption and production of species for this scheme. If the carbon dioxide and ammonia can be used for urea synthesis, and if the value of urea and ammonia is $65 and $60/ton, respectively, what is the maximum price at which the natural gas can be purchased?

14. **FIRE-RETARDANT PRODUCTION.** In Section 2.3 were presented the laboratory procedures for the production of dichloroneopentyl glycol for use as a fire-retardant coating. During full-scale production the reaction conditions can be adjusted so that little of the monochloro and trichloro products are formed. However, by-product hydrogen chloride is formed. What materials would be produced and consumed if the Kel–Chlor process were used to eliminate the waste hydrogen chloride?

15. **ROUTES TO PHENOL.** Figure 2.2–2 shows five routes to phenol. Determine the complete raw-material needs and by-product production for each reaction path. Assuming that there is no market for reaction products other than phenol, determine those paths which are economically viable at this early stage of process development. The solution to this problem will require data on current chemical prices (see, for example, *The Chemical Marketing Newspaper*).

16. **UREA SYNTHESIS.** How costly might it be to produce urea fertilizer from limestone and ammonia using the reaction of Figure 1.2–3. Limestone, $CaCO_3$, costs $14/ton and ammonia $60/ton. Fertilizer-grade urea is sold for $65/ton. Why the difference from your estimate?

17. **AMMONIUM NITRATE PRODUCTION.** Figure 1.2–3 shows a reaction path between ammonia and ammonium nitrate. If ammonia is worth $60/ton and ammonium nitrate is worth $45/ton, would you consider the engineering of a process based on this reaction economically feasible?

3

Material Balancing and Species Allocation

All problem solvers guess, but the sophisticated and the unsophisticated guess somewhat differently. . . . As soon as we are seriously concerned with our problem, we anticipate an outline of its solution . . . . We do not look for just any kind of a solution, but for a certain kind, a kind within a limited outline.

George Polya,
Mathematical Discovery

3.1 Introduction

Having selected the several reaction paths that seem on the surface to have potential for the mass production of a material, attention shifts to a whole new set of problems. At this point the process exists in the engineer's mind only as a sequence of chemical reactions that must be performed at rather restrictive conditions of temperature, pressure, feed purity, and reactive environment, and defined to a lesser extent by the separation and purification procedures found useful under laboratory conditions. Much remains to be done before economic production can be accomplished on the large scale.

The first problem to which the engineer ought to pay attention is the routing or *allocation* of the various materials throughout the emerging process. For example, the species required as chemical reactants must be found somewhere, perhaps in a raw material source or in the effluent of a reactor. Similarly, the reaction products and unconverted reactants must be disposed of, say, by recycle to a reactor, as a final product from the process, or as a waste material to be disposed.

These important allocation problems are particularly interesting, because they alter the nature of the separation problems that must be solved later. Suppose that we are designing a process to treat industrial waste by using pure oxygen, and we plan to use air as our source of oxygen. If we allocate the oxygen in the air to the waste-treatment reactor and the nitrogen in the air is vented as processing waste, this allocation introduces the problem of devising a means of separating the oxygen from the nitrogen. Nearly every allocation we make defines problems in separating materials from mixtures of other materials. For this reason we must keep our eye on the separation problems that arise as we map out the flow of material which ought to exist in our process.

Separations, especially complete separations, can be extremely expensive. These operating costs must be deducted from the gross profit computed from the production-consumption analysis of Chapter 2. Thus difficult and costly separations can reverse the economic attractiveness of a previously accepted reaction path.

However, before we can make species-allocation decisions or determine the difficulty of separation problems, it is necessary to understand the material flow in a process. Such an understanding is obtained by means of the very important engineering analysis technique, *material balancing*, which is based on the law of conservation of mass. The early part of this chapter deals with material balancing, and later the principles of species allocation are presented.

3.2 CONSERVATION OF MASS

Except in situations involving nuclear reactions, atoms are neither created nor destroyed. Atoms that enter any region in a process must either accumulate in the region or must leave. This observation leads to the material balance, Equation 3.2–1, which is valid for all atomic species in any region of a process.

$$\begin{pmatrix}\text{total atomic}\\ \text{species } i \text{ in}\end{pmatrix} - \begin{pmatrix}\text{total atomic}\\ \text{species } i \text{ out}\end{pmatrix} = \begin{pmatrix}\text{accumulation of}\\ \text{atomic species } i\end{pmatrix} \quad \text{for all species } i \qquad 3.2\text{–}1$$

By summing over all the atomic species entering and leaving a region, the total mass balance is obtained:

$$(\text{total mass in}) - (\text{total mass out}) = (\text{total accumulation}) \qquad 3.2\text{–}2$$

Equation 3.2–2 is a statement of the law of conservation of mass.

Conservation of mass also applies to molecular types, such as hydrochloric acid or methane, if it is known that that type of molecule is neither created nor destroyed within the region of the mass balance. In such situations, Equation 3.2–1 is applicable to the molecule as if it were an atom.

However, if the molecular species is known to undergo a chemical reaction, additional terms must be added to the conservation equation to account for the amount of that species formed or destroyed by chemical reaction:

$$\begin{pmatrix}\text{total molecular}\\ \text{species } i \text{ in}\end{pmatrix} - \begin{pmatrix}\text{total molecular}\\ \text{species } i \text{ out}\end{pmatrix} + \begin{pmatrix}\text{amount of species } i\\ \text{formed by chemical}\\ \text{reaction}\end{pmatrix} - \begin{pmatrix}\text{amount of species } i\\ \text{destroyed by chemical}\\ \text{reaction}\end{pmatrix} = \begin{pmatrix}\text{accumulation of}\\ \text{molecular species } i\end{pmatrix} \quad 3.2\text{–}3$$

The material balance may be performed over any convenient period of time. Especially in the case of continuous flow processes, it may be desirable to make continuous balance in every small increment of time. The result is a material flow rate balance:

$$\begin{pmatrix}\text{rate of species}\\ i \text{ in}\end{pmatrix} - \begin{pmatrix}\text{rate of species}\\ i \text{ out}\end{pmatrix} + \begin{pmatrix}\text{rate of formation}\\ \text{of species } i \text{ by}\\ \text{chemical reaction}\end{pmatrix} - \begin{pmatrix}\text{rate of destruction}\\ \text{of species } i \text{ by}\\ \text{chemical reaction}\end{pmatrix} = \begin{pmatrix}\text{rate of accumulation}\\ \text{of species } i\end{pmatrix} \quad 3.2\text{–}4$$

If the rate-of-accumulation term is negative, the inventory of the species in the part of the process around which the balance is made is being depleted. In the special case where the rate-of-accumulation term is zero, the region experiences neither a net gain nor loss of the species being balanced and is at *steady state.*

These are the basic ideas to be exploited in this section. However, there is more to material balancing than merely the basic principle of conservation of mass. Knowing when, where, and what to balance in a complex problem to obtain key information is an important part of the craft, which comes with experience. Section 3.3 offers the beginnings of that experience.

Example 3.2–1 Freezing Seawater. *As part of a process to obtain fresh water from the sea, seawater is partially frozen to form salt-free ice and a brine more concentrated in salt. If the seawater is 3.5 per cent salt by weight and if the brine is to be 7 per cent salt by weight, how much seawater must be brought into the process per pound of ice formed?*

In Figure 3.2–1 the ways in which salt and water enter and leave the process are identified. Assuming that none of the ice and brine are left in the process to accumulate,

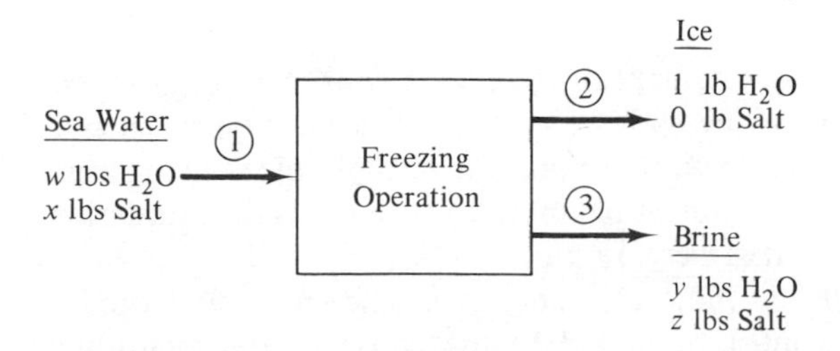

Figure 3.2–1

the following balances are valid on a basis of 1.0 lb of salt-free ice:

Salt balance:

$$x = z$$

Water balance:

$$w = 1 + y$$

The statement of seawater and brine salt concentrations gives the additional data that

$$\frac{x}{x + w} = 0.035$$

$$\frac{z}{z + y} = 0.07$$

Here we have four equations and four unknowns, the solution of which is

$$\left.\begin{aligned} w &= 2 \text{ lb } H_2O \\ x &= 0.07 \text{ lb salt} \end{aligned}\right\} \text{sea water}$$

$$\left.\begin{aligned} y &= 1 \text{ lb } H_2O \\ z &= 0.07 \text{ lb salt} \end{aligned}\right\} \text{brine}$$

Hence just over 2 lb of seawater must be brought into the process for each pound of salt-free ice formed. Often the results of the material balance are presented in tabular form (see Table 3.2–1).

TABLE 3.2–1 Seawater Freezing Material Balance

	Stream Amounts (lb)		
Species	Seawater, Feed 1	Ice, Stream 2	Brine, Stream 3
Water	2	1	1
Salt	0.07	0	0.07

Example 3.2–2 Inert-Gas Generation. *To provide an inert atmosphere to purge a reactor vessel, methane is burned in air to produce a mixture of CO_2, H_2O, and N_2. In what ratio should the methane and air be fed to the burner to provide such a gaseous product?*

Nitrogen passes through the process unaltered, so the material balance on it is particularly simple:

$$y = n$$

The carbon in the methane appears as the carbon in carbon dioxide, and hence the ratios of carbon in methane, 12:16, and in carbon dioxide, 12:44, enter the carbon balance:

$$1\left(\tfrac{12}{16}\right) = l\left(\tfrac{12}{44}\right)$$

The hydrogen in the methane appears in the water:

$$1(\tfrac{4}{16}) = m(\tfrac{2}{18})$$

The oxygen in the air appears in the carbon dioxide and the water combustion products:

$$x = l(\tfrac{32}{44}) + m(\tfrac{16}{18})$$

The atomic balances provide four equations, but there are five unknowns: l, m, n, x, *and* y *(see Figure 3.2–2). We must seek other sources of information to solve this problem.*

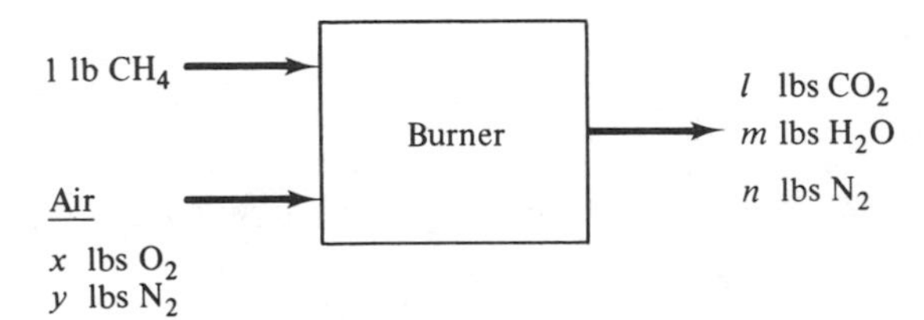

Figure 3.2–2

Air consists of 23 per cent O_2 by weight. This provides the needed additional equation:

$$\frac{x}{x + y} = 0.23$$

The solution is

$$\left.\begin{aligned} x &= 4 \text{ lb } O_2 \\ y &= 12 \text{ lb } N_2 \end{aligned}\right\} \quad \text{air}$$

$$\begin{aligned} l &= 2.75 \text{ lb } CO_2 \\ m &= 2.25 \text{ lb } H_2O \quad \text{inert gas} \\ n &= 12 \text{ lb } N_2 \end{aligned}$$

Hence to generate this inert atmosphere $x + y = 16$ lb of air must be provided for every pound of methane burned (see Table 3.2–2).

TABLE 3.2–2 Material Balance over Methane Burner

	STREAM AMOUNTS (LB)		
Species	Methane, Stream 1	Air Stream, Stream 2	Inert Gas, Stream 3
Methane	1	0	0
Oxygen	0	4	0
Nitrogen	0	12	12
Carbon dioxide	0	0	2.75
Water	0	0	2.25
Total mass	1	16	17

Example 3.2–3 Loss of Benzene in Process. *Benzene is soluble in water in the amount of 0.07 lb benzene/100 lb H_2O. In a process where water and benzene are in contact, 10,000 lb/hour (h) of water leave the process. How much benzene must be added to the process to make up for the resulting benzene loss? The benzene acts only as a solvent in the process and does not undergo any reaction.*

Here we can make a material balance over the benzene as if it were an atom, since it is not created or destroyed in the processing.

Benzene balance:

$$\text{benzene addition rate} = \text{benzene loss with water}$$

$$= (10{,}000)\frac{0.07}{100} = 7 \text{ lb/h}$$

These examples illustrate the most elementary applications of the principle of conservation of mass. In the next section we extend the usefulness of these ideas by examining matters of style in order to ease the formation and solution of material balances in more complex situations.

3.3 STYLE IN MATERIAL BALANCING

Every game is defined by certain rigid rules within which each player is free to develop his own style of play. The conservation of mass is the rigid rule of material balancing. We now consider certain matters of style to ease the formulation and solution of material balances in complex processing situations. The guidelines presented are flexible and should be modified as the engineer develops his own style.

Diagramming and Tabulation

Even in relatively simple problems it is worthwhile to draw a diagram, a schematic, or a flow sheet, which depicts the problem at hand. In complex problems diagramming is essential. Usually the flow sheet includes any units in which streams come together or are separated, or in which molecular species are generated or destroyed by reaction.

The streams connecting the units are usually named, such as reactor feed, recycle, purge, and so forth. The species known to be present in a stream are stated near the stream label, along with known flow rates and compositions. Pressures, temperatures, and other stream data not directly connected with the material balancing tend to clutter the flow sheet and should not be mentioned at this point in analysis. Only flow-rate data and compositions should appear on the material-balance diagram. The flow sheets in the following examples illustrate these points.

In addition to affixing flow rates to the appropriate streams in the flow sheet, it often is very useful to tabulate the results of the material balances. This provides an orderly way of checking the accuracy of the balance, and makes your results easily accessible to an interested outsider. This kind of tabulation is illustrated in several examples.

In many situations it is useful to choose a *basis* for the computations. That is, to select a single flow rate, usually a feed or product, upon which to base the material balance. It simplifies our thinking to specify a ton of product, one hundred pounds per hour of feed, or whatever is suitable for a basis.

Using Tie Components

Certain patterns that appear commonly in material balancing can be used with great effect. One pattern involves *tie components.* A tie component is an element, molecule, species, or even part of a molecule that goes through the balance region intact, in one place and out another place. Under steady-state conditions, if the flow rate of a tie component is known where it crosses the balance boundary in one place, it is known at the other place.

A pattern that is particularly useful consists of a tie component which passes through a part of the process where its flow rate is known, and through another part of the process where a complete stream composition is available. The flow rates of species involved in the stream of known composition can be computed quite easily.

In the following examples we show how tie components are used to expand the basic data originally appended on the flow sheet. The tie components are useful in these early stages of material balancing, and also in the later stages as more data emerge, exposing additional useful patterns.

Example 3.3–1 Leaching of Copper Ore. *An ore containing 7 per cent native copper is to be leached with sulfuric acid. All the copper in the ore is transferred to the acid phase, which is then extracted with an organic solvent. The organic solvent leaving the extractor is 20 per cent copper in solution, and this accounts for all the copper. The copper is then recovered from the solvent solution, and the solvent is recycled.*

If 800 tons/day of tailings leave the process, how much solvent must be recycled?

Figure 3.3–1 shows the flow sheet for the processing with the data affixed. The tailings flow rate and the composition of the ore are known, as well as the composition of the solvent extract. A tie exists between the ore and the tailings, which immediately indicates that 800 tons of tailings enter with the ore. Since the ore is 93 per cent tailings and 7 per cent copper, $(800)(\frac{7}{93}) = 60$ *tons/day of copper enters the process. A copper tie passes completely through the process (Figure 3.3–2), and at the solvent extraction effluent the composition is known to be 80 per cent solvent and 20 per cent copper. At that point the solvent flow rate must be* $(60)(\frac{80}{20}) = 240$ *tons/day.*

A solvent tie exists between the extractor effluent and the recycle solvent stream. Thus 240 tons/day of solvent is to be recycled (see Table 3.3–1).

To determine the solvent recycle rate in this case, we exploited three tie-component patterns. This was possible because of the unique sequence of given flow rates and compositions found in Figure 3.3–1.

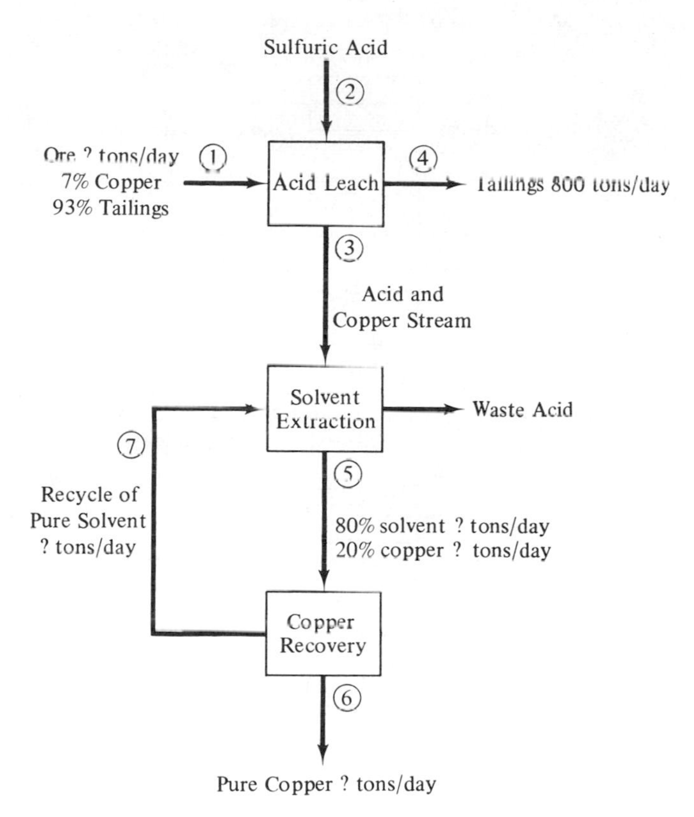

Figure 3.3–1. Preliminary Material-Balance Flow Sheet for Copper Recovery Process

Selecting Balance Boundaries

The law of conservation of mass applies to any and all regions in the process, and the engineer is free to select the boundaries of the region over which the mass balances are to be made. We now examine the effects of the location of balance boundaries.

If a boundary is selected which involves the crossing of only one unknown flow rate, then only that single unknown must be solved for. If several unknown flow rates enter into the balance, equations may have to be solved simultaneously, generally a more difficult task.

Initially, we attempt to select the component to be balanced and the balance boundaries to involve only the crossing of one unknown flow rate.

General Guidelines

There are additional considerations that may influence the material-flow analysis. Only sets of *independent* material-balance equations can be solved simultaneously. Equations are independent if no one of them can be derived by some algebraic

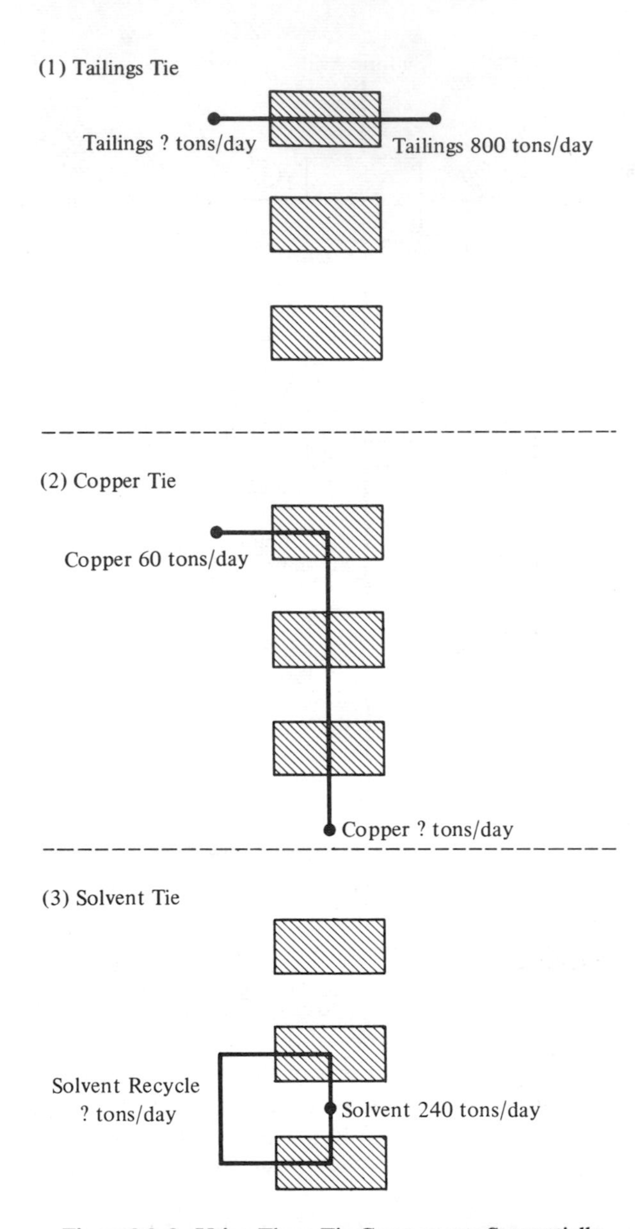

Figure 3.3–2. Using Three Tie Components Sequentially

combination of the remaining equations in the set. Thus if any two or more species exist in a fixed ratio in each stream in which they cross the balance boundaries, only one independent material-balance equation may be written for these species.

The number of unknown quantities to be calculated cannot exceed the number of independent material balances. Otherwise, the solution to the problem is inde-

TABLE 3.3–1 Material Balance in Copper Leaching Process of Figure 3.3–1

	STREAM FLOW RATES (TONS/DAY)						
Species	Feed 1	Acid 2	Leachate 3	Tailings 4	Extract 5	Product 6	Recycle 7
Copper	60	0	60	0	60	60	0
Tailings	800	0	0	800	0	0	0
Acid	0	*	*	0	0	0	0
Solvent	0	0	0	0	240	0	240

* Acid flow rate not specified in problem data.

terminate. If the number of independent equations exceeds the number of flow rates to be computed, it becomes a matter of judgment to determine which independent set of equations should be selected to solve the problem. If all the data in each equation were perfect, it would not matter which independent set of equations was selected. However, since such data are usually uncertain because of errors in sampling and measurement, it is generally best to use equations based on the species present in the greatest concentrations.

Example 3.3–2 Drying Solids with Hot Air. *One hundred pounds per hour of wet solids is to be dried by contact with hot air. To improve drying, the air is to be blown at high velocity over the solids and partially recycled through a heater. Bone-dry air is brought in to make up for the humid air removed along with the water taken from the solids. The solids are initially 30 per cent water and are to be dried to 5 per cent water. The humid air is to be 5 per cent water. How much dry air must be brought into the system? See Figure 3.3–3a.*

Figure 3.3–3b shows the balance flow sheet with the known data. Figure 3.3–4 is the three-step solution to this problem. The dry solids pass through the system unaltered, forming a tie between the wet-solids feed and the dry-solids product. Since the feed rate of dry solids is easily computed as $(0.70)(100) = 70$ *lb/h, and since the effluent dried solids composition is given as 5 per cent water, the water leaving with the dried solids is computed to be* $(70)(\frac{5}{95}) = 3.7$ *lb/h.*

If a water balance is made over the entire system, now only the rate at which water leaves with the humid air is unknown. That is computed to be 26.3 lb/h of water. Now, since the effluent humid air stream is to be 5 per cent water, the rate at which air leaves in that stream is $(26.3)(\frac{95}{5}) = 500$ *lb/h.*

An overall air balance indicates that air enters the system in the dry air stream and leaves in the humid air stream. Hence, the rate of addition of bone-dry air must be 500 lb/h (see Table 3.3–2).

In this problem we exploited a tie on the dry solids and two balances, which we selected to involve only one unknown each. This sequence led to the easy solution to this material-balancing problem.

Single conveyor dryer. The fans pull the air through the heater section located beneath them. The air then passes through the wet stock carried on the conveyor and back into the bottom of the heater. The conveyor belt is made of perforated-metal plate or woven wire, and the heater is usually of steam-heated finned tubes. Conveyor widths vary between 15 in. and 9 ft. and lengths vary up to 160 ft. Preforming of the feed to increase drying rate and obtain a more desirable product is often practised. It may be accomplished by extrusion or granulating steps or merely by scoring the cake removed from a continuous filter so that it breaks into reasonably uniform pieces. Thickness of material on the belt will depend on the material being processed, but it can vary between $\frac{1}{2}$ and 6 in. (Courtesy of Proctor & Schwartz, Inc.)

Figure 3.3–3a. Drying Solids with Hot Air. (Courtesy of Proctor & Schwartz, Inc., Philadelphia, Pennsylvania).

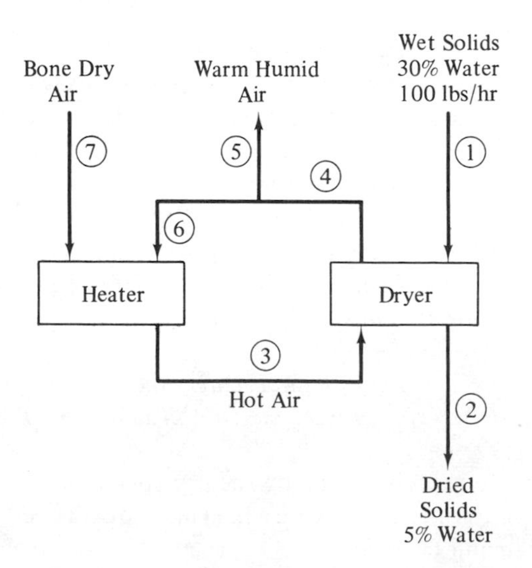

Figure 3.3–3b

TABLE 3.3–2 Material Balance on Solids Drier for Figure 3.3–3

	Stream Flow Rates (lb/h)						
	Wet Solids 1	Dry Solids 2	Recycle Air 3	Recycle Air 4	Air Effluent 5	Recycle Air 6	Dry Air 7
Solids	70	70	0	0	0	0	0
Water	30	3.7	*	*	26.3	*	0
Air	0	0	*	*	500	*	500

* Internal flow rates not determinable from data given.

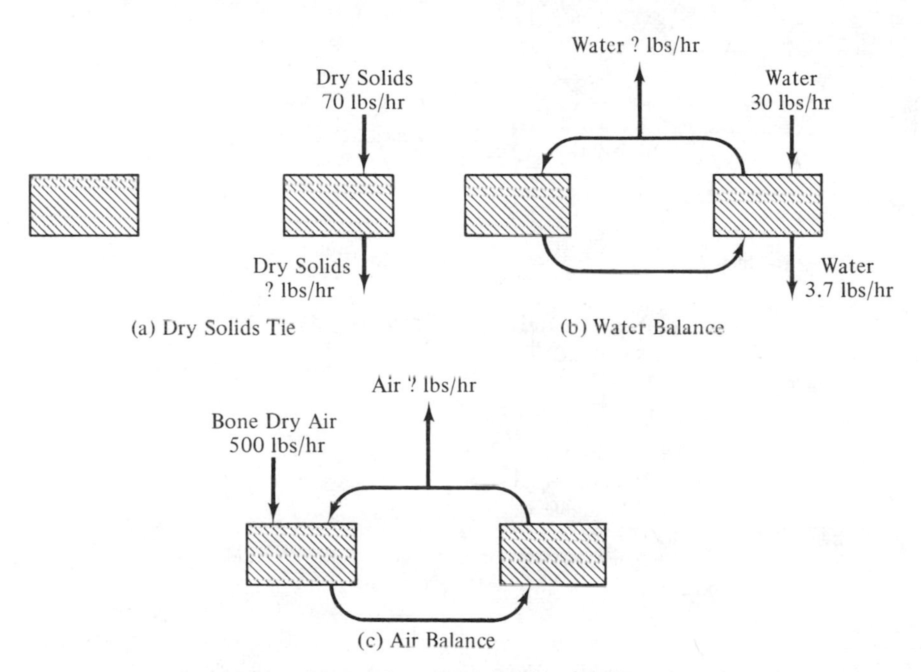

Figure 3.3–4. Three Steps in Material Balancing

Example 3.3–3 Soluble Coffee Powder. *The production of instant coffee is based on the flow sheet shown in Figure 3.3–5. Much of the water-soluble material is extracted from roasted and ground coffee beans in large percolators. The extract is spray dried to produce instant coffee powder, and the wet grounds are partially dried prior to disposal by incineration or as landfill. The flow rates of water, solubles, and insolubles for all the streams shown in Figure 3.3–5 can be constructed by the proper material balancing from the fragmentary data given.*

Figure 3.3–6 shows the routes of the three species and is useful in visualizing where material balances might be made. Attention is first attracted by the simple way the

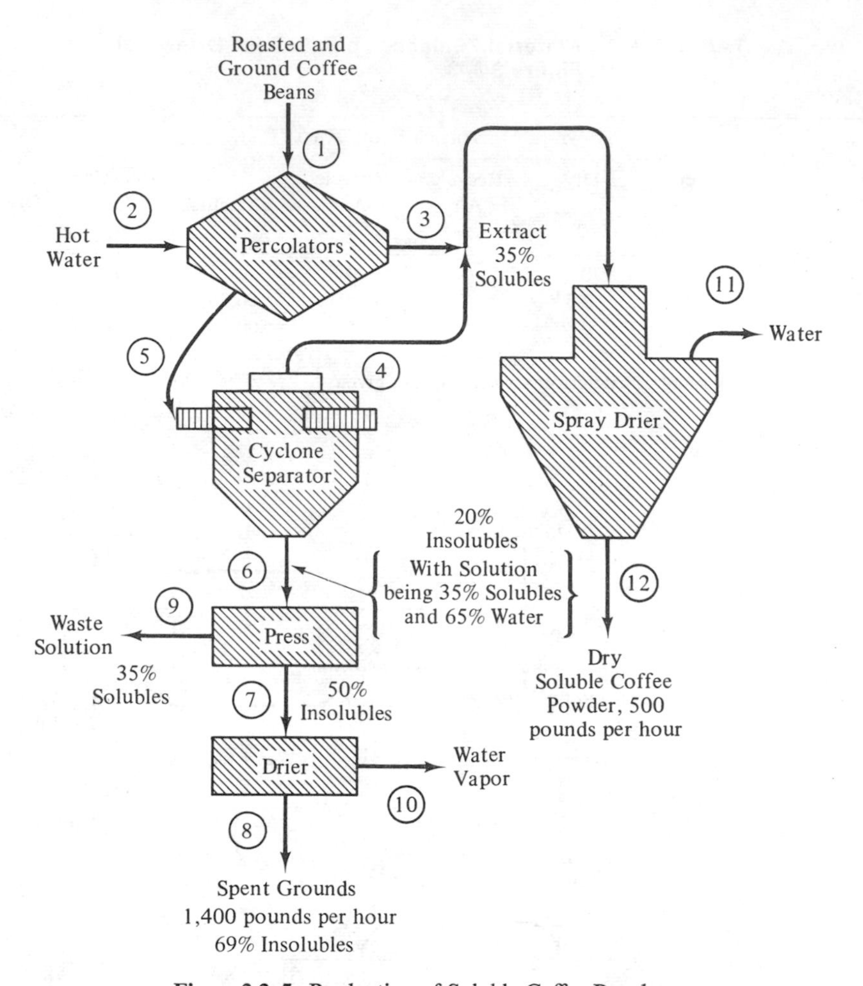

Figure 3.3–5. Production of Soluble Coffee Powder

insolubles pass through the process; this forms a tie component which enables the calculation of the flow rates of the other species along the path of insolubles.

- ***Insolubles Tie*** *The spent grounds stream is 69 per cent insolubles with a flow rate of 1400 lb/h; thus*

$$I_8 = (0.69)(1400) = 975 \text{ lb/h}$$

$$S_8 + W_8 = (0.31)(1400) = 425 \text{ lb/h}$$

where I, S, *and* W *denote insolubles, solubles, and water, with the subscript being the stream number. Because of the insolubles tie,*

$$I_8 = I_7 = I_6 = I_5 = I_1 = 975 \text{ lb/h}$$

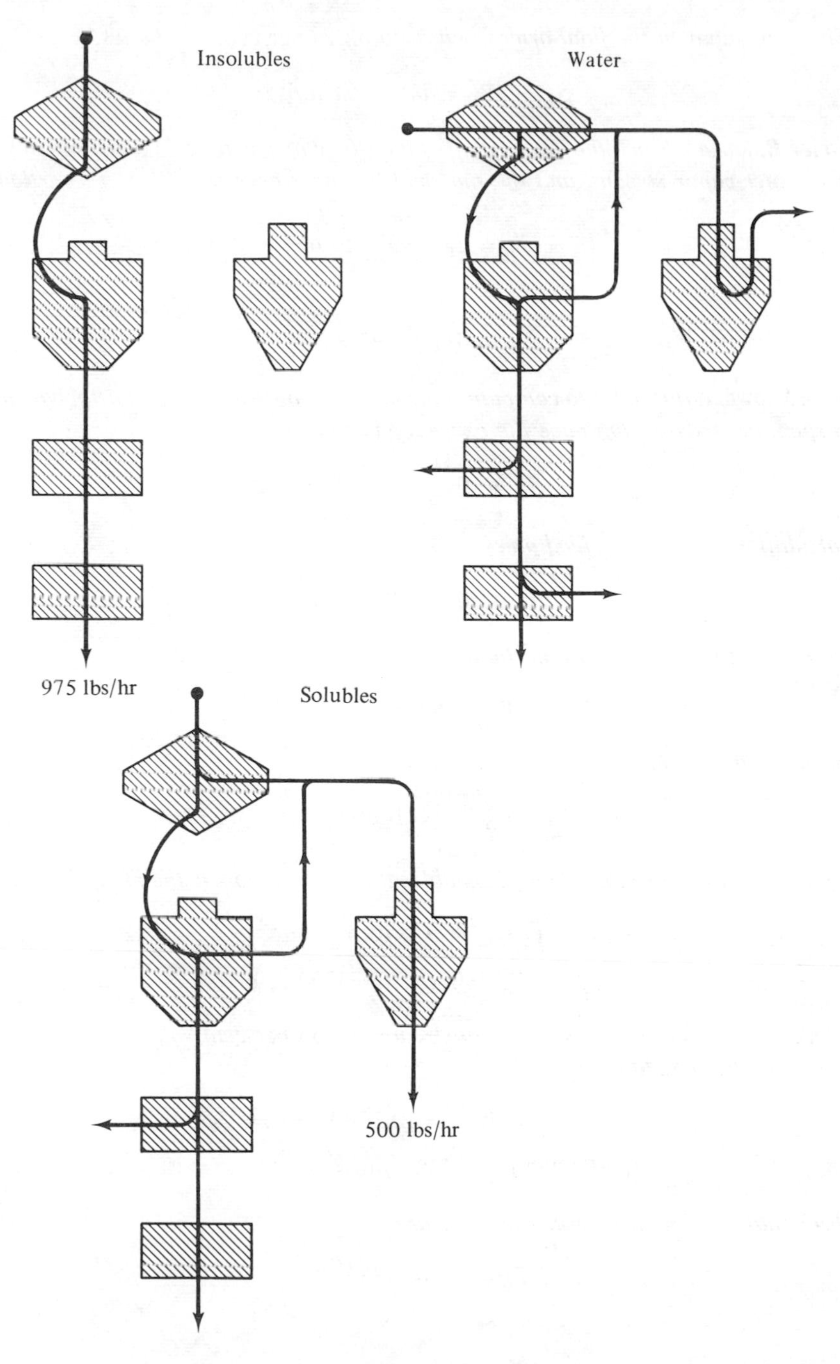

Figure 3.3–6. Routes of Insolubles, Solubles, and Water During Coffee-Powder Processing

Since the input to the final drier is given to be 50 per cent insolubles,

$$S_7 + W_7 = I_7 = 975 \text{ lb/h}$$

- ***Drier Balance*** *Since the drier removes the water by vaporization, no solubles leave with the water-vapor stream, and the material balances over the drier are particularly simple:*

$$I_7 = I_8 = 975 \text{ lb/h}$$

$$S_7 = S_8$$

$$W_7 = W_8 + W_{10}$$

Is enough known at this point to compute some of these flow rates? Our initial calculation on the spent grounds stream gave the expression

$$S_8 + W_8 = 425 \text{ lb/h}$$

The calculation on the drier feed gives

$$S_7 + W_7 = 975 \text{ lb/h}$$

Summing the last two balances on the drier gives

$$S_7 + W_7 = S_8 + W_8 + W_{10}$$

Solving these three equations gives

$$W_{10} = 550 \text{ lb/h}$$

- ***Press Balance*** *The insolubles tie enables the computation of the press feed rates at*

$$I_6 = 975 \text{ lb/h}$$

$$S_6 + W_6 = I_6(\tfrac{80}{20}) = 3900 \text{ lb/h}$$

Since it is given that the solution in stream 6 is 35 per cent solubles, the individual flow rates can be calculated as

$$S_6 = (0.35)(S_6 + W_6) = (0.35)(3900) = 1365 \text{ lb/h}$$

$$W_6 = (0.65)(3900) = 2535 \text{ lb/h}$$

The three material balances over the press are

$$I_6 = I_7 = 975 \text{ lb/h}$$

$$S_6 = S_9 + S_7$$

$$W_6 = W_9 + W_7$$

The press balance equations can be solved to yield

$$S_9 + W_9 = (S_6 + W_6) - (S_7 + W_7)$$
$$= 3900 - 975 = 2925 \text{ lb/h}$$

At 35 per cent solubles the waste solution from the press consists of

$$S_9 = (0.35)(S_9 + W_9) = (0.35)(2925) = 1025 \text{ lb/h}$$
$$W_9 = (S_9 + W_9) - S_9 = 2925 - 1025 = 1900 \text{ lb/h}$$

The remaining unknown flow rates at the press are now easily computed to be

$$S_7 = S_6 - S_9 = 1365 - 1025 = 340 \text{ lb/h}$$
$$W_7 = W_6 - W_9 = 2535 - 1900 = 635 \text{ lb/h}$$

• ***Spray Drier Balance*** *The product from the spray drier is 500 lb/h of pure dry soluble coffee. Thus*

$$I_{12} = 0$$
$$S_{12} = 500 \text{ lb/h}$$
$$W_{12} = 0$$

In Figure 3.3–6 we see that the solubles form a tie component passing through the spray drier, and in Figure 3.3–5 we see that the composition of the spray drier input is given as 35 per cent solubles and 65 per cent water. We compute

$$S_3 + S_4 = S_{12} = 500 \text{ lb/h}$$
$$W_3 + W_4 = (S_3 + S_4)(\tfrac{65}{35}) = 930 \text{ lb/h}$$

A water balance over the drier gives

$$W_3 + W_4 = W_{11} = 930 \text{ lb/h}$$

• ***Percolator–Cyclone Separator Balance*** *We do not know what fraction of the extract is drawn off directly from the percolator and what fraction is removed by the cyclone separator. For this reason we cannot expect to determine the detailed flow rates of streams 3, 4, and 5. However, balances made around the percolator–separator system give access to the flow rates of streams 1 and 2.*

A water balance, presuming the coffee beans are dry, gives

$$W_2 = W_6 + (W_3 + W_4)$$
$$= 2535 + 930 = 3465 \text{ lb/h}$$

and a solubles balance gives

$$S_1 = S_6 + (S_3 + S_4)$$
$$= 1365 + 500 = 1865 \text{ lb/h}$$

The insolubles in stream 1 are known from the initial tie to be

$$I_1 = 975 \text{ lb/h}$$

Table 3.3–3 summarizes all these material flows. It is interesting to notice where the soluble coffee goes. Of the 1865 lb/h entering, only 500 lb/h leave as product. The press waste stream contains 1025 lb/h of solubles. This suggests a tripling of the production rate by routing the press waste to the spray drier. If the pressing of the grounds does not release materials that alter the flavor of the coffee, this suggested change in the allocation of materials could make significant differences in the production.

TABLE 3.3–3 Material Balances for Soluble Coffee Processes

	COFFEE BEANS						SPENT GROUNDS	PRESS WASTE			PRODUCT
	1	2	3 + 4	5	6	7	8	9	10	11	12
Solubles	1865	0	500	?	1365	340	340	1025	0	0	500
Insolubles	975	0	0	975	975	975	975	0	0	0	0
Water	0	3465	930	?	2535	635	85	1900	550	930	0

Handling Recycle

A continual source of interesting material-balance problems is processes in which material is recycled back from a downstream position to an upstream position. In such a situation we cannot solve the upstream part of the process until the downstream part is solved, and vice versa. This generally leads to the need for solving the material balances simultaneously for several parts of the process.

Example 3.3–4 Production–Consumption Analysis with Recycle. *In Example 2.5–1 the production–consumption analysis of an equilibrium reaction with partial recycle of the unconverted reactants was discussed. Now we show by material balances that the practice of deducting the amount of recycle from both the reactor effluent and the original feed of the unconverted reactant is correct.*

Figure 3.3–7 illustrates again Example 2.5–1. Isopropyl alcohol and acetic acid react to form isopropyl acetate and water at only 60 per cent conversion:

$$\text{alcohol} + \text{acid} \rightarrow 0.6 \text{ acetate} + 0.6 \text{ H}_2\text{O} + 0.4 \text{ alcohol} + 0.4 \text{ acid}$$

How much raw acid and alcohol must be brought into the system to produce the same 0.6 mole of ester if 80 per cent of the unconverted reactants is recovered and recycled? How much acid and alcohol are lost as waste?

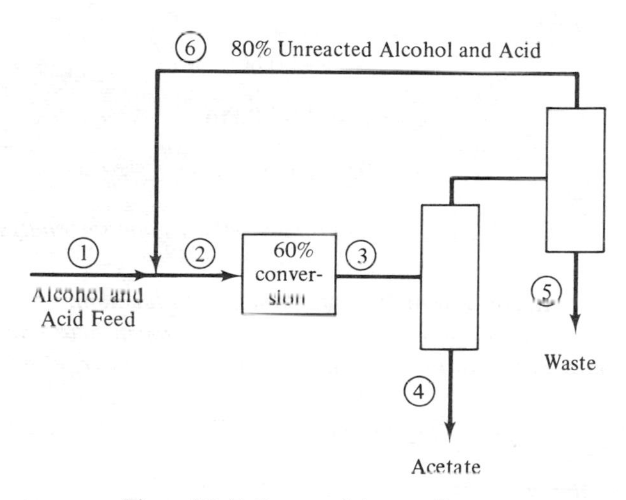

Figure 3.3–7. Isopropyl Acetate Process

Table 3.3–4 includes all the known flow rates in the process for each 0.6 mole of acetate produced. Since the acid and alcohol everywhere appear in the same ratio, only one equation for these two species can be written.

Since the reactor operates at 60 per cent conversion, an acetate production of 0.60 mole implies that the acid and alcohol feed to the reactor was 1.00 mole. The

TABLE 3.3–4 Data for Process

Species	Raw Feed	Reactor Feed	Reactor Effluent	Acetate Product	Waste	Recycle
Alcohol	X	1.00	0.40	0	0.2(0.40)	0.8(0.40)
Acid	X	1.00	0.40	0	0.2(0.40)	0.8(0.40)
Acetate	0	0	0.60	0.60	0	0
H_2O	0	0	0.60	0	0.60	0

reactor then also produces 0.40 mole each of unconverted acid and alcohol. Since 80 per cent of this is recovered and recycled, the raw acid and alcohol requirements are then obtained from a single balance around the reactor premixer:

$$\text{acid or alcohol,} \quad X + 0.80(0.40) = 1.00$$
$$X = 0.68 \text{ mole}$$

This raw-material requirement is the stoichiometric coefficient reduced by the amount of recycle. Similarly, the amount of acid or alcohol lost is obtained from a balance around the waste separator:

$$\text{loss} = 0.40 - 0.80(0.40)$$
$$= 0.40(1 - 0.80)$$
$$= 0.08 \text{ mole}$$

This again is the same as obtained by the production–consumption analysis procedure.

Example 3.3–5 Argon Purge During Ammonia Production. *Figure 3.3–8 shows the flow sheet for the production of ammonia from nitrogen and hydrogen. The feed stream is a nitrogen and hydrogen mixture in the 1:3 ratio required of the reaction chemistry:*

$$N_2 + 3\,H_2 \rightarrow 2\,NH_3$$

Trace amounts of the inert gas argon occur in the feed.

In the converter, 10 per cent of the nitrogen and hydrogen react to form ammonia. The ammonia product is removed by the condenser in which only ammonia is liquified, and the unreacted nitrogen and hydrogen, along with the argon, recycle to the converter.

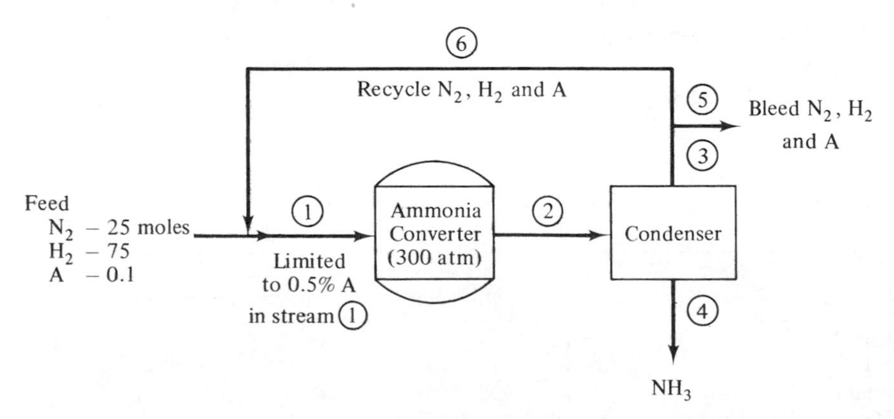

Figure 3.3–8. Ammonia Production

Without a bleed stream, argon would only enter the process and would accumulate to dominate the recycle. The bleed stream vents part of the recycle stream to prevent the buildup of argon.

If, for the proper performance of the converter, its feed is limited to 0.5 per cent argon, what fraction of the recycle stream must be vented?

Table 3.3–5 includes all the known flow rates in the process, and in Table 3.3–6 we begin to build up the simultaneous equations that must be solved to handle the recycle. A flow rate of X moles of nitrogen is postulated to enter the converter. Since hydrogen

TABLE 3.3–5 Initial Data on Ammonia Process

Species	Fresh Feed	Converter Feed 1	Converter Effluent 2	Condenser Effluent 3	Ammonia Product 4	Bleed 5	Recycle 6
N_2	25				0		
H_2	75				0		
A	0.1				0		
NH_3	0	0		0		0	0

TABLE 3.3–6 Data Generated by Postulating That *X* Moles of Nitrogen Enter Converter

Species	Fresh Feed	Converter Feed 1	Converter Effluent 2	Condenser Effluent 3	Ammonia Product 4	Bleed 5	Recycle 6
N_2	25	X	$0.90X$	$0.90X$	0	$f0.90X$	$(1-f)0.90X$
H_2	75	$3X$	$2.70X$	$2.70X$	0	$f2.70X$	$(1-f)2.70X$
A	0.1	$0.02X$	$0.02X$	$0.02X$	0	$f0.02X$	$(1-f)0.02X$
NH_3	0	0	$0.20X$	0	$0.20X$	0	0

is to be in a stoichiometric ratio, 3X moles of hydrogen enter. The argon flow rate is 0.5 per cent of the total flow, or 0.02X moles. Ten per cent of the feed nitrogen and hydrogen are converted to ammonia; thus the nitrogen, hydrogen, and ammonia flows leaving the converter are 0.90X, 0.27X, and 0.20X, respectively. The argon passes through the converter unaltered. All the ammonia leaves in the product, and the remainder of the materials in the converter effluent appear as the condenser effluent stream. Of this stream a fraction f *is bled out of the process, and the remainder recycles to the converter.*

In forming Table 3.3–6 we have performed a good fraction of the material balancing, and at this point several key balances are to be made. A balance over the mixing junction indicates that the converter feed stream 1 is the sum of the fresh feed stream and the recycle stream. Balances on the nitrogen and hydrogen give the same equation. The argon balance gives new information.

Nitrogen:

$$25 + (1 - f)0.90X = X$$

Hydrogen:

$$75 + (1 - f)2.70X = 3X$$

Argon:

$$0.1 + (1 - f)0.02X = 0.02X$$

Multiplying the argon balance by 0.90/0.02 and subtracting the result from the nitrogen balance eliminates the (1 − f) term, giving

$$20.5 = 0.1X$$

or

$$X = 205 \text{ moles}$$

Substituting this result into the nitrogen balance gives one equation in f:

$$25 + (1 - f)184.5 = 205$$

or

$$f = 1 - \frac{180}{184.5} = 0.025$$

Thus 2.5 per cent of the recycle must be bled off to maintain the low argon concentration. Table 3.3–7 summarizes the entire balance.

TABLE 3.3–7 Complete Balance on Ammonia Process

Species	Fresh Feed	Converter Feed 1	Converter Effluent 2	Condenser Effluent 3	Ammonia Product 4	Bleed 5	Recycle 6
N_2	25	205	184	184	0	4.7	179
H_2	75	615	553	553	0	15	538
A	0.1	4.1	4.1	4.1	0	0.1	4.0
NH_3	0	0	41	0	41	0	0

It is interesting to notice the amount of ammonia production that must be sacrificed to remove the argon from the recycle. The fresh feed contains 25 moles of N_2 and 75 moles of H_2, which if all converted would produce 50 moles of NH_3. The actual production from the process is 41 moles, which is nearly 20 per cent lower. There is no way of getting around this; the trace argon in the feed, 0.1 per cent, has a significant effect on the process.

In the previous examples the recycle material balances were formed in terms of algebraic expressions in the unknown flow rates, leading to the simultaneous equations in several unknown variables. Recycle material balances also can be solved numerically by assuming trial values of an unknown flow rate and detecting imbalances in the material balances. The direction of the imbalance suggests the direction in which the assumed flow rate must be changed. Such trial-and-correction procedures are often the easiest way of handling recycle. The next example shows how this is done.

Example 3.3–6 Caustic Manufacture. *In the process shown in Figure 3.3–9, 21,200 tons/day of soda ash (Na_2CO_3) is reacted with 14,800 tons/day of slaked lime [$Ca(OH)_2$] to produce 16,000 tons/day of caustic ($NaOH$) and 20,000 tons/day of an insoluble carbonate ($CaCO_3$):*

$$Na_2CO_3 + Ca(OH)_2 \rightarrow 2\ NaOH + CaCO_3$$

$$21{,}200 \text{ tons/day} + 14{,}800 \text{ tons/day} \rightarrow 16{,}000 \text{ tons/day} + 20{,}000 \text{ tons/day}$$

The reaction occurs in a dilute solution in the reactor and the carbonate precipitates out. The main product is an 8 per cent aqueous caustic solution. However, some of this solution is entrained in the carbonate sludge, which is 30 per cent insoluble $CaCO_3$ and 70 per cent solution. From the first thickener the sludge, containing an 8 per cent caustic

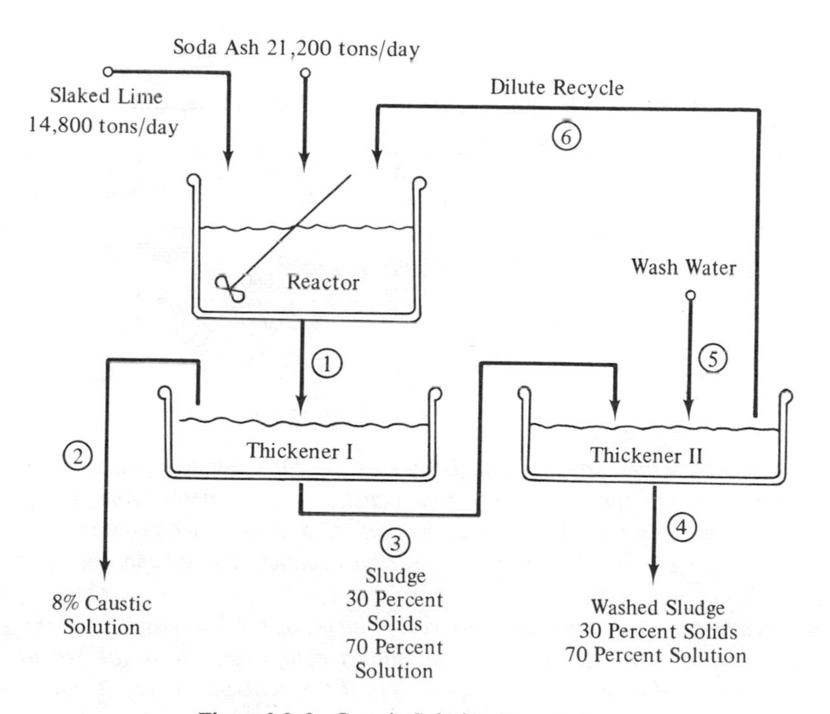

Figure 3.3–9. Caustic Solution Manufacture

solution, is sent to the second thickener where wash water is added. The caustic solution, now diluted with wash water, is recycled to the reactor. However, some of the diluted solution is lost with the washed sludge. The sludge is 30 per cent insoluble carbonate and 70 per cent solution.

How much wash water must be added to balance the material flows and to produce the caustic solution of 8 per cent composition?

We begin by tabulating all the known and easily calculated flow rates. The 20,000 tons/day of $CaCO_3$ originates in the reactor and passes through streams 1, 3, and 4. The sludge leaving thickener I is 30 per cent $CaCO_3$ solids and 70 per cent the 8 per cent caustic solution. Thus at stream 3 the total solution flow rate is $(20{,}000)(\frac{70}{30}) =$ 46,666 tons/day. Of that solution 8 per cent is caustic and the remaining 92 per cent water. Hence the caustic flow in stream 3 (circled subscript numbers indicate the stream number) is

$$NaOH_{③} = (46{,}666)(0.08) = 3733 \text{ tons/day}$$

and the water flow is

$$H_2O_{③} = 46{,}666 - 3733 = 42{,}933 \text{ tons/day}$$

This easily obtained initial information is tabulated in Table 3.3–8. Notice that the 46,666 tons/day solution lost with the washed sludge is calculable, but composition of the solution is not known.

TABLE 3.3–8 Basic Data on Flow Rates (Tons/Day)

	Reactor Effluent 1	Stream Product 2	First Sludge 3	Washed Sludge 4	Wash Water 5	Recycle 6
H_2O			42,933	46,660 (H₂O + NaOH)		
NaOH			3,733		0	
$CaCO_3$	20,000	0	20,000	20,000	0	0
		8% caustic				

The difficulty which now faces us is that we do not know how much dilute solution should be recycled to form the 8 per cent caustic in the reactor. The recycle stream carries with it the caustic washed from the sludge, and we must know the composition of this recycle stream to adjust to the 8 per cent product. Let us approach this by trial and error.

We assume a wash-water addition rate, calculate the composition of the solution in the thickener II, and then calculate the effluent composition from the reactor formed by the recycle of solution of that composition. If the reactor effluent is not 8 per cent, the initial wash-water addition-rate assumption was in error. Should that be the case, a better guess should be made.

Step 1. Postulate $H_2O_{⑤} = 100{,}000$ tons/day.

Step 2. Calculate composition of solution in thickener II:

$$\text{total NaOH in} = NaOH_{③} = 3733 \text{ tons/day}$$

$$\text{total } H_2O \text{ in} = H_2O_{③} + H_2O_{⑤} = 42{,}933 + 100{,}000$$

$$= 142{,}933 \text{ tons/day}$$

$$\% \text{ NaOH} = \frac{3733}{142{,}933} \times 100 = 2.7\%$$

Step 3. Compute effluent streams from thickener II:

46,666 tons/day of solution, 2.7% NaOH

leave with the washed sludge:

$$NaOH_{④} = (46{,}666)(0.027) = 1210 \text{ tons/day}$$

$$H_2O_{④} = 46{,}666 - 1210 = 45{,}456 \text{ tons/day}$$

The material recycled is, therefore,

$$NaOH_{⑥} = NaOH_{③} - NaOH_{④}$$

$$= 3733 - 1210 = 2523 \text{ tons/day}$$

$$H_2O_{⑥} = H_2O_{③} + H_2O_{⑤} - H_2O_{④}$$

$$= 42{,}933 + 100{,}000 - 45{,}456$$

$$= 92{,}477 \text{ tons/day}$$

Step 4. Calculate reactor effluent composition and compare to 8 per cent:

$$\text{total NaOH in reactor} = 16{,}000 \text{ tons/day} + NaOH_{⑥}$$

$$= 16{,}000 + 2523 = 18{,}523$$

$$\text{total } H_2O \text{ into reactor} = H_2O_{⑥}$$

$$= 96{,}477$$

$$\text{caustic composition} = \frac{18{,}523}{96{,}477} = 19.3\%$$

which is much greater than 8 per cent. The amount of wash water assumed added (100,000 tons/day) must be too low. A trial calculation with a higher wash-water rate is in order. Repeating the calculations for wash-water rates of 200,000 and 250,000 tons/day generates the numbers in Table 3.3–9.

We see that 200,000 tons/day of wash water generated a solution too concentrated, 9.7 per cent, and that 250,000 tons per day generated a solution slightly too dilute, 7.7 per cent. This suggests a trial wash rate of, say, 230,000 tons/day, a calculation sequence left as an exercise to the reader.

TABLE 3.3–9 Converging to Product Composition

Trial	Assumed Wash-Water Rate $H_2O_{⑤}$ (tons/day)	Composition of Reactor Effluent Solution (%)
1	100,000	19.3
2	200,000	9.7
3	250,000	7.7

3.4 SYNTHESIS OF MATERIAL FLOW

The *analysis* of material flow through an existing processing plan is a passive activity, which merely identifies more accurately the quantitative features of the process. The *synthesis* of the material flow that ought to exist in a process is a creative activity leading to processing plans which may not have existed before. Synthesis is the end to which we must direct our skill in material balancing. We need to create, in a yet nonexistent process, the flow of material that has a good chance of leading to an efficient and economic process.

The experienced engineer allocates the species using *exact* techniques of conservation of mass and the *heuristic* techniques of engineering judgment. In this section we show how these techniques each solve part of the species-allocation problem. We begin with a simple example that arises in the synthesis of a sulfur dioxide process. Following that we look at a more complex allocation problem, the exploitation of a two-step reaction path to ethyl acetate. In these examples material balances are applied to the definition of certain material flows that must occur. We then illustrate how the engineer must add his intuition on the material flows that *ought* to occur.

Example 3.4–1 Sulfur Dioxide by Direct Oxidation. *Sulfur dioxide often is manufactured by the direct oxidation of sulfur:*

$$S + O_2 \rightarrow SO_2$$

However, to maintain a reasonably low temperature in the burner, sufficient amounts of cool and inert gas must dilute the oxygen before it is fed to the burner. The oxygen mixture should be about 70 mole per cent inert and 30 per cent oxygen.

What species allocations suggest themselves by the use of nitrogen and sulfur dioxide as the diluent gases? Air is to be used as the source of oxygen.

Figure 3.4–1 shows two allocations of nitrogen, oxygen, sulfur, and sulfur dioxide. In the first allocation, air and sulfur are reacted directly to produce a mixture of sulfur dioxide and nitrogen. The nitrogen and sulfur dioxide are separated to recover the sulfur dioxide product.

In the second allocation nitrogen and oxygen are separated initially, and the pure oxygen is mixed with cool recycled sulfur dioxide to provide the diluted oxygen feed to the burner. This concept eliminates the sulfur dioxide purification step required of the first allocation.

Both allocations satisfy the technical requirements that sufficient amount of diluent be present in the burner feed and that all the material balances be satisfied. What other dominant differences exist with which we could distinguish the better allocation?

These allocations differ in two important ways, the difficulty of the separation problems encountered and the tolerance to uncontrolled feed changes.

- ***Differences in Separations*** *Allocation 1 requires the separation of a mixture of*

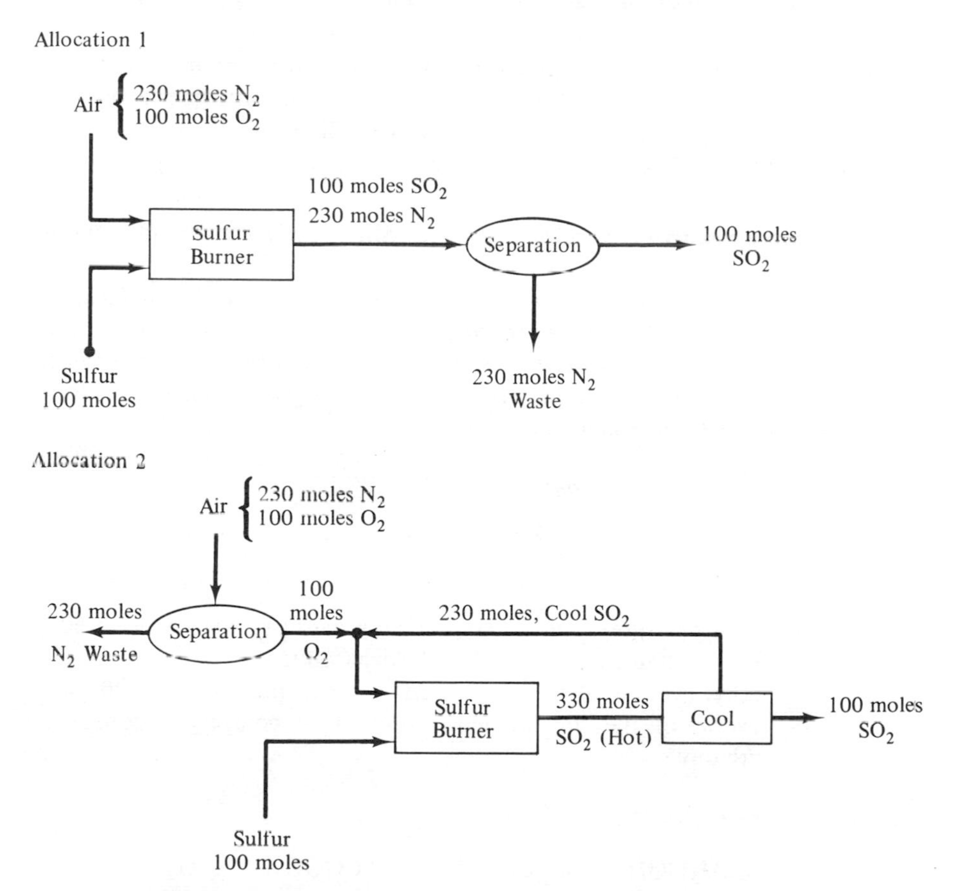

Figure 3.4–1. Alternative Species Allocations During Sulfur Oxidation

nitrogen and sulfur dioxide into sulfur dioxide (boiling point 14°F) and nitrogen (boiling point −320°F). These species differ greatly in boiling point, the one tending toward a liquid and the other a gas. This is a relatively easy separation.

Allocation 2 requires the separation of a mixture of nitrogen (boiling point −320°F) and oxygen (boiling point −297°F) into pure oxygen and nitrogen. Both of these are very much gases, a much more difficult separation than in allocation 1, yet there are commercially available processes to perform N_2–O_2 separation.

Since in both cases 330 moles of material must be separated, we favor allocation 1, which ought to lead to the easier separation.

- ***Tolerance to Feed Disturbances*** *Sudden changes in the feed flow rates and quality are always possible, leading either to an excess of oxygen or sulfur from the burner. Allocation 2 requires perfect stoichiometric feed rates to obtain pure sulfur dioxide product, whereas allocation 1 has its separation units in a position to catch these disturbances. This then favors allocation 1, as it leads toward a more flexible process.*

It is our judgment that the first allocation in Figure 3.4–1 is far superior.

In this simple example we witnessed the role that engineering judgment plays in assessing alternative allocations. The material balances were satisfied in both cases, and we had to look at other factors to carry process discovery further. In the next example we extend these ideas to a more complex situation.

Example 3.4–2 Ethyl Acetate from Ethanol. *Ethyl acetate, a common commercial solvent, can be manufactured from ethanol (ethyl alcohol) by a two-step reaction path. The first reaction is the controlled partial oxidation of ethanol to acetic acid and water. In the second reaction, the acetic acid is reacted with additional ethanol to form ethyl acetate and water. In this example, the reaction conditions are fixed rigidly to focus attention on species allocations; in practice, the engineer might exploit flexibility in the reaction conditions and conversions.*

Reaction 1. Oxidation of Ethanol

$$\underset{\text{(ethanol)}}{C_2H_5OH} + \underset{\text{(oxygen)}}{O_2} \rightarrow \underset{\text{(acetic acid)}}{CH_3COOH} + \underset{\text{(water)}}{H_2O}$$

Conditions: high pressure, vapor phase, reaction over catalyst. At least 50 mole per cent nitrogen needed in feed as diluent. Prohibited in feed is ethyl acetate. Water is allowed. Oxygen must be in a 20 per cent excess of the stoichiometric amount to completely consume the ethanol.

Reaction 2. Esterification

$$\underset{\text{(ethanol)}}{C_2H_5OOH} + \underset{\text{(acetic acid)}}{CH_3COOH} \rightleftharpoons \underset{\text{(ethyl acetate)}}{CH_3COOC_2H_5} + \underset{\text{(water)}}{H_2O}$$

Conditions: reaction occurs in solution at ambient conditions, at a conversion of 60 per cent. Oxygen is prohibited in feed. Water and nitrogen are allowed contaminants.

- ***Raw Materials*** *A 70 mole per cent water solution of ethanol is the source of ethanol, and air (80 mole per cent nitrogen, 20 mole per cent oxygen) is the source of oxygen.*

- ***Products*** *One thousand moles per hour of pure ethyl acetate is the main product. No ethanol, acetic acid, or ethyl acetate is to be allowed to leave the process; this is a pollution-control constraint.*

What concrete information is contained in this problem statement? How much of the process is uniquely determined?

Reaction 2 is the only source of ethyl acetate, and for this reason it must generate 1000 moles/h of ethyl acetate. Since the conversion is only 60 per cent, more than the stoichiometric amounts of ethanol and acetic acid must be fed to the reactor: that is, 1667 moles/h of ethanol and of acetic acid (1000/0.60). The reaction will generate 1000 moles/h of water as a by-product. The reactor effluent will contain 1000 moles/h of ethyl acetate, 1000 moles/h of water plus whatever comes in with the feed, and 667 moles/h each of unreacted ethanol and acetic acid.

Reaction 1 is only one source of the 1667 moles/h of acetic acid fed to reaction 2; the effluent from reaction 2 is a potential source of 667 moles/h of unreacted acetic acid. For each mole per hour of acetic acid X produced from reaction 1, X moles of ethanol are consumed, 1.20X moles/h of oxygen must be supplied to provide the 20 per cent excess oxygen, and at least 2.20X moles/h of nitrogen must be supplied as diluent. The effluents from this reactor will be X moles/h of acid, X moles of water plus any water that enters the reactor, 0.20X moles of unreacted oxygen, and 2.20X moles of nitrogen. Moreover, we can state that X will not be less than 1000 moles/h, and not more than 1667 moles/h: the first if all the acetic acid is recycled in reaction 2, and the second if none is recycled.

However, if we have no means of disposing of excess acetic acid (say, by neutralization, recovery, or combustion), all of it must be recycled. The recycle can either be directly to reactor 2 or indirectly via reactor 1. In any case the net result is a recycle to reactor 2.

Acetic acid is one of the three pollutants that must not leave the process. This additional fact, unique to this problem, enables us to fix X at 1000 moles/h; that is, only the amount of acetic acid needed to make the ethyl acetate is to be produced by reaction 1.

Figure 3.4–2 shows that portion of the material flow which is uniquely determined. The species allocations are only partially defined. We must go beyond that which is uniquely determinable.

From this point on materials are allocated on the basis of plausible arguments that have a good chance of leading to an economic process.

We need 1000 moles/h of ethanol for reaction 1, but there are two sources of ethanol, the ethanol solution (which is only 70 per cent ethanol) and the unreacted

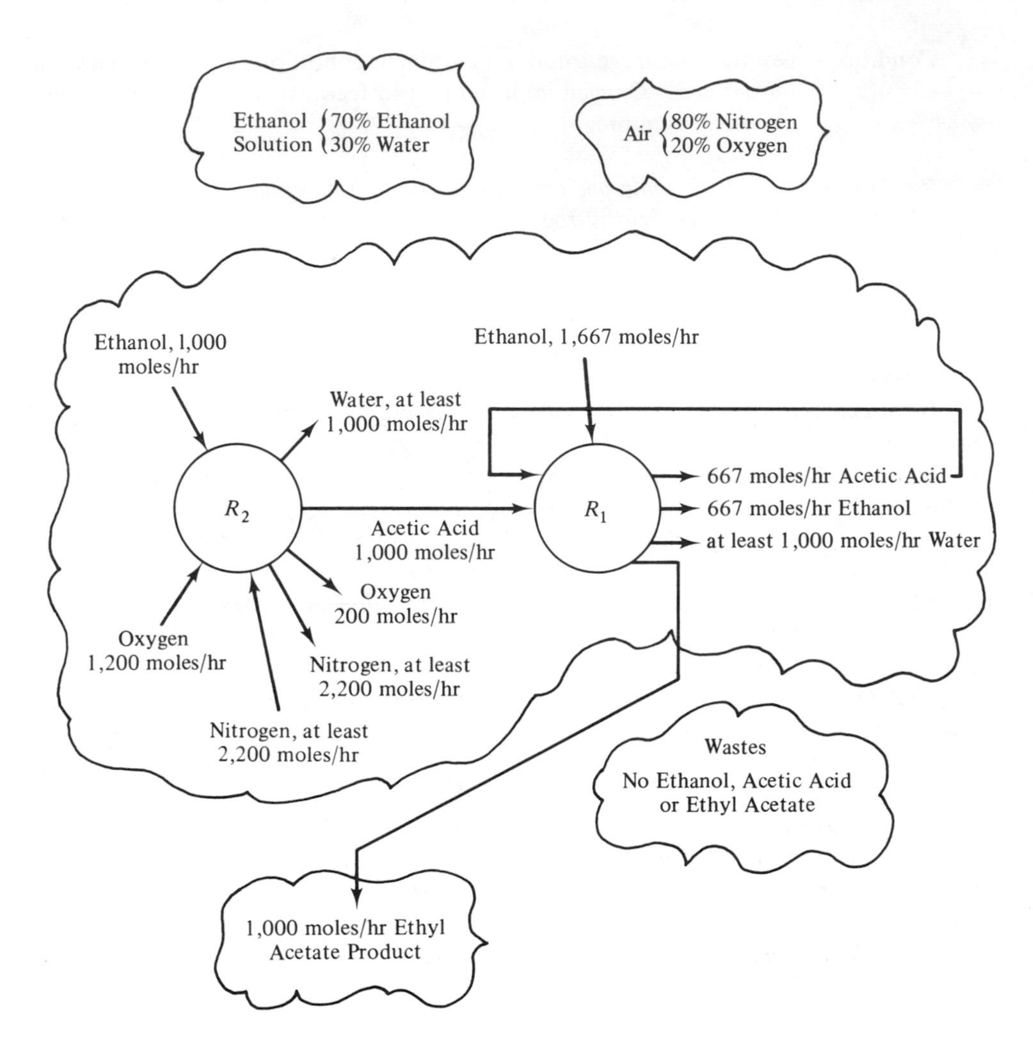

Figure 3.4–2. Basic Fragmentary Species Allocations That Must Occur During Ethyl Acetate Production

ethanol from reaction 2. What is a plausible source of the ethanol? Water is an allowed contaminant in the feed to reactor 1, but ethyl acetate is prohibited. If we used the recycle ethanol in reaction 1, care must be taken to remove the ethyl acetate; but if we recycled it to reaction 2, any contaminant of ethyl acetate would be no problem. The ethanol solution looks like a good source of the ethanol for reaction 1. Should we remove the water before using the ethanol as the feed to reaction 1? Water is going to be produced by the reaction anyway, so perhaps it is simplest merely to feed the solution to the reactor. With these plausible arguments, 1000 moles/h of ethanol and 430 moles/h of water will be fed to reactor 1.

There are two sources of oxygen for reactor 1 to meet the need of 1200 moles/h. An unlimited supply is available from the air, and 200 moles/h is available as the effluent from reaction 1. What is a reasonable allocation? At least 1000 moles/h of oxygen must be obtained from the air; it seems simplest to bring in all 1200 moles/h from that source. Will there be enough diluent nitrogen? Twelve hundred moles/h of oxygen from the air will bring with it 4800 moles of nitrogen, which is well above the 2200 moles/h minimum. Is it too far above? A check with reactor design experts might be in order.

These plausible decisions lead to the influx of 1200 moles/h of oxygen and 4800 moles/h of nitrogen into reactor 1.

A material balance over reactor 1 indicates that the effluents are 1000 moles/h of acetic acid, 200 moles/h of oxygen, 4800 moles/h of nitrogen, and 1430 moles/h of water. What should be done with these effluents?

The acetic acid and water are normally liquids, and the oxygen and nitrogen are normally gases. The separation of the two phases would be easy. Furthermore, the acetic acid is required in reaction 2 and the oxygen is prohibited; they must *be separated. It seems reasonable to attempt a separation of gas and liquid phases at this point. What about a water–acetic acid separation? Water is going to be produced by reaction 2 and the reaction occurs in solution, so it seems to be easiest to leave the two unseparated.*

These plausible decisions lead to the transfer of 1000 moles/h of acetic acid and 1430 moles/h of water from reactor 1 to reactor 2. The 200 moles/h of oxygen and 4800 moles/h of nitrogen are sent out as wastes, there being no purpose to trying the difficult separation of nitrogen and oxygen.

What is the best source of the 1667 moles/h of ethanol and the extra 667 moles/h of acetic acid needed as feeds to the reactor? We will recycle the 667 moles/h excess of ethanol and acetic acid, which appears as effluent from reactor 2, and bring in from the raw-materials source 1000 moles/h of ethanol. The additional ethanol solution will introduce 430 moles/h of water.

These reasonable decisions lead to a reactor 2 effluent of 1000 moles/h of ethyl acetate, 667 moles/h of acetic acid and of ethanol, and 2860 moles/h of water. This stream must be separated into a pure product stream of 1000 moles/h, a waste-water stream of 2860 moles/h, and a stream to recycle back to reactor 2.

This complete material allocation, shown in Figure 3.4–3, is the result of exact material balances and the plausible judgment of the process inventor. This allocation may well be the basis for the discovery of an economic process design.

This example shows vividly how the species allocation, used to connect the raw materials, the chemical reactors, and the process effluents, determines the separation problems to be solved. In Figure 3.4–3 two separations occur: first, the separation of oxygen and nitrogen from a mixture of oxygen, nitrogen, acetic acid, and water, and, second, the separation of ethanol and acetic acid and of water from a mixture of ethanol, acetic acid, water, and ethyl acetate. The allocations were purposely chosen to lead to these simpler separations. The choice was guided by material balances and judgment.

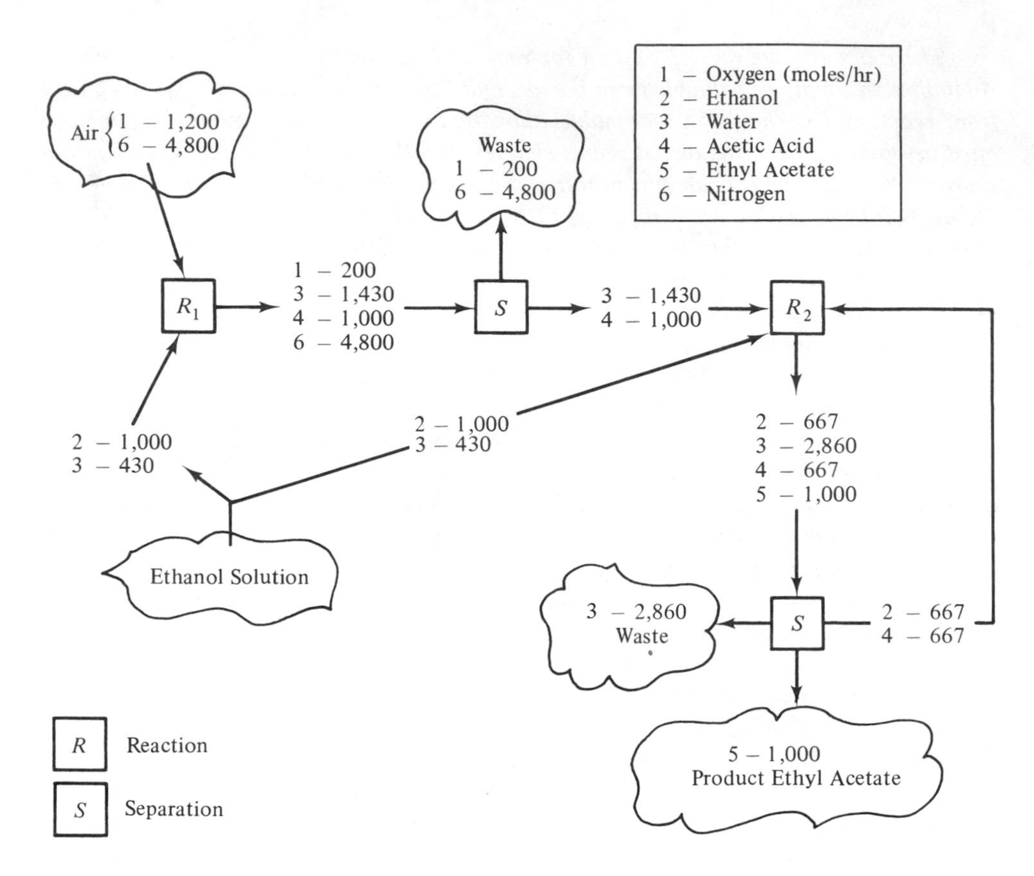

Figure 3.4–3. One of Many Plausible Complete Species Allocations

3.5 ALTERNATIVE ALLOCATIONS

We have seen how material balances, reaction stoichiometry, and engineering judgment determine the allocation of species that ought to lead to efficient processing. While the reaction stoichiometry and the material balances lead to single and exact results, engineering judgment adds the flexibility required of processing innovation. A significant portion of the material flow is subject to change and alteration to improve processing efficiency.

This section serves to emphasize the alternative allocations that must be considered. Commonly, species allocations will lead to several possible processing concepts, which cannot be distinguished; the separations involved may be different, but it may not be clear which of several is the easiest. In such a situation the engineer must keep several of the alternative allocations for further study. Several candidate processing schemes exist to be engineered further.

Example 3.5–1 Octanes from Butanes. *It is proposed to manufacture octanes (C_8H_{18}) for use as motor fuel from a feed stock of 40 per cent isobutane (iC_4H_{10}) and 60 per cent normal butane (nC_4H_{10}). Such molecular modifications of hydrocarbons are the heart of modern petroleum processing in which low-value material is transformed into high-grade products.*

The alkylation reaction 1 is to be used to generate the octane molecules. This reacts isobutane and normal butylene (nC_4H_8) in the presence of hydrofluoric acid (HF) at low temperatures. The iC_4H_{10}, nC_4H_8 feed ratio must be 8:1 to prevent the polymerization of nC_4H_8. All the nC_4H_8 is consumed. In this reaction any nC_4H_{10} is not altered and is an allowed contaminant.

1. Alkylation

$$\underset{\text{(isobutane)}}{iC_4H_{10}} + \underset{\text{(normal butylene)}}{nC_4H_8} \xrightarrow[\text{low temp.}]{\text{HF}} \underset{\text{(octane)}}{C_8H_{18}} \quad \text{plus small amounts of tars}$$

There is no butylene in the feed stock, so it is necessary to dehydrogenate the normal butane to generate normal butylene required, reaction 2. Dehydrogenation occurs at 1000°F in the pores of a Cr_2O_3/Al_2O_3 catalyst in a reactor operated at low pressure. To prevent the coking of the catalyst, the feed must be free of hydrocarbons heavier than the butane. In addition to hydrogen, small amounts of lighter hydrocarbons are produced. Thirty per cent of the feed is converted to product in one pass through the bed of catalyst.

2. Dehydrogenation

$$\underset{\text{(normal butane)}}{nC_4H_{10}} \xrightarrow[\text{catalyst}]{1000^\circ\text{F}} \underset{\text{(normal butylene)}}{nC_4H_8} + \underset{\text{(hydrogen)}}{H_2} \quad \text{plus small amounts of lighter hydrocarbons}$$

A reaction production and consumption analysis leads to the overall balance shown in Figure 3.5–1. The production of 100 moles of C_8H_{18} is taken as the basis, and required for this production is 100 moles of iC_4H_{10} and nC_4H_{10}. The unequal feed ratio

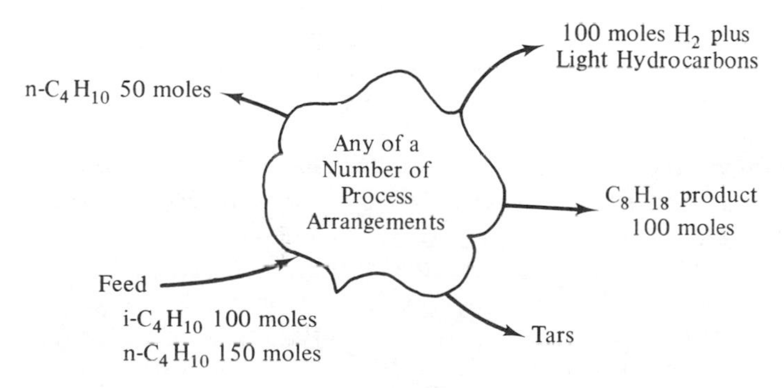

Figure 3.5–1. Overall Balance on Butane-to-Octane Process

requires that 50 moles of nC_4H_{10} be not used for want of an iC_4H_{10} mate. The by-product of 100 moles of H_2, tars, and light hydrocarbons leaves. The small amounts of tar and light hydrocarbons were not included in the balance. Regardless of the internal detail of the process, Figure 3.5–1 holds.

The central problem in species allocation is to provide for the following:

> *333 moles nC_4H_{10} to feed the dehydrogenation reactor. At 30 per cent conversion this reactor produces 100 moles nC_4H_8, 100 moles H_2, 233 moles unreacted nC_4H_{10}, plus light hydrocarbons.*
>
> *800 moles iC_4H_{10} plus 100 moles nC_4H_8 for the alkylation reactor. This reactor generates 100 moles C_8H_{18}, 700 moles unreacted iC_4H_{10}, plus tars.*

Figure 3.5–2 shows three species allocations which accomplish these ends. We examine these separately, realizing that still other allocations exist which might lead to successful processing plans.

In allocation 1 the feed stock is split into iC_4H_{10}, which goes to the alkylation reactor, and into nC_4H_{10}, part of which leaves the process and part of which goes to the dehydrogenation reactor. Recycled nC_4H_{10} from the dehydrogenator is added to feed stock nC_4H_{10} to complete the dehydrogenator feed requirements. The excess iC_4H_{10} found in the alkylation reactor effluent is used to complete the feed requirements to the alkylation reactor. This allocation leads to these three separation problems:

Feed stock:

100 iC_4H_{10}
150 nC_4H_{10}
→ 100 iC_4H_{10}
→ 150 nC_4H_{10}

Dehydrogenator effluent:

100 H_2
. . . light hydrocarbon
233 nC_4H_{10}
100 nC_4H_8
→ {100 H_2, . . . light hydrocarbon}
→ 233 nC_4H_{10}
→ 100 nC_4H_8

Alkylator effluent:

700 iC_4H_{10}
100 C_8H_{18}
. . . tars
→ 700 iC_4H_{10}
→ 100 C_8H_{18}
→ . . . tars

In allocation 2 the feed stock is fed directly to the alkylator. This is possible because the nC_4H_{10} is an allowed contaminant there and passes through the reactor unharmed. This leads to these separations:

Dehydrogenator effluent:

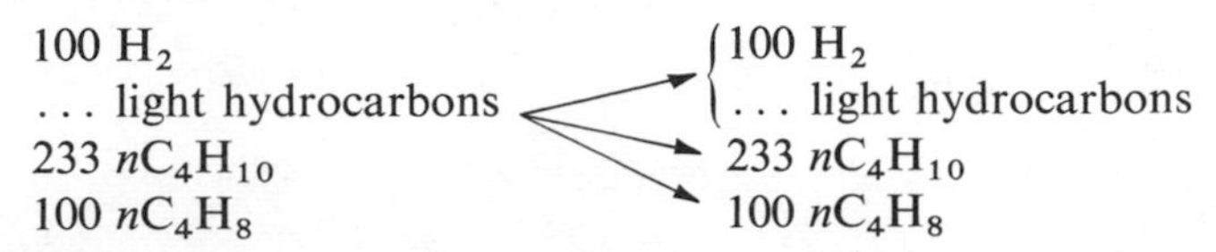

Alkylator effluent:

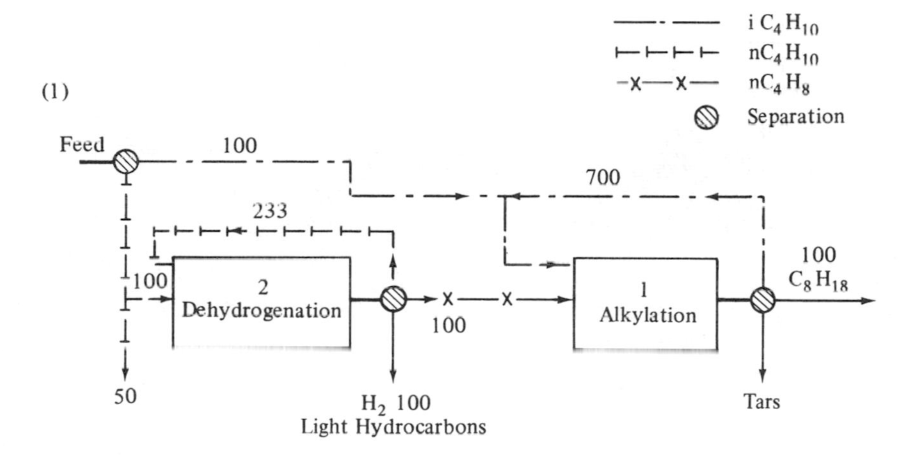

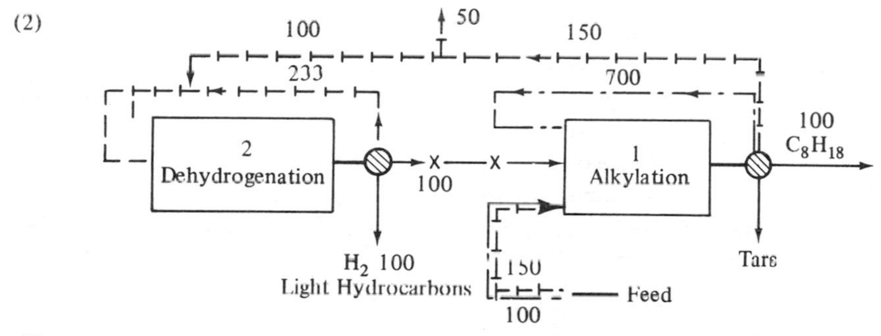

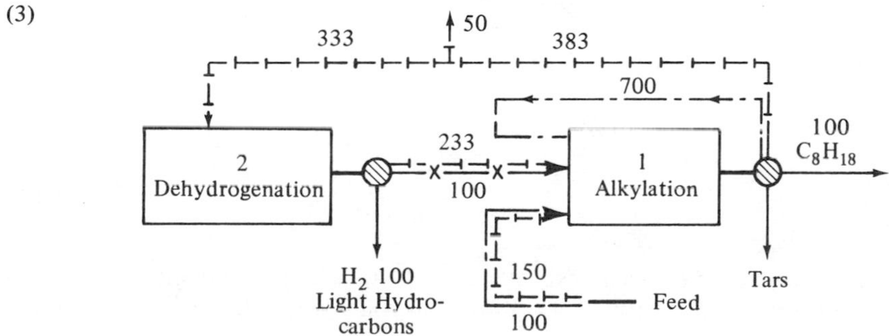

Figure 3.5–2. Three Alternative Species Allocations to Support Dehydrogenation and Alkylation

Finally, in allocation 3, allocation 2 is modified to have the unreacted nC_4H_{10} *in the dehydrogenator effluent pass through the alkylator before recycle. This leads to this set of separation problems:*

Dehydrogenator effluent:

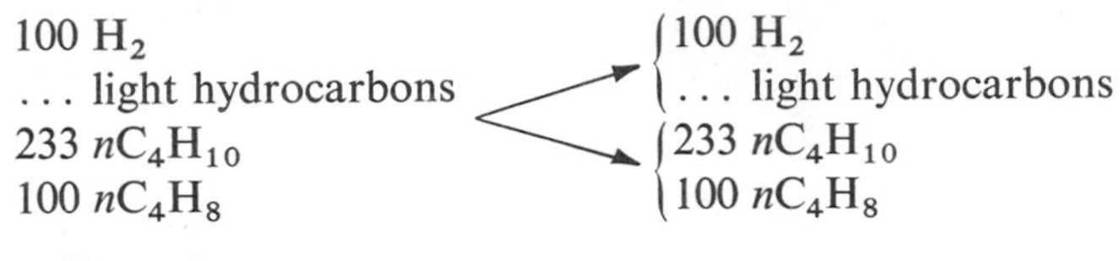

Alkylator effluent:

700 iC_4H_{10}	→	700 iC_4H_{10}
100 C_8H_{18}	→	100 C_8H_{18}
. . . tars	→	. . . tars
383 nC_4H_{10}	→	383 nC_4H_{10}

Which of these and other allocations lead to simple separation problems? These separation problems are so similar that we must learn more about separation technology to isolate the best allocation.

3.6 CONCLUSIONS

At the beginning of this chapter only the chemistry of processing was in mind. We discovered that the conservation of mass led directly to certain material flows that must exist within the emerging process. The techniques of material balancing were developed to analyze these flows.

However, it became clear that material balancing only tied down uniquely a certain fraction of the total material flow which might exist in the process. The species allocations that are not unique must be determined by judgment, and this flexibility leads to processing innovations.

One major effect of the flexibility in the species allocations is to offer the possibility of alternative separation problems. The major element of judgment then became that of identifying relatively easy species separations. The information needed to assess the difficulty of separations is discussed in the next chapter.

REFERENCES

There are a number of excellent texts on material balancing; our favorites include:

J. C. WHITWELL and R. K. TONER, *Material and Energy Balances*, McGraw-Hill, New York, N.Y., 1973.

D. M. HIMMELBLAU, *Basic Principles and Calculations in Chemical Engineering*, 2nd ed., Prentice-Hall, Inc., Englewood Cliffs, N.J., 1967.

For a more advanced discussion of problems in species allocation, see:

J. J. SIIROLA and D. F. RUDD, *Industrial and Engineering Chemistry Fundamentals*, **10**, 353, Aug. 1971.

PROBLEMS

These problems are divided into three groups: material balances that involve no recycle, material balances that involve recycle, and species-allocation problems. It has been our experience that the jump in required problem-solving sophistication between successive problem groups is significant, and that the initiate does well to progress gradually from the simple to the more sophisticated through the sequence of problems laid out here.

1. **MINERALS FROM THE SEA.** The composition of seawater is, in parts per million (ppm),

Chloride	Cl^-	18,980
Sodium	Na^+	10,561
Sulfate	SO_4^{2-}	2,649
Magnesium	Mg^{2+}	1,272
Calcium	Ca^{2+}	400
Potassium	K^+	380
Bicarbonate	HCO_3^-	142
Bromide	Br^-	65
Other salts		34

How many cubic feet of seawater, density 60 lb/ft^3, must be processed to recover 1 ton of sodium chloride? Of magnesium? Of bromine?

2. **CRYSTALLIZER OPERATION.** One ton per hour of a hot solution containing 35 per cent solids is cooled to cause solids precipitation. At the cool outlet temperature of the crystallizer the solubility of the solid is 10 per cent. How many tons per hour of solids are formed?

3. **HYDROGEN SULFIDE REMOVAL.** Hydrogen sulfide gas is to be removed from a hydrogen gas stream by an absorber–stripper process. The hydrogen sulfide is absorbed in a solvent in the absorber, and the loaded solvent is sent to the stripper, where the hydrogen sulfide is removed. The stripped solvent is recycled to the absorber to pick up more hydrogen sulfide. How much hydrogen sulfide leaves the stripper?

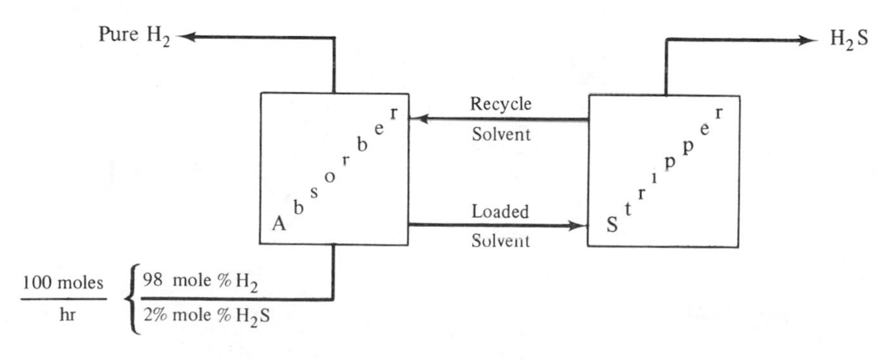

4. **SUGAR REFINING.** An evaporator is to produce a 70 weight per cent sugar solution from a 30 weight per cent solution. How many pounds of water must be evaporated per ton of sugar solution produced?

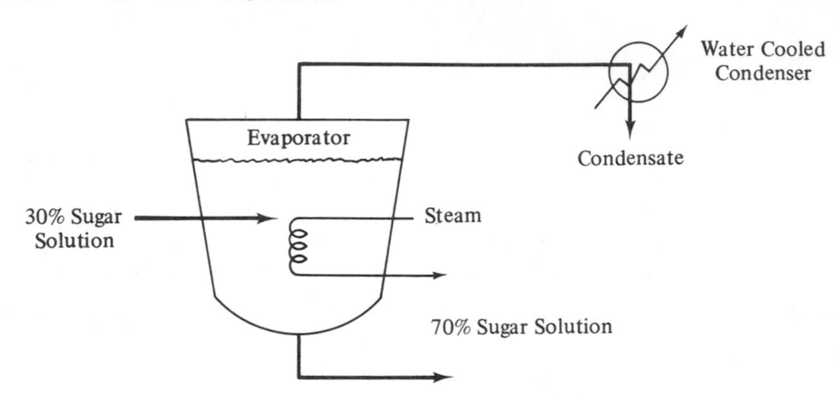

5. **PRECIPITATION OF BENZENE FROM CHLOROFORM.** What percentage of the benzene, contained in a 40 mole per cent solution of benzene in chloroform, will precipitate out if the solution is cooled to −70°F? A saturated solution at −70°F is 30 mole per cent benzene, and the solid benzene contains no chloroform.

6. **MOISTURE ADDITION TO SOAP.** In the manufacture of soap, a 30-ton batch is found to contain 10 per cent moisture. However, to meet quality-control standards, the soap must contain 25 per cent moisture. How much water must be added to bring the batch to quality?

7. **OIL AND MEAL FROM SEED.** Oilseed protein sources include soybean, cotton seed, peanut, sunflower, copra, rapeseed, sesame, safflower, castor, and flax. Commonly, the separation of the oil and the protein meal is performed by solvent extraction.

   The analysis of cottonseed is 4 per cent hull, 10 per cent linter, 37 per cent meal, and 49 per cent oil. During the extraction step about 2 lb of solvent, hexane, must be

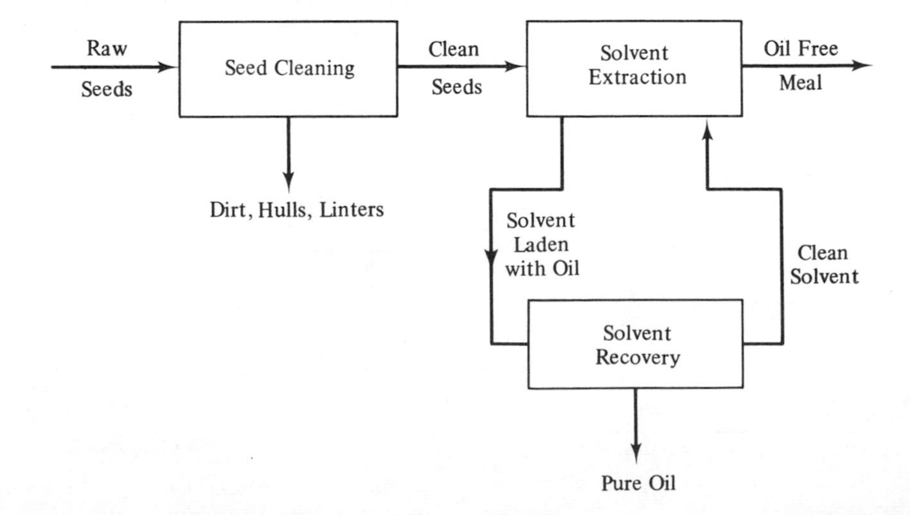

used per pound of clean seeds processed. For each ton of raw seeds to be processed, determine the amount of oil and oil-free meal produced and the amount of hexane that must be recycled through the solvent extraction unit.

8. **REDUCTION OF IRON ORE.** Six hundred pounds of carbon are reacted with 1 ton of pure iron ore (Fe_2O_3) to produce iron metal:

$$Fe_2O_3 + 3\,C \rightarrow 2\,Fe + 3\,CO_2$$

If the reaction goes to completion, determine the pounds of
   (a) Fe_2O_3 or carbon left unreacted.
   (b) Pure iron formed.
   (c) Carbon monoxide formed.

9. **DRYING SOLIDS.** Two tons per hour of a sludge containing 30 per cent solids is centrifuged to form a thickened sludge of 60 per cent solids. This is dryed to a 5 per cent moisture-content powder. How much water is removed per hour by the centrifuge and by the dryer?

10. **HYDROGEN SULFIDE REMOVAL.** Hydrogen sulfide is stripped by an absorber from a stream of 25 mole per cent H_2S and 75 per cent N_2. How many pounds of H_2S can be recovered per pound of gas processed?

11. **SOLIDS SCREENING.** Solids of $\frac{1}{4}$ in. diameter are to be recovered from a mixture of solids of various sizes by passing the mixture over a screen with $\frac{3}{8}$-in. openings. Most of the $\frac{1}{4}$-in. solids pass through the screen, but some remain in the oversize material. The initial mixture contains 60 weight per cent $\frac{1}{4}$-in. solids, 30 per cent $\frac{1}{2}$-in. solids, and 10 per cent $\frac{3}{4}$-in. solids. The screening operation gives 40 tons of $\frac{1}{4}$-in. solids and 60 tons of oversize (including the entrained $\frac{1}{4}$-in. solids). Calculate the composition of the oversize material.

12. **PHOSPHATE ROCK BENEFICIATION.** Low-grade phosphate rock containing 5 per cent P_2O_5 is to be upgraded to a 20 per cent concentrate. How much low-grade ore must be processed to produce 1 ton of the concentrate?

13. **CITY REFUSE TO SOIL CONDITIONER BY COMPOSTING.** One sixth of Houston's city refuse is turned into an organic soil conditioner by Lone Star Organics, a subsidiary of Metropolitan Waste Conversion Corporation. The unit handles 300 tons/day of garbage and refuse, which consists of 75 per cent paper cardboard and rags, 13 per cent garbage and garden refuse, 7 per cent tin cans and light ferrous metal, 1.5 per cent tramp metal (large pieces of iron and steel), 2 per cent noncompostibles (rubber, heavy plastic, cement), 1 per cent glass, and 0.5 per cent aluminum and nonferrous metals.

    First, the glass, aluminum, tramp metals, and other noncompostibles are removed by manual sorting and picking. Then the tin cans and light ferrous materials are removed by magnetic separation. The remaining material is milled to less than 1 in. and then mixed with sludge from the sewage-treatment plant. The sludge provides the bacteria to compost the waste and the organic nitrogen to maintain the proper carbon and nitrogen ratio. After 6 days of composting the refuse is ground and dried, stopping bacterial action.

    The processed compost is gray and has a texture similar to peat moss. As a soil conditioner, the compost is slightly better than steer manure.

    The flow sheet shown is a simplified version of the actual process (*Chemical Engineering*, Nov. 7, 1967, pp. 232–234).

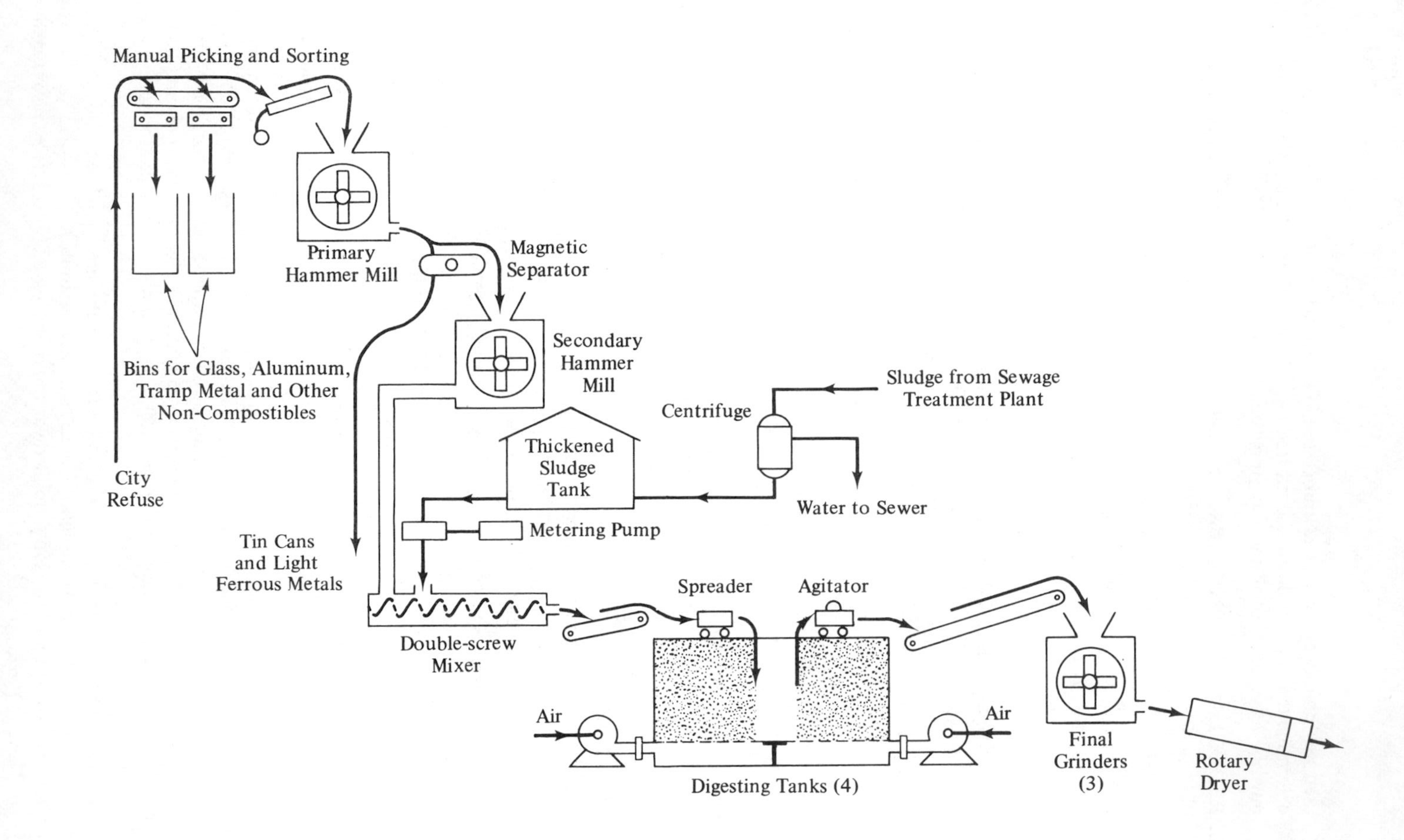

Manual Picking and Sorting
Primary Hammer Mill
Magnetic Separator
Secondary Hammer Mill
Bins for Glass, Aluminum, Tramp Metal and Other Non-Compostibles
City Refuse
Sludge from Sewage Treatment Plant
Centrifuge
Thickened Sludge Tank
Water to Sewer
Metering Pump
Tin Cans and Light Ferrous Metals
Spreader
Agitator
Double-screw Mixer
Air
Air
Final Grinders (3)
Rotary Dryer
Digesting Tanks (4)

(a) How many tons per day are recovered of glass, aluminum, tramp metal, and other noncompostibles?

(b) It has been reported that 50,000 lb/day of tin cans and other light ferrous metals are recovered. Is this reasonable?

(c) The sludge from sewage treatment is 30 per cent solids and is thickened to 50 per cent solids. If the thickened sludge and milled refuse are mixed in equal amounts in the double-screw mixer, how much thickened sludge must be added per day?

(d) How many tons per day of sludge must be brought in for thickening, and how much water leaves the centrifuge for the sewer?

(e) If paper, cardboard, and rags are removed for sale before the secondary hammer mill, how much less sludge must be brought in for the composting?

14. **WATER REMOVAL IN WASTE PROCESSING.** Figure 5.7–3 is the flow sheet for a process we shall develop for obtaining a high-protein dry concentrate and a high-lactose dry concentrate from the wastes of a cheese factory. Determine the water removal rates from the evaporator and spray drier that dry the protein concentrate. How much water is removed from the lactose spray drier?

15. **DRIED MILK.** Skim milk contains 3.0 to 3.5 per cent casein, 0.5 to 0.8 per cent whey protein, 0.1 to 0.2 per cent lower-molecular-weight polypeptides, amino acids, and urea, 4 to 5 per cent lactose, and 0.5 to 1.0 per cent mineral salts and various amounts of citric acid, lactic acid, and vitamins. Both the casein and whey protein are proteins. What percentage protein is dried skim milk?

16. **COD-LIVER EXTRACTION.** The following data are available on a multistage cod-liver-oil extraction unit, in which the valuable oil is being extracted from the livers, using ether.

Entering livers:
1000 lb/h
33 mass per cent oil
66 mass per cent inerts

Entering solvent:
2000 lb/h
1 mass per cent oil
99 mass per cent ether

Leaving extract:
1820 lb/h
19 mass per cent oil
81 mass per cent ether

Leaving processed livers:
1180 lb/h

Determine the composition of the processed livers.

17. **UPPER AND LOWER BOUND ON WATER USAGE.** Throughout this chapter we analyze a variety of reactor–thickener configurations for the production of a caustic solution from soda ash and lime (Example 3.3–6 and Problems 23 and 29). In this problem we establish an upper and lower bound on the water required by such a process.

This illustrates the use of material balances to isolate the gross operating characteristics of a process before the processing details are known.

Using overall material balances, show that the concentration (x tons caustic/ton solution) of caustic in the washed sludge solution is related to the wash-water rate (W tons/day) as follows:

$$W = 230{,}666 - \frac{x}{0.08}(46{,}666)$$

The wash-water rate for any design must then be somewhere between 230,666 tons/day and 184,000 tons/day. What are the bounds on the caustic production?

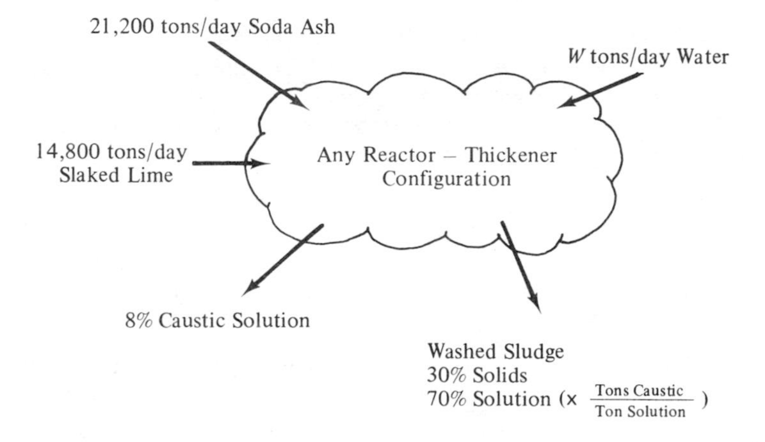

18. BENEFICIATION OF PHOSPHATE ROCK. Low-grade phosphate rock can be upgraded by froth flotation processing. The low-grade rock is ground into a powder and suspended as a water slurry. The addition of small amounts of oil and other surface-active agents causes the phosphate particles to become adherent to air bubbles, and by frothing the slurry with air a phosphate-rich foam floats off the top of a froth flotation unit.

Illustrated is a sequence of rougher and cleaner flotation units arranged to produce a 22 per cent P_2O_5 concentrate from a 5 per cent low-grade rock. Laboratory data give the following stream analyses:

	WT % P_2O_5 ON DRY BASIS
Feed F	5
Waste W	1
Recycle R	10
Middling M	14
Product P	22

All the streams are a slurry of approximately 30 per cent solids. The weight of the frothing agents is negligible to the weight of the ore processed.

(a) What percentage of the phosphate in the feed material is recovered as the product?

(b) Determine the flow rates of all the streams if 100 tons/h of feed enters the process.

(c) How much water leaves with the waste rock?

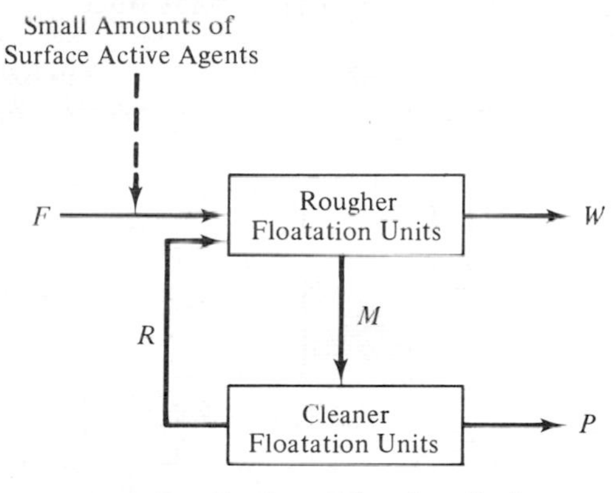

Benefication of Phosphate Rock
by Froth Floatation

19. DEGREASING OF MACHINE PARTS. One hundred tons per day of machine parts are to be degreased in a kerosene vat according to the plan shown. The parts contain 5 lb of grease/100 lb of parts. How much fresh kerosene must be added per day?

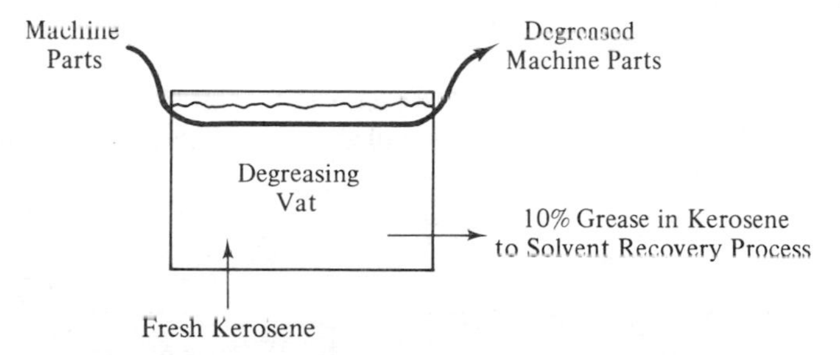

20. WATER TREATMENT. A water source contains 90 ppm Ca^{2+} and 80 ppm Mg^{2+} that must be removed before the water can be used as a boiler feed. It is proposed to add Na_3PO_4 in 10 per cent excess to ensure the removal of the calcium and magnesium ions as insoluble $Ca_3(PO_4)_2$ and $Mg_3(PO_4)_2$. How many pounds of sodium phosphate must be added to each 100 lb of water? The notation (ppm) refers to weight of solid per million weight of solution.

21. BENZENE–TOLUENE DISTILLATION. The feed to a distillation column contains 36 per cent benzene by weight, the remainder being toluene. The overhead distillate is to contain 52 per cent benzene by weight, while the bottoms are to contain 5 per cent benzene by weight. Calculate

(a) The percentage of the benzene feed contained in the distillate.

(b) The percentage of the total feed that leaves as distillate.

22. GLYCERINE DESALTING. (From Whitwell and Toner; see references.) A glycerol plant is treating a solution (that is 87 per cent water, 10 per cent glycerine, and 3 per cent

NaCl) with butyl alcohol in a solvent extraction tower. The alcohol fed to the tower contains 2 per cent water. The raffinate leaving the tower contains all the original salt and has a composition of 1.0 per cent glycerine and 1.0 per cent alcohol. The extract from the tower is sent to a distillation column. The distillate from this column is the alcohol, containing 5 per cent water. The bottoms from the column is 25 per cent glycerine and the rest water.

The two feed streams to the extraction column amount to 1000 lb/h each. Calculate the output of glycerine per hour from the distillation column.

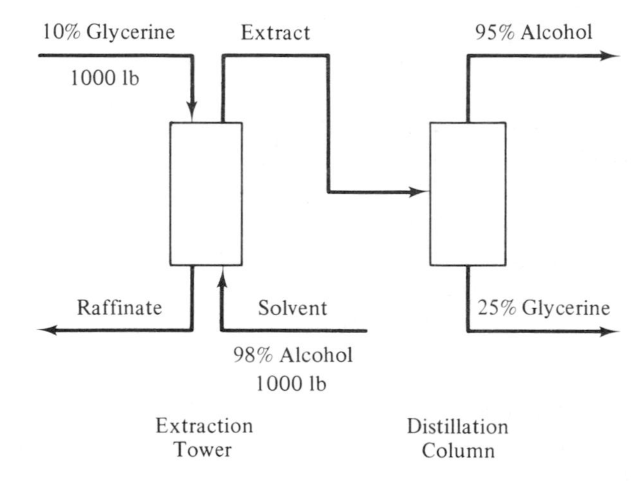

23. CAUSTIC PROCESS WITH ONE THICKENER. This is a simplified version of the process in Example 3.3–6 for the production of an 8 per cent caustic solution. The second

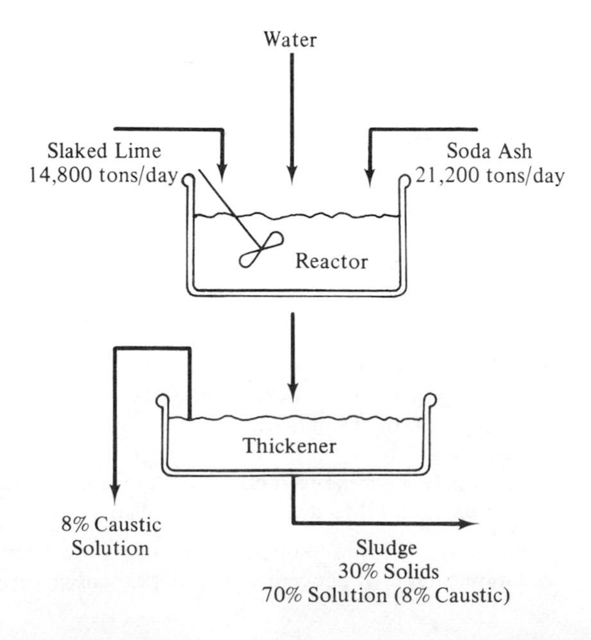

thickener has been removed and the water added directly to the reactor. Determine the water addition rate and the rate of production of the 8 per cent caustic. Comparing this production to that of the process in Example 3.3–6, determine the percentage of increase in production caused by the use of the second thickener.

24. **BLOWDOWN OF BOILER.** Boiler water feed, containing 750 ppm solids, is used to manufacture steam for circulation to a process system. Thirty per cent of the steam formed is lost and the remaining 70 per cent is condensed and returns to the boiler as additional boiler feed. To prevent the buildup of solids in the boiler water to beyond 3000 ppm solids, water is drawn off in the blowdown stream.

How many pounds of blowdown water must be withdrawn per pound of feed water added?

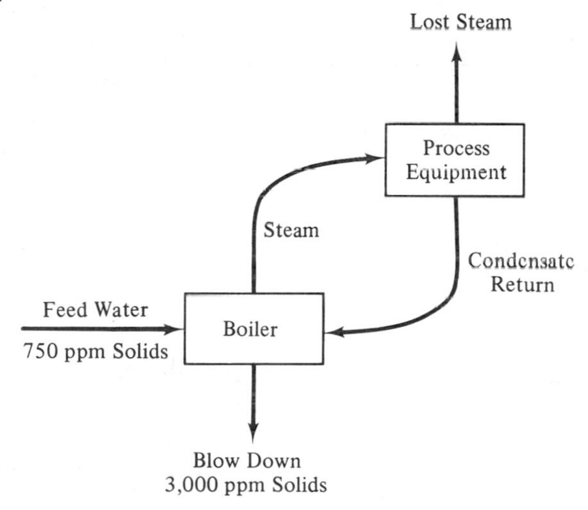

25. **ACID SLUDGE WASHING.** One ton per hour of an acid sludge containing 50 per cent by weight of inert solids in a 30 per cent H_2SO_4 solution is to be washed with fresh water to recover the acid. Tests show that the retention of the liquid on the solid is independent of acid concentration and is 1 lb of liquid/1 lb of dry solid.

(a) How much water is required per hour to recover 90 per cent of the acid if only one settling tank is used?

(b) If the same amount of water is used with two settling tanks employing countercurrent flow of solids and liquid, what percentage of the acid would be recovered?

26. **DRYING WITH AIR RECYCLE.** A solid containing 15 per cent water by weight is dried with air to a moisture content of 7 per cent. The fresh-air feed contains 0.01 lb of H_2O/lb of dry air, the recycle stream, 0.1, and the air to the drier at point *A*, 0.03. How many pounds of air are recycled per 100 lb of solid feed, and how much fresh air is required?

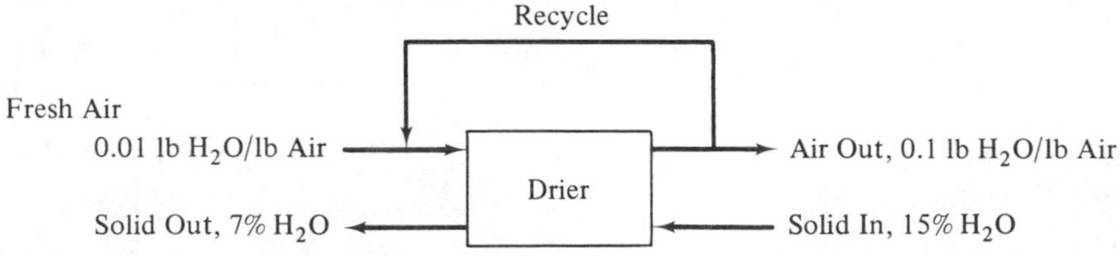

27. BUTANE TO BUTENE. Butane is to be dehydrogenated to butene in a catalytic reactor with 5 per cent of the butane entering the reactor being converted to butene and the other products shown. For what feed rate should the reactor be designed?

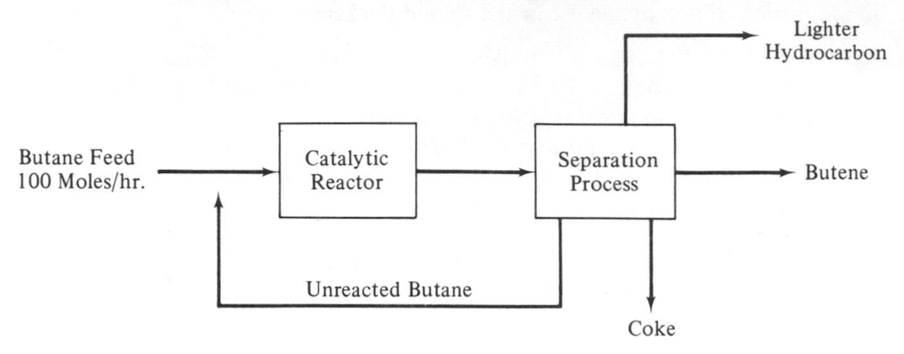

Answer 2,000 Moles/hr.

28. MODIFICATION OF INSTANT-COFFEE PROCESS. In Figure 3.3–5 the waste solution from the press leaves the process for waste disposal. It is proposed to increase processing efficiency by recycling this stream back to the percolators. Determine all the new flow rates in Table 3.3–3 resulting from this design modification.

29. THIRD THICKENER INTRODUCED. In Problem 23 and in Example 3.3–6 the effects of different sludge-thickening operations were investigated for processes to manufacture a caustic solution. It was found that caustic recovery was increased by the use of the second thickener. Determine the water addition rate and the production of 8 per cent caustic when a third thickener is used, as illustrated.

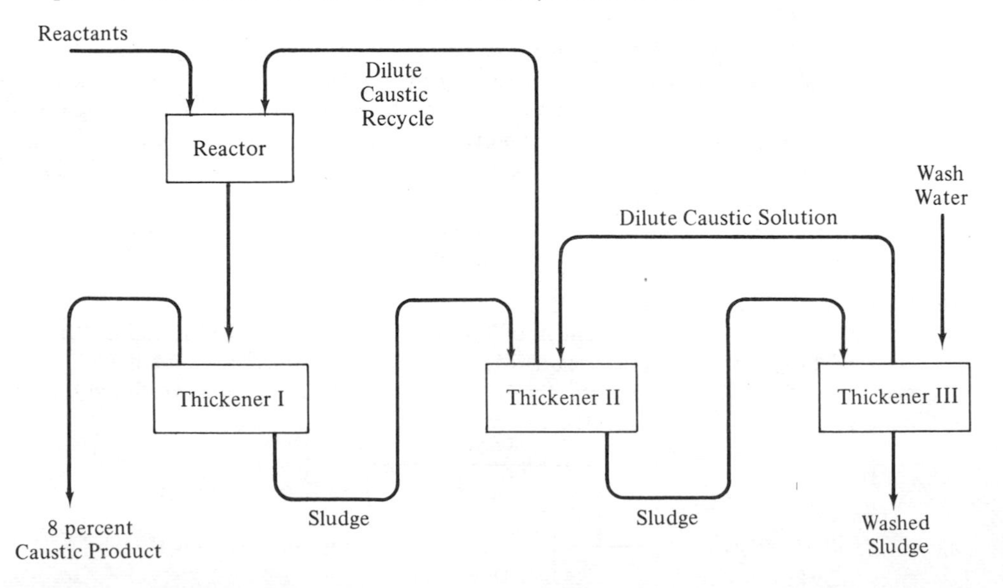

30. MANUFACTURE OF PHENOL. Phenol is to be manufactured from benzene, chlorine, caustic, and hydrogen chloride by the following three reactions:

1. $C_6H_6 + Cl_2 \xrightarrow[3\text{ atm}, FeCl_3]{60°C} C_6H_5Cl + HCl$ 90% conversion
(benzene) (chlorine) (chlorobenzene) (hydrogen chloride)

2. $C_6H_5Cl + NaOH + 2H_2O \xrightarrow[400\text{ atm}]{360°C} C_6H_5ONa + NaCl + 2H_2O$ 90% conversion
(chlorobenzene) (caustic)

3. $C_6H_5ONa + HCl + H_2O \xrightarrow[1\text{ atm}]{25°C} C_6H_5OH + NaCl + H_2O$ 100% conversion
(phenol)

Important data include:
- (a) Only benzene and chlorine allowed in reaction 1 feed.
- (b) Only chlorobenzene, caustic, and water allowed in reaction 2 feed.
- (c) C_6H_5ONa, HCl, H_2O, NaCl, C_6H_5Cl, and NaOH allowed in reaction 3 feed.

Determine three alternative species allocations to support these reactions, and compare the difficulty of material separations.

31. MANUFACTURE OF STYRENE. Styrene is to be manufactured from ethylene and benzene by the following reactions. Synthesize the three species allocations you feel would lead to economic processing.

1. $C_6H_6 + C_2H_4 \xrightarrow[3\text{ atm}]{95°C} C_6H_5\text{—}(C_2H_5) + \text{trace} \begin{Bmatrix}\text{toluene}\\ \text{diethylbenzene}\\ \text{tar}\end{Bmatrix}$ 40% conversion
(benzene) (ethylene) (ethylbenzene)

2. $C_6H_5\text{—}(C_2H_5) + 15\,H_2O \xrightarrow[1\text{ atm}]{500°C} C_6H_5(C_2H_3) + H_2 + 15\,H_2O + \text{trace} \begin{Bmatrix}\text{methane}\\ \text{toluene}\end{Bmatrix}$ 60% conversion
(ethylbenzene) (styrene) (hydrogen)

Additional information:
- (a) Raw benzene contains 2 per cent water.
- (b) Only benzene and ethylene are allowed in reaction 1.
- (c) Only benzene, water, and trace toluene are allowed in reaction 2.
- (d) No impurities are allowed in product styrene.

32. MANUFACTURE OF VINYL CHLORIDE. Prepare the two best species allocations to support these reactions for the manufacture of vinyl chloride from ethylene, air, and chlorine.

1. $C_2H_4 + Cl_2 \xrightarrow[3\text{ atm}]{95°C} C_2H_4Cl_2$ 90% conversion
(ethylene) (chlorine) (ethylene dichloride)

2. $C_2H_4Cl_2 \xrightarrow[20\text{ atm}]{400°C} C_2H_3Cl + HCl$ 80% conversion
(ethylene dichloride) (vinyl chloride) (hydrogen chloride)

3. $C_2H_4 + \frac{1}{2}O_2 + 2\,HCl \xrightarrow[5\text{ atm}]{300°C} C_2H_4Cl_2 + H_2O$ 70% conversion
(ethylene) (hydrogen chloride) (ethylene dichloride)

Additional information:

(a) No impurities are allowed in vinyl chloride product.
(b) Raw material ethylene contains 2 per cent carbon particles.
(c) Only pure Cl_2 and C_2H_4 are allowed in reaction 1 feed.
(d) Only pure $C_2H_4Cl_2$ is allowed as reaction 2 feed.
(e) Feed to reaction 3 must be pure.

33. DEACON PROCESS FOR WASTE RECYCLE. The Deacon process is used to convert by-product hydrogen chloride to chlorine by the reaction path

$$\underset{\text{(hydrogen chloride)}}{4\,HCl} + \underset{\text{(oxygen)}}{O_2} \rightleftharpoons \underset{\text{(chlorine)}}{2\,Cl_2} + \underset{\text{(water)}}{2\,H_2O}$$

This reaction occurs with a 60 per cent conversion of the feed HCl to Cl_2, at the temperature and pressure proposed. Air is the source of oxygen, and oxygen should be present in the required stoichiometric amount.

The Deacon process is to be used to recover and recycle the chlorine in waste HCl produced in the manufacture of 100 tons/day of chlorobenzene by the reaction path

$$\underset{\text{(benzene)}}{C_6H_6} + \underset{\text{(chlorine)}}{Cl_2} \rightarrow \underset{\text{(chlorobenzene)}}{C_6H_5Cl} + \underset{\text{(hydrogen chloride)}}{HCl}$$

It is proposed to keep the chlorine concentration below 10 mole per cent in the feed to this reaction to prevent the formation of higher chlorinated components. Also, no water or oxygen is allowed in the feed. The reaction goes to completion in the reactor.

Propose two species allocations for these two processes operating together, and rank the allocations according to ease of material separations. How would the processing scheme change if oxygen and water were allowed in the feed to the chlorobenzene reactor? The following boiling-point data should be used to estimate difficulties of separation.

SPECIES	BP (°K AT 1 ATM)
Oxygen	90
Hydrogen chloride	188
Chlorine	239
Water	373
Nitrogen	77
Benzene	353
Chlorobenzene	405

4

Separation Technology

There appears to be no limit in the variations that can be applied to the development of new separation processes. . . . The field has been called an inventor's paradise.

J. E. Powers

4.1 INTRODUCTION

Having established the reaction-path chemistry and having outlined the flow of material through the proposed process, attention naturally focuses on the problems of material separation. To separate materials economically, the engineer must create an environment in which the materials behave in a drastically different way. Such an environment is detected by the examination of the chemical and physical properties of the materials, looking for property differences that cause drastic differences in behavior. One species in the mixture might float while the others sink; some species might change to a gas or solid while the others remain a liquid; one species might dissolve in a solvent while the others remain insoluble; small molecules might pass through a plastic membrane while others remain behind.

Table 4.1–1 summarizes a wide variety of separation phenomena in common use. These operations are divided into three broad categories, depending on whether the separation is a result of phase equilibria, a difference in rate of some process, or a mechanical process. The table also includes the separating agent that creates the environment in which the species behave differently.

TABLE 4.1–1 Classes of Separation Processes*

Name	Feed Phase	Separating Agent	Product Phases	Property Exploited	Practical Example
Equilibration Separation Processes					
1. Evaporation	Liquid	Heat	Liquid + vapor	Difference in volatilities (vapor pressure)	Concentration of fruit juices
2. Flash expansion	Liquid	Pressure reduction (energy)	Liquid + vapor	Same	Flash process for seawater desalination
3. Distillation	Liquid and/or vapor	Heat	Liquid + vapor	Same (repeated internally)	Crude-oil separation
4. Stripping	Liquid	Noncondensable gas	Liquid + vapor	Same	Removal of light hydrocarbon from crude-oil fractions
5. Absorption	Gas	Nonvolatile liquid	Liquid + vapor	Preferential solubility	Removal of CO_2 and H_2S from natural gas by absorption into ethanolamines
6. Extraction	Liquid	Immiscible liquid	Two liquids	Different solubilities of different species in the two liquid phases	Recovery of antibiotics
7. Crystallization	Liquid	Cooling, or else heat causing simultaneous evaporation	Liquid and solids	Difference in freezing tendencies; preferential participation in crystal structure	Freezing process for seawater desalination
8. Adsorption	Gas or liquid	Solid adsorbent	Fluid and solid	Difference in adsorption potentials	Drying of gases by solid desiccants
9. Ion exchange	Liquid	Solid resin	Liquid + solid resin	Law of mass action applied to available anions or cations	Water softening
10. Drying of solids	Moist solid	Heat	Dry solid + humid vapor	Evaporation of water	Food dehydration
11. Leaching or washing	Solids	Solvent	Liquid + solid	Preferential solubility	Leaching of $CuSO_4$ from calcined ore

* From C. J. King, *Separation Processes*, McGraw-Hill Book Company, New York, 1970.

TABLE 4.1–1 **Classes of Separation Processes (*Cont.*)**

Name	Feed Phase	Separating Agent	Product Phases	Property Exploited	Practical Example
12. Clathation	Liquid	Clathrating molecule and cooling	Liquid + solid	Preferential participation in crystal structure	Hydrate process for seawater desalination
13. Osmosis	Salt solution	More concentrated salt solution; membrane	Two liquids	Tendency to achieve uniform osmotic pressures removes water from more dilute solution	Suggested for food dehydration
14. Bubble fractionation; foam fractionation	Liquid	Rising air bubbles; sometimes also complexing surfactants	Two liquids	Tendency of surfactant molecules to accumulate at gas–liquid interface and rise with air bubbles	Removal of detergents from laundry wastes; ore flotation
15. Flotation	Mixed powdered solids	Added surfactants; rising air bubbles	Two solids	Tendency of surfactants to adsorb preferentially on one solid species	Ore flotation; recovery of zinc sulfide from carbonate gangue
16. Magnetic separation	Mixed powdered solids	Magnetic field	Two solids	Attraction of materials in magnetic field	Concentration of ferrous ores
17. Paper chromatography	Liquid	Capillarity; paper or gel phase	Regions of moistened paper	Preferential solubilities and adsorption potentials in two phases	Protein separation for analytical purposes
18. Freeze drying	Frozen water-containing solid	Heat	Dry solid and water vapor	Sublimation of water	Food dehydration
19. Desublimation	Vapor	Cooling	Solid and vapor	Preferential condensation (desublimation); preferential participation in	Purification of phthalic anhydride
20. Gel filtration	Liquid	Solid gel (e.g., cross-linked dextran)	Gel phase and liquid	Difference in molecular size and hence in ability to penetrate swollen gel matrix	Purification of pharmaceuticals; separation of proteins

TABLE 4.1–1 Classes of Separation Processes (*Cont.*)

Name	Feed Phase	Separating Agent	Product Phases	Property Exploited	Practical Example
21. Dual-temperature exchange reactions	Fluid	Heating and cooling	Two fluids	Difference in reaction equilibrium constant at two different temperatures	Separation of hydrogen and deuterium
22. Zone melting	Solid	Heat	Solid of nonuniform composition	Same as crystallization	Ultrapurification of metals
Rate-Governed Separation Processes					
1. Gaseous diffusion	Gas	Pressure gradient (compressor work)	Gases	Difference in rates of Knudsen or surface diffusion through a porous barrier	Concentration of $U^{235}F_6$ from natural UF_6
2. Sweep diffusion	Gas	Condensable vapor	Gases	Different diffusivities against sweeping motion of cross-flowing vapor	Suggested for isotope separation, helium from methane, etc.
3. Thermal diffusion	Gas or liquid	Temperature gradient	Gases or liquids	Different rates of thermal diffusion	Suggested for isotope separation, etc.
4. Mass spectrometry	Gas	Magnetic field	Gases	Different charges per unit mass	Isotope separation
5. Dialysis	Liquid	Selective membrane; solvent	Liquids	Different rates of diffusional transport through membrane (no bulk flow)	Recovery of NaOH in rayon manufacture; artificial kidneys
6. Electrodialysis	Liquid	Anionic and cationic membranes; electric field	Liquids	Tendency of anionic membranes to pass only anions, etc.	Desalination of brackish waters
7. Gas permeation	Gas	Selective membrane; pressure gradient	Gases	Different solubilities and transport rates through membrane	Purification of hydrogen by means of palladium barriers

TABLE 4.1–1 **Classes of Separation Processes (*Cont.*)**

NAME	FEED PHASE	SEPARATING AGENT	PRODUCT PHASES	PROPERTY EXPLOITED	PRACTICAL EXAMPLE
8. Electrophoresis	Liquid containing colloids	Electric field	Liquids	Different ionic mobilities of colloids	Protein separation
9. Electrolysis plus reaction	Liquid	Electrical energy	Liquids	Different rates of discharge of ions at electrode	Concentration of HDO in H_2O
10. Ultracentrifuge	Liquid	Centrifugal force	Two liquids	Pressure diffusion	Separation of large, polymeric molecules according to molecular weight
11. Reverse osmosis	Liquid solution	Pressure gradient (pumping power) + membrane	Two liquid solutions	Different combined solubilities and diffusivities of species in membrane	Seawater desalination
12. Ultrafiltration	Liquid solution containing large molecules or colloids	Pressure gradient (pumping power) + membrane	Two liquid phases	Different permeabilities through membrane (molecular size)	Waste-water treatment; protein concentration; artificial kidney
13. Molecular distillation	Liquid mixtures	Heat and vacuum	Liquid and vapor	Difference in kinetic theory maximum rate of vaporization, proportional to vapor pressure	Separation of vitamin A esters and intermediates
Mechanical Separation Processes					
1. Filtration	Liquid + solid	Pressure reduction (energy); filter medium	Liquid + solid	Size of solid greater than pore size of filter medium	Recovery of slurried catalysts
2. Mesh demister	Gas + solid or liquid	Pressure reduction (energy); wire mesh	Gas + solid or liquid	Same	Removal of H_2SO_4 mists from stack gases
3. Settling	Liquid + solid or another immiscible liquid	Gravity	Liquid + solid or another immiscible liquid	Density and size difference	Clarification of murky solutions

TABLE 4.1–1 Classes of Separation Processes (*Cont.*)

Name	Feed Phase	Separating Agent	Product Phases	Property Exploited	Practical Example
4. Centrifuge (sedimentation type)	Liquid + solid or another immiscible liquid	Centrifugal force	Liquid + solid or another immiscible liquid	Density and size difference	Recovery of insoluble reaction products
5. Centrifuge (filtration type)	Liquid + solid	Centrifugal force	Liquid + solid	Size of solid greater than pore size of filter medium	Dewatering sludges
6. Cyclone	Gas + solid or liquid	Flow (inertia)	Gas + solid or liquid	Density and size difference	Recovery of fluidized catalyst fines
7. Electrostatic precipitation	Gas + fine solids	Electric field	Gas + fine solids	Charge on fine solid particles	Dust removal from stack gases

The selection of an exploitable chemical and/or physical property difference is a major problem in process synthesis, since generally the material to be separated will differ in a variety of ways. The factors that enter into the selection of a particular property difference include the physical property itself, the magnitude of the property difference, the amount of material to handle, the relative proportions of the different species in the material, the purity required, the need for recovering other species from the material, the chemical behavior of the material during separation, the phases involved, the corrosiveness of the material, and the general know-how and experience available. For this reason, it cannot be stated unequivocally that one particular property difference is best to be exploited for one particular class of separation problems. The separations must be tailored to the problem currently at hand.

In this chapter we examine in some detail a small part of the field of separation technology. We see how solids are separated from other solids, how differences in volatility are exploited to separate liquid mixtures, and how differences in solubility are used by absorbers and extractors.

4.2 Separation of Solids from Solids

Figure 4.2–1 shows some of the property differences useful in engineering on the separation of valuable solids from other solids. Usually, equipment designed to exploit these property differences operates most effectively on mixtures of solids within a given size range, and for this reason a separator may be preceded by crushing, grinding, and/or other size-changing operations. In addition to size adjustment, the

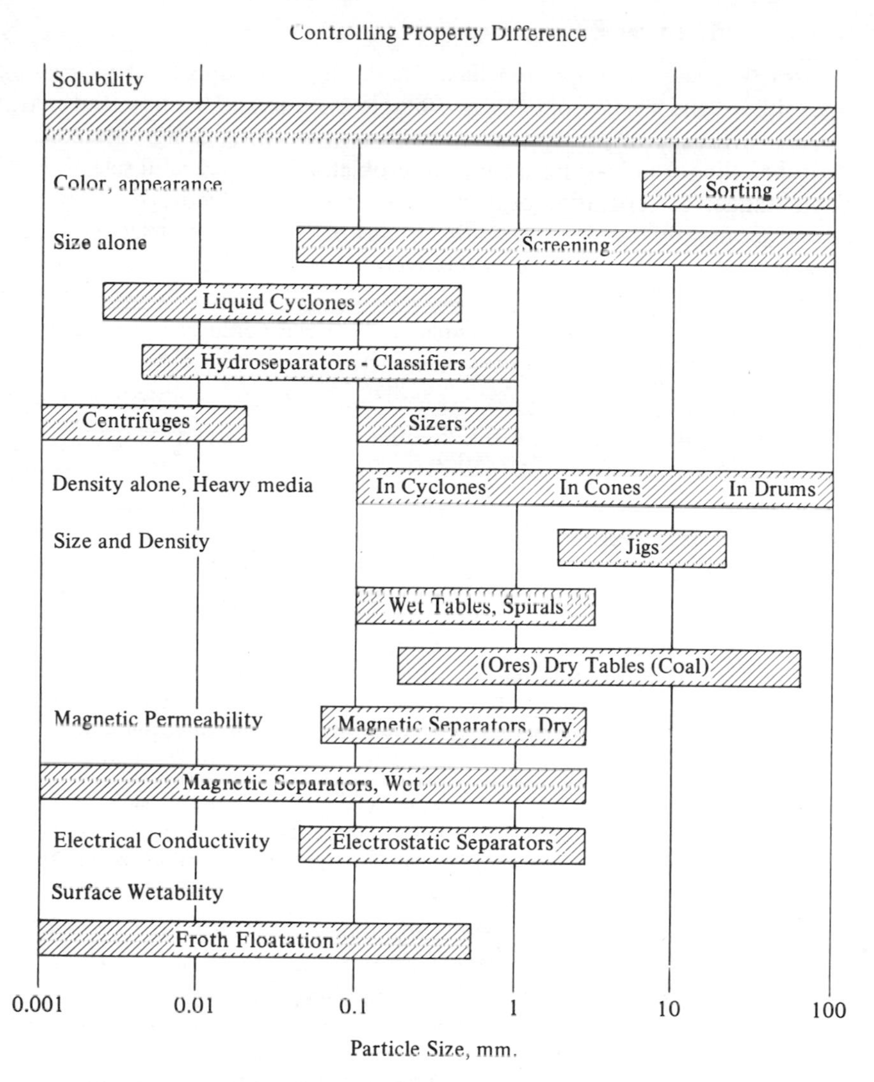

Figure 4.2–1. Property Differences Exploited by Common Methods of Solids from Solids Separation. (Adapted from E. J. Roberts et al., "Solids Concentration," *Chemical Engineering*, June 29, 1970.)

grinding operation may be needed to liberate solids caught mechanically by or loosely bound chemically to the waste solids. Frequently, the grinding and separating operations are performed sequentially, each aiding the other. For example, magnetic taconite ores are ground in at least three stages with intermediate magnetic separation of the valuable material, thereby reducing the grinding costs vastly.

Solubility Differences

Solvent extraction is commonly used in mineral, chemical, and food processing to remove a soluble solid or oil from other insoluble materials. The dissolved material, or solute, must then be separated from the solvent, which by design must be easier to do than the original solid–solid separation problem. By the careful selection of the solvent this tradeoff in separation problems can be used to advantage.

Normally, one would seek a solvent that is cheap, has relatively selective extractive properties, is nonhazardous, leads to a relatively simple solvent recovery and recycle problem, and, especially in the processing of foods and medicines, does not impart undesirable properties to the product. Table 4.2–1 shows some of the solvents found to meet these constraints in several processing areas.

TABLE 4.2–1 Typical Commercial Liquid–Solid Extraction Systems*

EXTRACTANT	FEED	SOLVENT AND REMARKS
Diffusional Extraction		
Soluble coffee	Coffee beans	Aqueous
Fish oils	Fish meal, whole fish	Alcohols, chlorinated hydrocarbons, hexane, etc., for the production of fish protein concentrate
Undeveloped silver	Photographic film	Aqueous solution containing a silver halide solvent (sodium or ammonium thiosulfate), a hardening agent (acetic or citric acid), etc.
Sugar	Sugar beets	Aqueous solutions containing lime
Animal sugars	Hides	Sugars are a result of action between skin and tannin to produce leather and sugar. The sugars are extracted and oxidized to acids
Washing Extraction		
Vegetable oils	Oilseeds	Furfural, propane, or hexane
Potassium chloride	Sylvinite	Sodium chloride solutions used to recover potash from deep formations
Sugar	Sugar cane	Aqueous solutions containing lime
Vanilla	Vanilla beans	35% ethanol—65% water
Flavors, odors, etc.	Natural products	Hexane, benzene, petroleum ether
Flavors, odors	Flowers	Fat—alcohol
Salts, etc.	Corn hulls	Aqueous solutions (during corn starch production)
Turpentine and wood rosin	Stump wood	Naphtha
Leaching		
Activated carbon	Pulp wastes	Sulfuric acid
Phosphoric acid	Phosphate rock	Sulfuric acid, nitric acid, or hydrochloric acid

* From, R. N. Rickles, *Chemical Engineering*, Mar. 15, p. 157, 1965.

Table 4.2–1 Typical Commercial Liquid–Solid Extraction Systems (*Contd.*)

Extractant	Feed	Solvent and Remarks
Sodium aluminate	Bauxite	Caustic solution (Bayer process)
Sodium uranyl carbonate	Uranium ore concentrate	Acid or sodium carbonate extraction
Uranyl nitrate	Uranium fuel elements	Nitric acid, nitric and hydrofluoric acid, or nitric and sulfuric acid
Rare earths	Rare earth concentrates	Hot sulfuric acid
Sodium dichromate	Chromite ore	Hot sulfuric acid
Titanyl sulfate	Ilmenite	Sulfuric acid
Copper sulfate	Copper ore	Sulfuric acid—ferric sulphate
Zinc	Zinc ore	Sulfuric acid
Gold, silver	Ores	Cyanide solution
Nickel	Ores	Aqueous ammonia or acids
Chemical Extraction		
Gelatin	Bones and skins	Aqueous solutions of pH 3–4
Glue	Bones and skins	Degreasing solvent followed by hot water
Lignine	Wood	Preparation of pulp by dissolving lignins, which are hydrolyzed to alcohols and acids
	Sulfate	12.5% solution of caustic Na_2S and soda ash
	Sulfite	7% by weight SO_2 of which 4.5% is as sulfurous acid and 2.5% as $Ca(HSO_3)_2$
	Soda	12.5% NaOH, Na_2CO_3
Iodine	Seaweed	Sulfuric acid

The equipment used to contact the solid and the solvent varies greatly from industry to industry. At one extreme, the solvent may percolate through an undisturbed ore deposit, or, at the other extreme, elaborate equipment may be needed to effect extraction with expensive and dangerous solvents. Figure 4.2–2 illustrates several typical industrial extractors. In the Bonotto extractor, for example, the solids are scraped over horizontal plates to openings where the solids drop to the next lower plate, and the solvent passes upward through the tower, contacting the solids countercurrently. The Bonotto extractor would work well only on solids that settle quickly through the solvent. Special mechanical designs, including screw conveyors and moving buckets, are needed for other solid-solvent-handling jobs.

Example 4.2–1 Ore Leaching. *One hundred tons per hour of an ore, 15 per cent of which is soluble, is to be washed with 100 tons/h of acid. The underflow from each of the two stages, in the two-stage countercurrent extractor shown in Figure 4.2–3, contains 0.3 lb of solution per pound of the insoluble ore. How much of the soluble material is leached out of the ore and appears in the wash water leaving the process?*

A material balance around each extractor unit in Figure 4.2–3 provides sufficient information. We begin with the first extractor.

- ***Inerts*** *The inerts do not dissolve in the water and can leave only with the underflow slurry:*

$$I_1 = 85 \text{ tons inert/h}$$

- ***Soluble Material*** *The soluble material enters this extractor in the fresh ore (S_0 = 15 tons/h) and in the wash water from stage 2 (w_2y_2 tons soluble/h). The soluble material leaves with the effluent acid (w_1y_1 tons soluble/h) and in the slurry going to stage 2 [$I_1(0.3)y_1$ tons soluble/h].*

$$15 + w_2y_2 = w_1y_1 + 85(0.3)y_1 \quad \text{tons soluble/h.}$$

- ***Acid*** *Acid enters stage 1 from stage 2 [$w_2(1 - y_2)$ tons/h], and leaves as the effluent from the process [$w_1(1 - y_1)$] and in the slurry going to stage 2 [$I_1(0.3)(1 - y_1)$].*

$$w_2(1 - y_2) = w_1(1 - y_1) + 0.85(0.3)(1 - y_1) \text{ tons acid/h.}$$

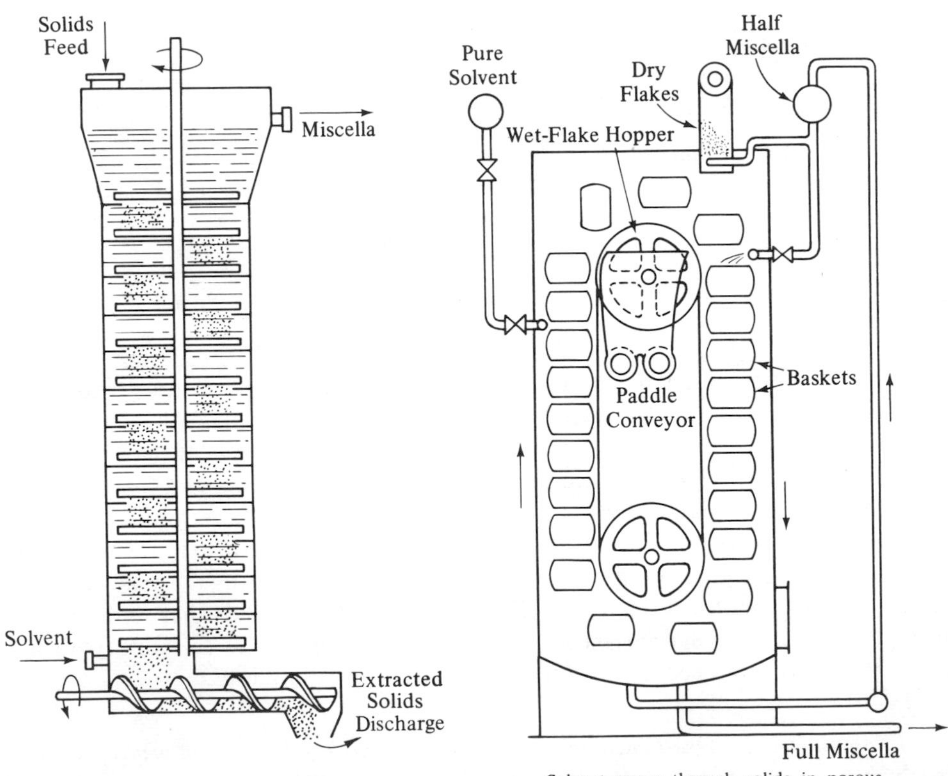

Bonotto extractor scrapes solids over horizontal plates to alternate openings.

Solvent pours through solids in porous bottom baskets. Baskets move vertically, dumping solids after one cycle.

Figure 4.2–2

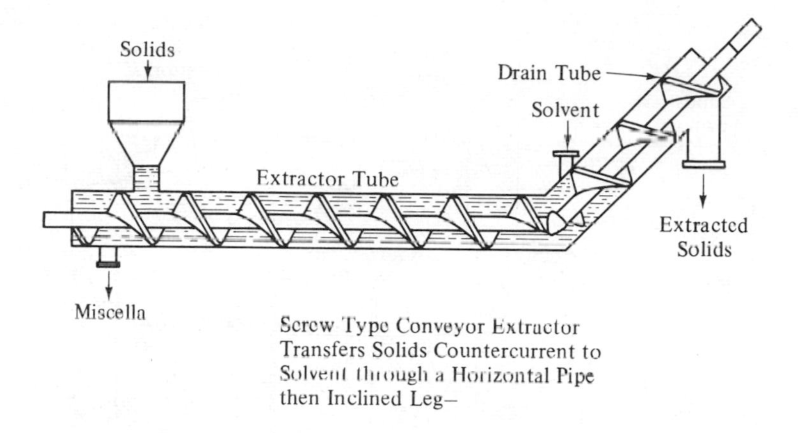

Screw Type Conveyor Extractor Transfers Solids Countercurrent to Solvent through a Horizontal Pipe then Inclined Leg–

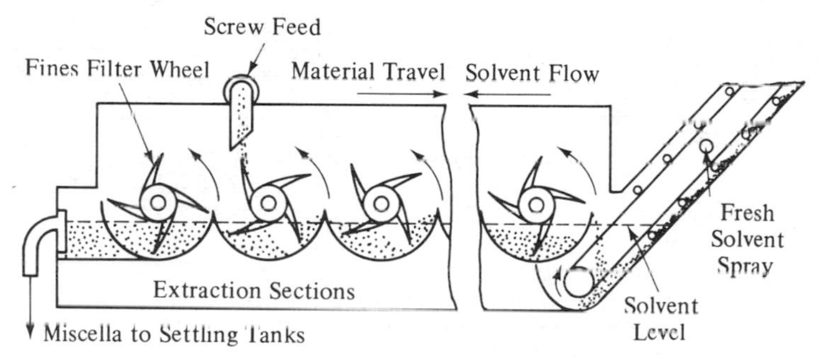

Kennedy Extractor Uses Impellers to Lift Solids from Solvent through a Series of Horizontal Compartments.

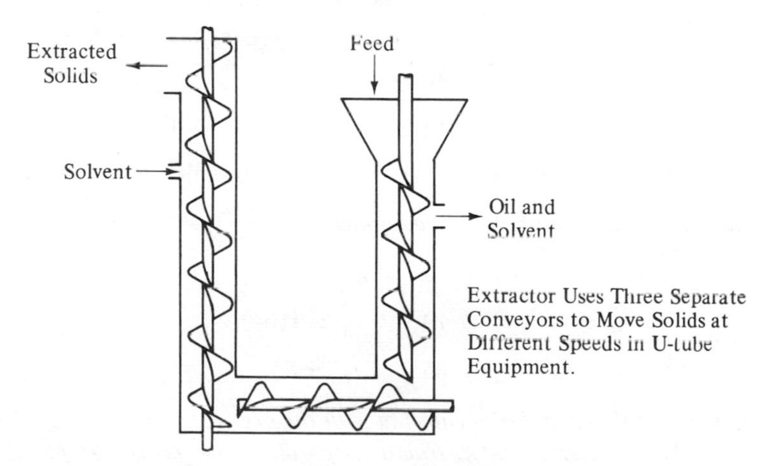

Extractor Uses Three Separate Conveyors to Move Solids at Different Speeds in U-tube Equipment.

Figure 4.2–2 (*continued*)

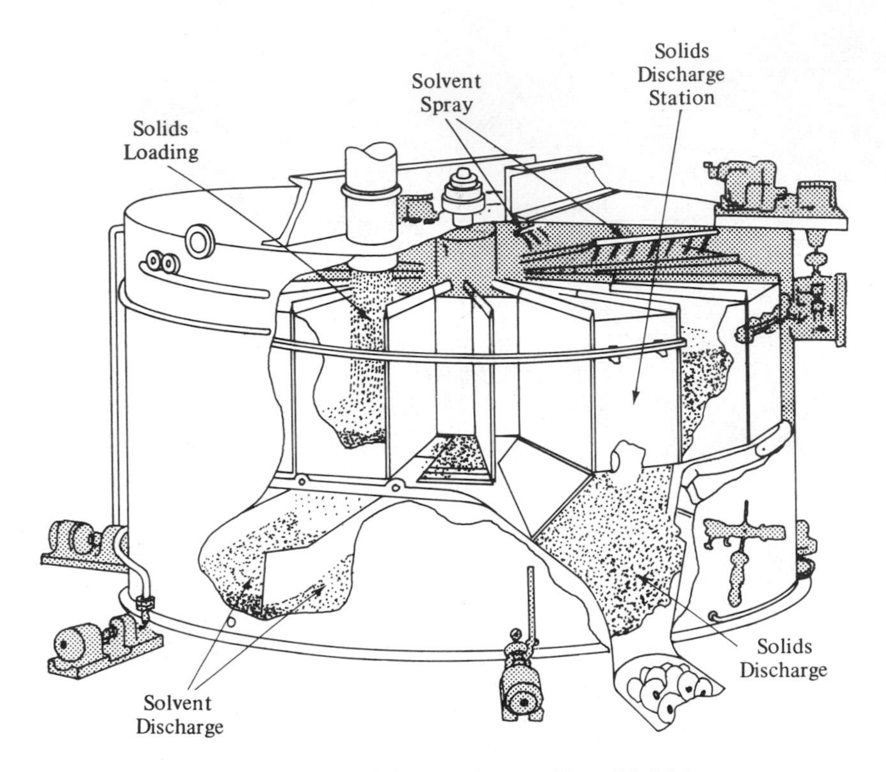

Rotocal Extractor Feeds Solvent to Top of Solids in Screen-Bottomed Baskets at Various Points as They Move Around Cycle–

Fig. 4.2–2 (*continued*)

Similar balances over stage 2 yield

$$I_2 = 85 \quad \text{inert}$$

$$(85)(0.3)y_1 = w_2 y_2 + (85)(0.3)y_2 \quad \text{soluble}$$

$$100 + (85)(0.3)(1 - y_1) = w_2(1 - y_2) + (85)(0.3)(1 - y_2) \quad \text{water}$$

The solution of these material-balance equations is

$$w_1 \simeq 75 \qquad w_2 \simeq 100$$

$$y_1 \simeq 19 \qquad y_2 \simeq 0.04$$

$$I_1 \simeq 85 \qquad I_2 \simeq 85$$

Of the 15 tons/h of soluble material entering with the ore, $w_1 y_1 = 14$ *tons/h leaves with the acid. Thus this countercurrent extractor is predicted to recover 94 per cent of the soluble ore.*

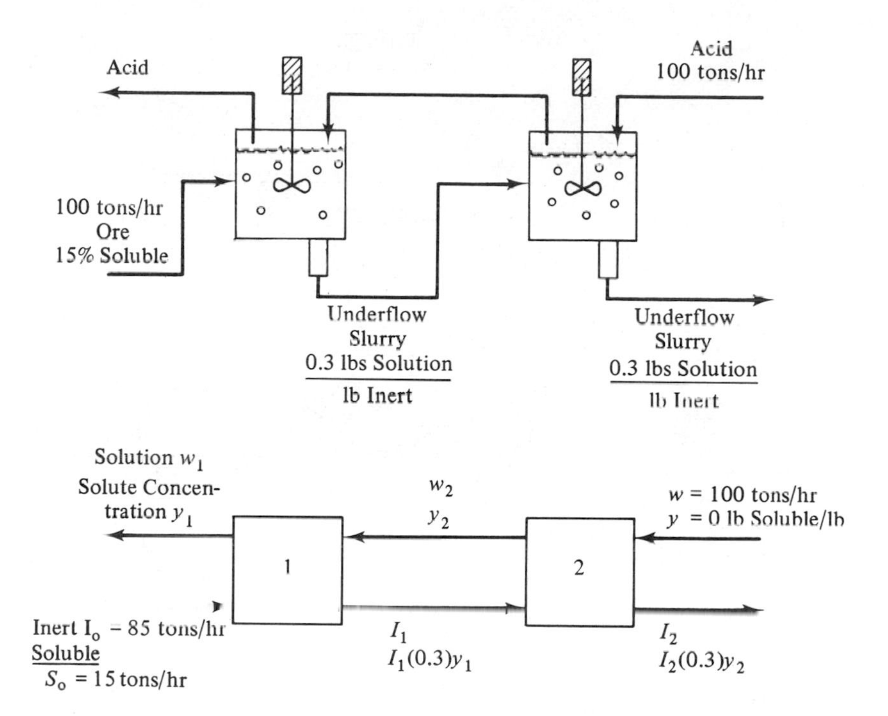

Figure 4.2–3. Ore-Washing Example

Appearance Differences

Figure 4.2–4 illustrates the use of appearance differences in the separation of diamonds from waste materials. In the method developed by the DeBeer's Diamond Research Laboratory, X-ray detectors detect the dark and discolored industrial diamonds, and a control device activates a blast of compressed air, which ejects the diamonds from a falling stream of ore. In a similar separation based on appearance differences, premium halite, which is 99.98 per cent soluble, is separated from waste in equipment that takes advantage automatically of the crystal's appearance under diffuse light illumination. Unless the separations based on appearance can be automated, the separations become much too costly for practical use.

Size and Density Differences

Gravitational forces acting on a particle are proportional to its mass, buoyant forces proportional to volume, and surface forces to surface area. The careful adjustment of these forces in the proper equipment can lead to the economic separation of solids that differ in size and density.

Figure 4.2–5 shows the common heavy-media circuit used to separate materials that differ in density. Materials that are more dense than a fluid sink and those less dense float. If a liquid can be found that is of intermediate density, solids can be

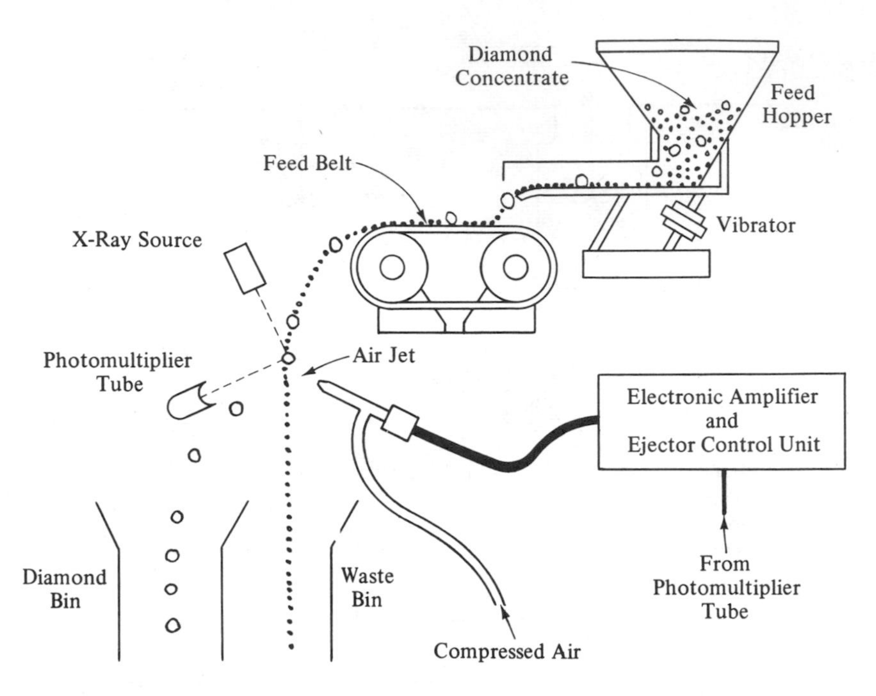

Figure 4.2–4. Diamond Separation Process Uses X-Rays to Detect Diamonds in Ore, Triggering Device for Making the Separation
From E. J. Roberts et al., "Solids Concentration," *Chemical Engineering*, June 29, 1970.

separated by this sink–float method. Since most solids are more dense than water, other liquids are necessary. Heavy organic liquids, such as carbon tetrachloride or acetylene tetrabromide, have been proposed, but are toxic, costly, and difficult to recover. Saturated calcium chloride solutions have been used to process coal, but this leads to corrosion and salt-recovery problems. In the 1930s researchers found that water suspensions of finely divided solids behave like a heavy liquid. Suspensions of magnetite and ferrosilicon in water can attain specific gravities between 1.25 and 3.7; furthermore, these solids can be recovered easily by magnetic separation.

The heavy-media circuit in Figure 4.2–5 is used extensively in the recovery of iron, copper, fluorspar, coal, tin, gravel, diamonds, ilmenite, and manganese. The feed is first prepared to eliminate very fine particles, the solids are separated in the heavy media, the medium is removed from the separated products, and the medium is recovered and cleaned for reuse.

Example 4.2–2 Separation of Sodium and Potassium Salts. *Potassium is an essential element in the production of many agricultural products. It aids in the production of sugar in fruits and vegetables, starches in tubers and grains, and fibrous material in plants. The presence of potassium in the soil helps prevent certain plant diseases and lessens the effects of excessive nitrogen application.*

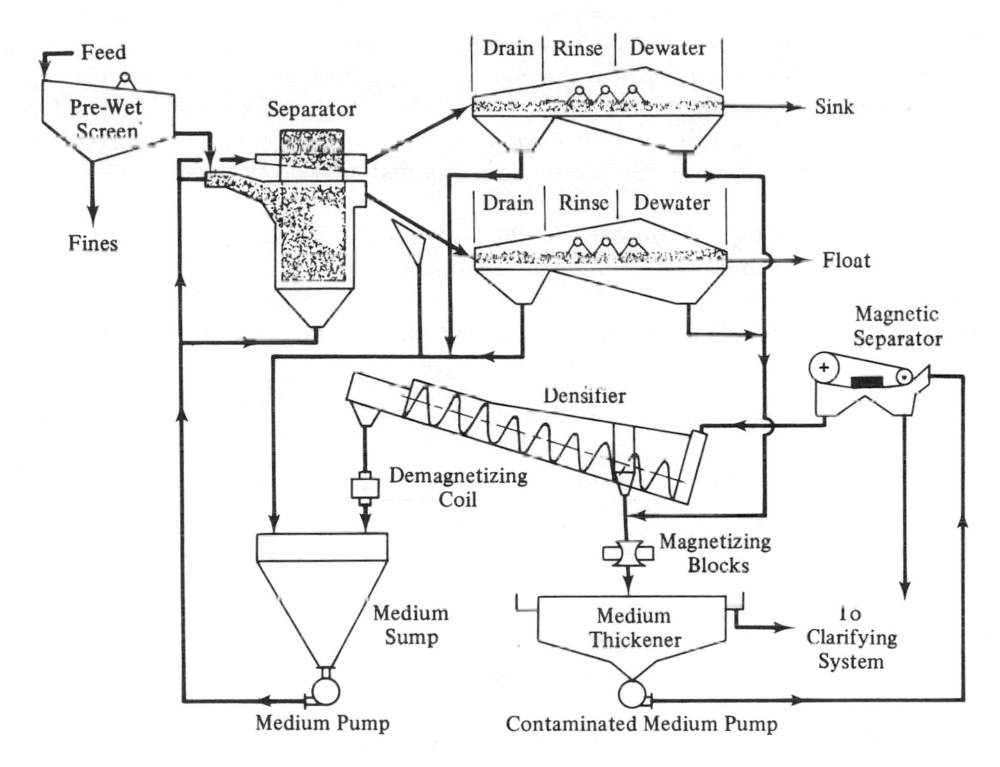

Figure 4.2–5. Heavy-Media Separation: Typical Process Involves Feed Preparation, Treatment with Heavy Media, Separation, Recovery of the Media
From E. J. Roberts et al., "Solids Concentration," *Chemical Engineering*, June 29, 1970.

Potassium is often found in deposits of sylvinite, a mixture of KCl and NaCl. The separation of these two salts is a major problem in the production of fertilizers. Unfortunately, the similar solubilities in water hinder the selective crystallization processing one would be immediately attracted to in salts separation. However, the specific gravity of KCl is 1.98 and of NaCl, 2.16; this difference suggests the possibility of a dense-media separation. The proper mixture of the organic liquids acetylene tetrabromide (specific gravity: 2.96) and methylene chloride (specific gravity: 1.34) can be formed with a specific gravity between that of the sodium and potassium chlorides.

The dry salt mixture is suspended in the dense organic media and fed into a cyclone separator, in which the gravitational forces that cause separation are accentuated by the swirling of the fluid. The heavier sodium salt settles to the outer wall of the cyclone, and the ligher potassium salts leave through the central outlet.

Several schemes are used to recover and recycle the dense media. In one scheme the suspensions are filtered and the solvent is evaporated, or stripped away with steam, leaving behind the solid. The condensed steam and organic are then separated by decantation. A second scheme, shown in Figure 4.2–6, reduces the steam costs by using a brine solution to wash the salts. In this way all the salts do not need to be heated to the steam temperature. It is interesting that in this dense-media separation, the recovery of the dense media is the major processing problem that must be overcome.

The Humphrey spiral concentrator illustrated in Figure 4.2–7 has found use in the recovery of ilmenite, rutile, zircon, and monazite from beach sand, the recovery of iron ore, the separation of coal from slate, and other separations based on shape and density. As a feed slurry of 20 to 40 per cent solids flows down through the spiral

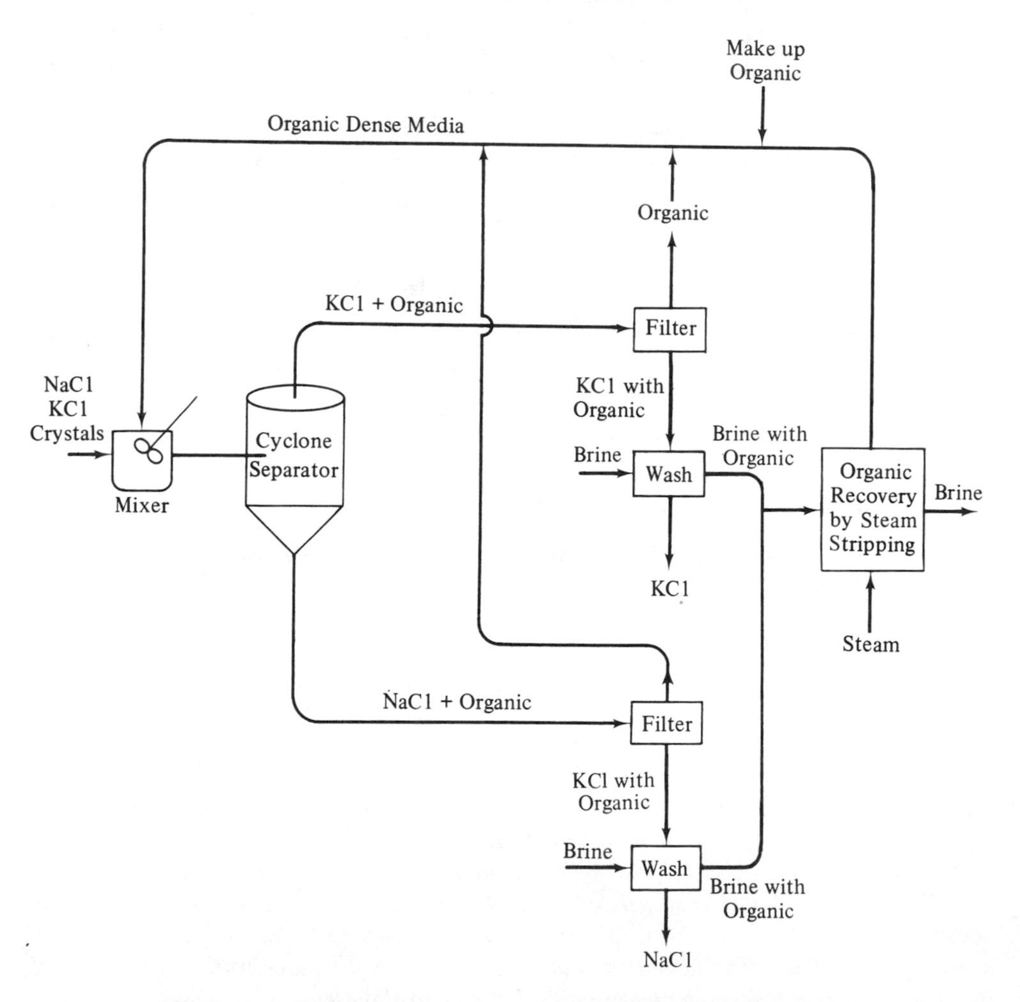

Figure 4.2–6. Separation of Sodium and Potassium Chlorides Using a Dense Media of Acetylene Tetrabromide and Methylene Chloride. Dense Media Is Recovered by Filtration, Brine Washing, and Steam Stripping

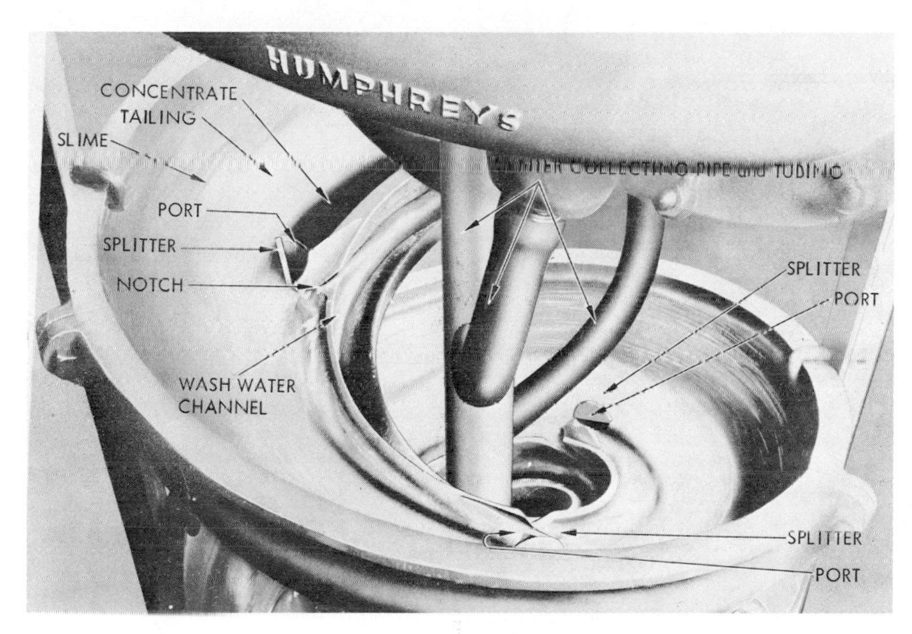

Figure 4.2–7. Spiral Concentrator Has Take-Off Ports for Various Fractions in a Slurry. (Courtesy of *Chemical Engineering*, June 29, 1970, p. 62, McGraw-Hill.)

trough, several forces act on the particles: gravity, the drag of the water, friction against the trough, and centrifugal forces. The heavy particles settle to the bottom of the fluid where the fluid velocity is least, and tend not to centrifugate. Light and flaky materials are swept by centrifugal forces outward. These spirals have no moving parts, are inexpensive and trouble free, and handle 1 to 2 tons/h.

A variety of other separation devices exploits differences in shape, size, and density, and the reader interested in mechanically ingenious devices will find the literature cited at the end of the chapter a source of entertaining reading.

Magnetic and Electrostatic Differences

A limited number of materials are influenced by strong magnetic fields, and a larger number by strong electrostatic fields. Figure 4.2–8 illustrates a variety of devices used to separate magnetic from less magnetic materials. Similar machines perform the electrostatic separation of dielectric from conducting materials. Machines are available that handle up to 120 tons/h of material, and electrostatic separation has been applied to the cleaning of tobacco and cocoa beans, the separation of potassium feldspar from sodium feldspar, and the concentration of iron, silver, graphite, zircon, ilmenite, rutile, tin, and phosphate rock.

Example 4.2–3 Processing with Immobilized Enzymes. *Enzymes, immobilized on the surface of glass, ceramic, or plastic particles, are finding increasing use in the process*

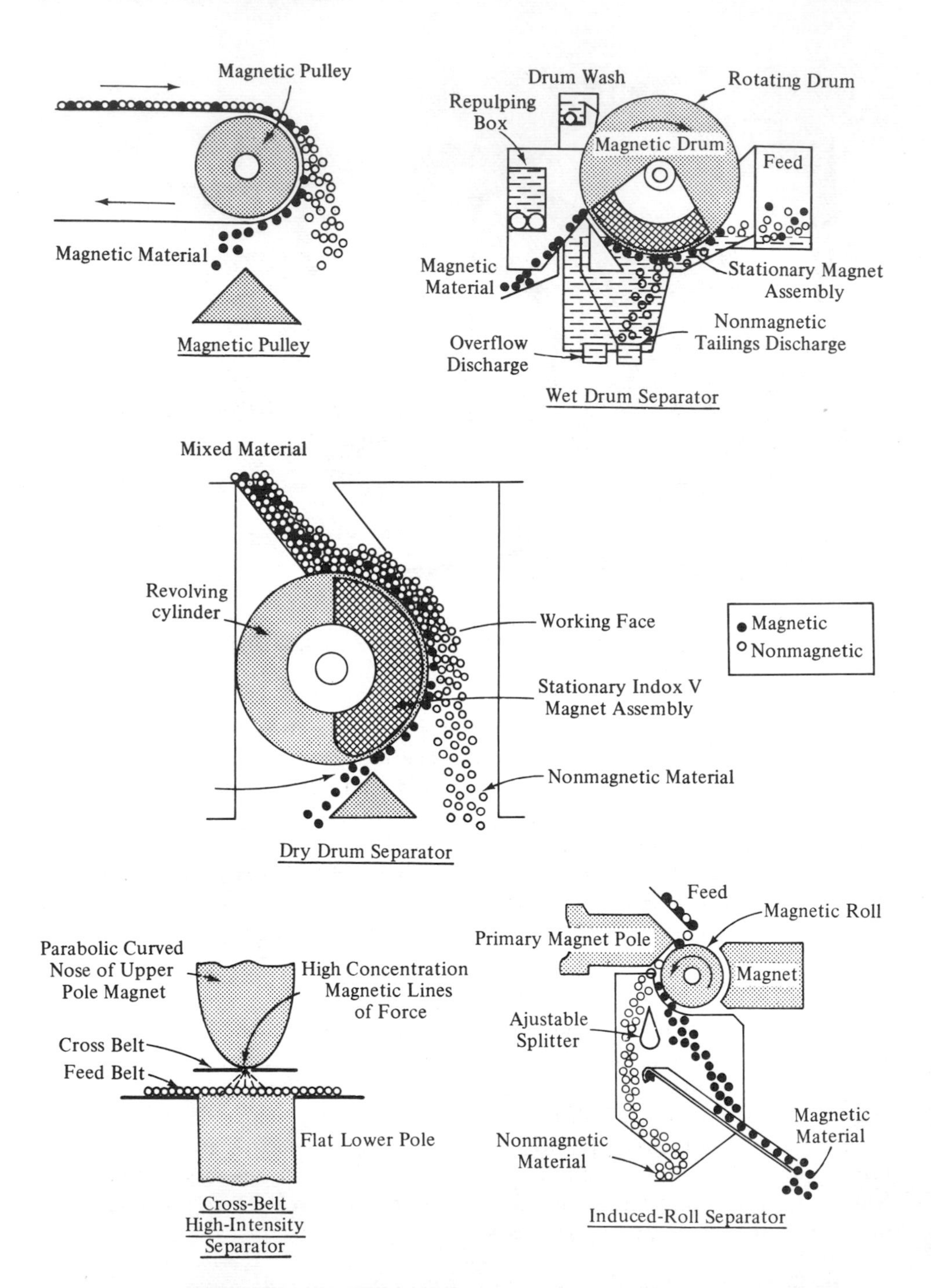

Figure 4.2–8. Magnetic Separators: A Variety of Devices Is Available for Handling Strong and Weak Magnetic Materials

industries. The protein in fish scrap can be solubilized by some enzymes, sewage treatment is possible by other enzymes, and beer can be clarified by still other enzymes. However, a major obstacle is the recovery of the enzyme, for often it is caught up in the pulp of the material being processed.

In 1971 R. A. Grieger and G. J. Powers submitted the following invention as a patent disclosure. By immobilizing the enzymes on particles acted on by a magnetic field, enzyme recovery may be practical. For example, iron particles might be encapsuled in the material upon which the enzyme is normally immobilized, and the wet drum separator shown in Figure 4.2–9 used to recover the particles from the waste process pulp.*

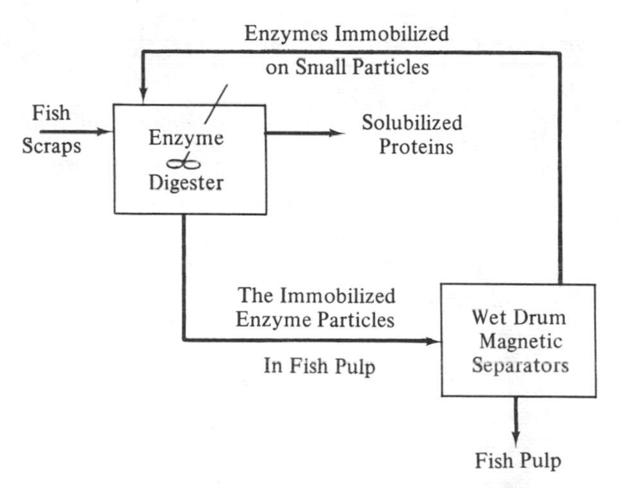

Figure 4.2–9. Processing of Fish Scraps by Enzymes Immobilized on Magnetically Susceptible Particles

Surface Wettability Differences

Over 90 per cent of the world's copper, lead, zinc, molybdenum, antimony, and nickel are produced from ores by froth flotation. Other applications occur in treating industrial and municipal wastes, and in biological and food processing. Flotation is possible whenever differences in chemical or crystal structure create surface property differences, causing the selective adherence of particles to air bubbles rising through a slurry. Nonpolar substances such as sulfur, graphite, diamond, and coal naturally tend to adhere to air bubbles; other materials require surface modification.

The materials to be separated must be finely ground, surface treated to enhance surface wettability differences, and frothed in a device similar to that shown in Figure 4.2–10. The concentrate rises as a froth to be skimmed off the top, and the tailings remain behind to be removed. As an example, fluorite (40 per cent CaF_2 in

* April 6, 1971, Department of Chemical Engineering, University of Wisconsin.

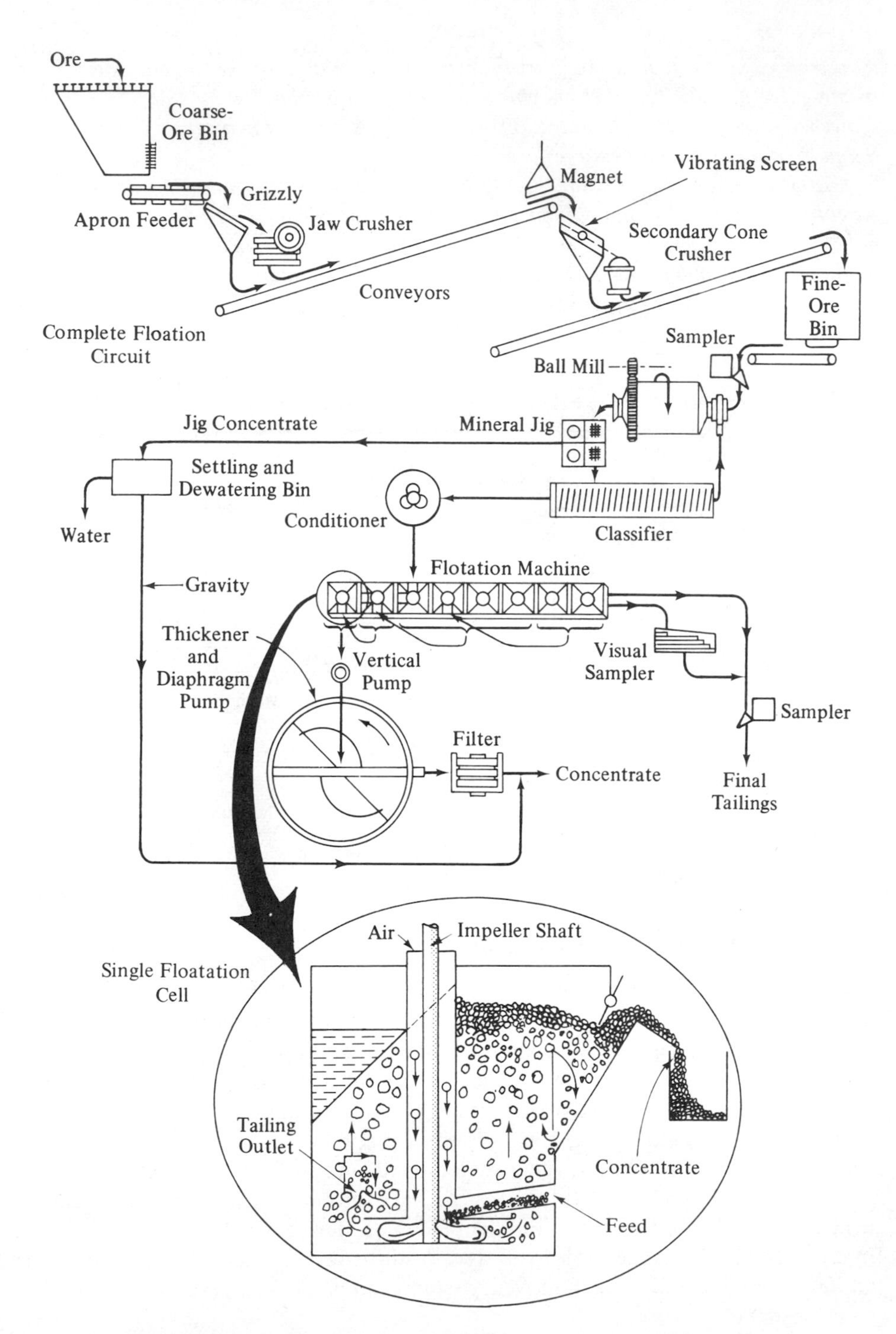

Figure 4.2–10. Flotation Is a Popular Method for Separating Solids. Machinery Involves Use of an Aeration Device

$CaCO_3$, SiO_2, and iron oxide) can be purified to 99 per cent CaF_2 using a surface promoting mixture of the following for each ton of ore: 7 lb of Na_2CO_3, 1.5 lb of $NaSiO_3$, 0.8 lb of refined fatty acid, 0.1 lb of alcohol frother, and 0.15 lb of quebracho.

4.3 EXPLOITING VOLATILITY DIFFERENCES

So effective is the separation technology which exploits volatility differences that the first response of the engineer when confronted with a separation problem is to search for boiling-point differences. Differences in the boiling point of the pure substances are a good indication of differences in the volatility of mixtures. Should these differences exist, there is an excellent chance that separation by vaporization and/or condensation would be effective. When boiling points are similar, a clean separation may not occur, only an enrichment.

Boiling points of the pure substances, their vapor pressures, and relative volatilities between substances are all quantitative measures of the enrichment that might occur should a liquid mixture be partially vaporized or a gas mixture partially condensed. The relative volatility between two components in solution is defined as

$$\alpha_{12} = \frac{Y_1/X_1}{Y_2/X_2}$$

where Y and X are the mole fractions in the vapor and liquid phases in equilibrium. If $\alpha_{12} > 1$, enrichment of component 1 in the vapor would occur if the liquid were partially vaporized, and the remaining liquid would be richer in component 2. For ideal solutions, the relative volatility equals the ratio of the vapor pressures of the pure substances:

$$\alpha_{12} = \frac{P_1}{P_2}, \qquad \text{for ideal solutions}$$

Table 4.3–1 gives values of P_1/P_2 at different temperatures for a variety of ideal solutions. The greater the deviation from unity, the easier the separation by distillation. Since the normal boiling point of a pure substance is the temperature at which its vapor pressure equals atmospheric pressure, differences in boiling points indicate differences in vapor pressure and hence the possibility of enrichment by partial vaporization and condensation.

Example 4.3–1 Enrichment of a Benzene–Toluene Mixture. *Suppose that an equimolar liquid mixture of benzene and toluene were heated at 1 atm pressure until half the liquid is vaporized. What would be the composition of the vapor and liquid?*

The relative volatility of the benzene–toluene mixture is given in Table 4.3–1 as

$$\alpha_{BT} = \frac{y_B/x_B}{y_T/x_T} \cong 2.5$$

TABLE 4.3–1 **Relative Volatilities of Ideal Mixtures**

MIXTURE 1 AND 2	BP OF 1, °C	P_1/P_2[a]	BP OF 2, °C	P_1/P_2
Benzene–ethylene dichloride	80.1	1.113	83.48	1.109
Benzene–toluene	80.1	2.61	110.7	2.315
a-Butyl chloride–*n*-butyl bromide	77.5	2.08	101.6	1.87
Chloroform–carbon tetrachloride	61.1	1.71	76.6	1.60
Ethanol–isopropanol	78.3	1.18	82.3	1.17
Ethanol–propanol	78.3	2.18	97.2	2.03
Ethyl chloride–ethyl bromide	12.5	3.23	38.4	2.79
Ethyl ether–benzene	34.6	5.16	80.2	3.95
Ethylene dibromide–propylene dibromide	131.7	1.30	141.5	1.30
Ethylene dichloride–trichloroethane	83.5	2.52	113.7	2.33
a-Heptane–methylcyclobexane	98.4	1.058	100.3	1.056
a-Hexane–*n*-heptane	69.0	2.613	98.4	2.33
Methanol–ethanol	64.7	1.73	78.1	1.64
Methanol–isobutanol	64.6	6.1	107.5	4.4
Methanol–propanol	64.6	3.89	97.2	3.15
Methyl acetone–ethyl acetate	56.8	2.036	77.1	1.923
Phenol–*o*-cresol	181.2	1.30	190.6	1.275
Phenol–*m*-cresol	181.2	1.768	201.5	1.699
Phenol–*p*-cresol	181.2	1.793	202.2	1.728
Toluene–benzyl chloride	110.7	7.75	178.0	4.45
Toluene–chlorotoluene	110.7	4.76	162.0	3.65
Water–ethylene glycol	100.0	49.8	197.0	13.2
Water–ethylene glycol[b]	60.1	98	150.2	21
Water–glycerol[c]	38.1	76.400	202.0	244

[a] Values of P_1/P_2 (approximately equal to α) are given at the boiling points of components 1 and 2, respectively.

[b] Pressure 150 mm Hg.

[c] Pressure 50 mm Hg.

Since the system consists of only benzene and toluene, the mole fractions in the vapor phase and the liquid phase must add to unity:

$$y_B + y_T = 1$$

$$x_B + x_T = 1$$

If we consider 1 mole of the original liquid, $L_o = 1$*, to be changed into half vapor,* $V = 0.5$*, and half liquid,* $L = 0.5$*, a material balance over the benzene gives*

$$L_o x_{Bo} = L_B x_B + V_B y_B$$

or

$$0.5 = 0.5x_B + 0.5y_B$$

We have four equations and four unknowns, the solution of which is

$$y_B = 0.61, \qquad y_T = 0.39$$

$$x_B = 0.39, \qquad x_T = 0.61$$

A liquid 50 per cent benzene and 50 per cent toluene by partial vaporization can be separated into a vapor 61 per cent benzene and a liquid 39 per cent benzene. Further enrichment can be accomplished by partially condensing the separated vapor, and by partially vaporizing the separated liquid (multistage distillation).

A partial vaporization and partial condensation leads to the *enrichment* of the mixtures, rather than the complete separation. If the relative volatility is large, the enrichment may be a complete separation, but generally multiple enrichments are required for complete separation. Vapors are partially condensed and liquids are partially vaporized, requiring heat removal and heat addition, respectively. Heat addition and heat removal can be costly operations, and it is an economic advantage if the condensing of a vapor can vaporize another liquid, the one operation driving the other. This integration of heat usage leads to multistage distillation equipment or a fractionator, commonly used for difficult separations.

Figure 4.3–1 illustrates the integration of heat usage during multistage distillation. The sequence of operations to be implemented involves (1) the partial vaporization of the feed liquid, (2) the partial condensation of the vapor resulting from 1, and (3) the partial vaporization of the liquid resulting from 1. The liquid resulting from operation 2 is doubly enriched in the more volatile component, and the liquid resulting from operation 3 is doubly enriched in the less volatile component. Liquid 2 and the vapor 3 are not as enriched and should be sent back to 1 for reprocessing, and vapor 3 will be sufficiently hot to provide some of the heat needed to operate the vaporization 1. This integration of energy and material usage leads to the continuous multistage distillation equipment commonly used for large-scale separations.

Crude Oil Distillation

In this subsection we show how elaborate systems of distillation towers are used to produce petroleum products from crude oil. Figure 4.3 2 is a photograph of the petroleum distillation process here described.

After preheating by heat exchange with the hot distillates, the crude oil is heated in a tubular heater and partly vaporized. The heater outlet temperature is adjusted in accordance with the characteristics of the crude and lies within the range of 660 and 715°F (350 and 380°C). The mixture of vapor and liquid enters a fractionation tower. Direct steam is blown into the lower section of the tower to remove the light ends from the residue. Gasoline goes overhead as light material and is condensed. Part of the gasoline is pumped as reflux to the uppermost tower tray, the balance being delivered as straight-run gasoline for further processing. Low-boiling hydrocarbons, such as propane and butane, which might be contained in the crude, can be recovered from the overhead gas. The higher-boiling fractions, such as kerosene and diesel oil, become concentrated on the lower tower trays and are withdrawn as sidestreams for stripping with steam in sidestream columns, if necessary. The stripped light ends, together with the steam from the sidestream columns, are returned to the main tower. Kerosene and diesel oil are withdrawn as bottoms from the sidestream columns. The residue from the main tower is either used as fuel oil or further treated by vacuum distillation.

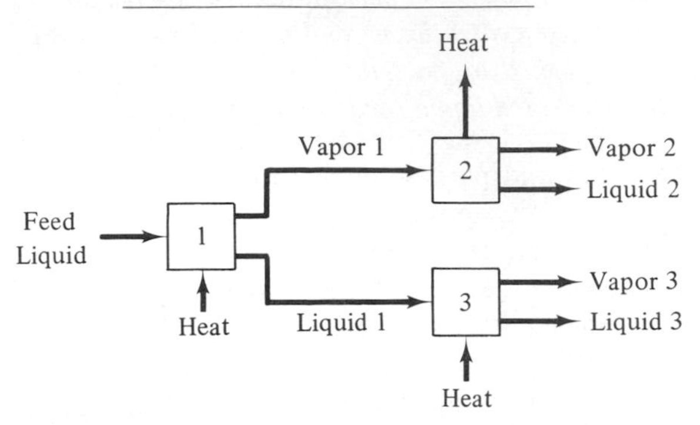

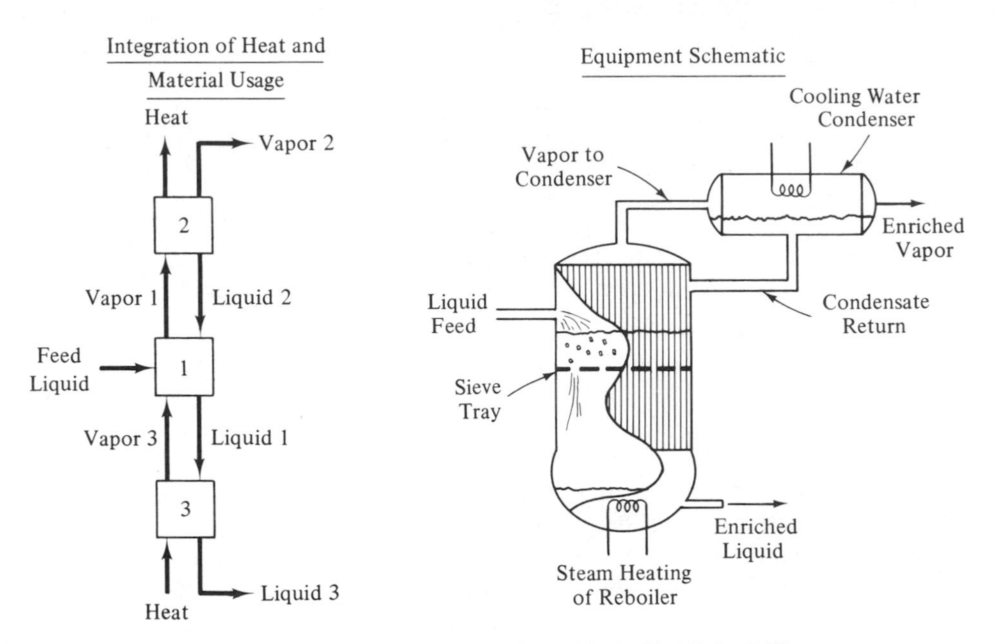

Figure 4.3–1. Exploiting Enrichments Caused by Boiling-Point Differences

The residue from atmospheric distillation is heated in a tubular heater to 660 to 715°F (350 to 380°C), depending upon the type of crude, and distilled under a vacuum. The necessary vacuum is created by steam jet ejectors. Higher-molecular-weight components evaporate in the vacuum without cracking. Distillates are withdrawn from the top of the tower and from the upper trays; these distillates are either used as feedstocks for catalytic cracking plants or, provided that the properties of the crude permit it, further processed into lube oils; these distillates are generally stripped in sidestream columns. Part of the distillate is returned to the top of the column as reflux.

As the boiling point and the vapor pressure of the straight-run gasoline from atmospheric distillation frequently do not meet with the market requirements,

Figure 4.3–2. Distillation of Petroleum. (Courtesy of Lurgi Gesellschaften, Frankfurt, Germany.)

stabilization or redistillation of the gasoline is required. For the adjustment of the desired gasoline vapor pressure and initial boiling point, C_3 hydrocarbons and part of the C_4 hydrocarbons are removed by distillation under pressure in a stabilization column. The distillate, together with the uncondensed fractions from crude distillation, is further processed and separated into its various components.

Gasolines with a boiling end point of above 390°F (200°C) are redistilled. The gasoline is removed overhead from a redistillation column. The residue is added to the kerosene or gas oil fraction.

Azeotropic Distillation

Not all materials form ideal solutions in which the relative volatilities are constant over wide ranges of composition. Should the relative volatility of a binary solution become unity, no enrichment occurs during vaporization or condensation. If

$$\alpha_{12} = \frac{y_1/x_1}{y_2/x_2} = 1$$

then $y_1/y_2 = x_1/x_2$. In such a case, the vapor and the liquid are of the same composition.

Solutions in which the relative volatility varies with composition are called nonideal solutions, and should the variable relative volatility reach unity, an azeotropic mixture results. An azeotrope cannot be enriched by vaporization–condensation operations. Figure 4.3–3 shows vapor–liquid equilibrium data for chloroform–acetone and ethyl acetate–ethanol systems, both of which form an azeotropic mixture. For example, in chloroform–acetone mixtures, the chloroform is less volatile than the acetone up to a concentration of 66 per cent chloroform; beyond that the acetone is the more volatile component. At 66 per cent both components have the same volatility, and the vapor and liquid have the same composition. Ordinary multistage distillation cannot affect separations beyond an azeotrope composition.

However, the introduction of a third component may alter the relative volatilities sufficiently to create a different azeotropic mixture. Figure 4.3–4 shows a scheme for producing absolute alcohol by introducing benzene to break the water–ethanol azeotrope of 96 per cent ethanol. The benzene forms a ternary minimum azeotrope with ethanol and water, which boils at a lower temperature and contains a higher ratio of water to ethanol than the ethanol–water binary azeotrope. Ninety six per cent alcohol is fed to column A. The ternary azeotrope is taken overhead in this column, and absolute alcohol is obtained as bottoms product. The overhead vapors are condensed and passed to decanter B, in which two liquid layers form. The upper layer, rich in benzene, is returned to column A as reflux, and the lower layer is fed

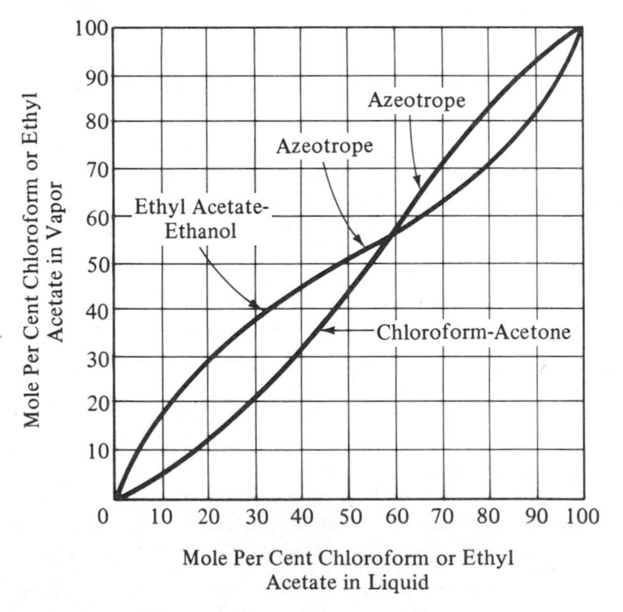

Figure 4.3–3. Vapor–Liquid Equilibrium Data at 760 mm Hg for Chloroform–Acetone and Ethyl Acetate–Ethanol Systems

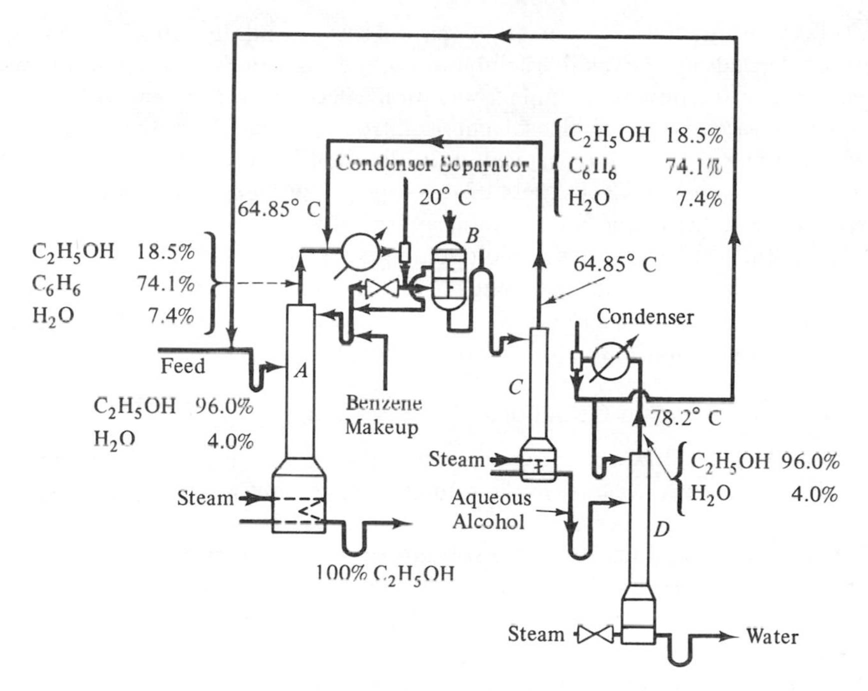

Separator Equilibrium

	Top Layer	Bottom Layer
Vol. % Overhead	84.0	16.0
	Compositions	
C_2H_5OH	14.5%	53.0%
C_6H_6	84.5	11.0
H_2O	1.0	36.0

Figure 4.3–4. Dehydration of 96 Per Cent Ethanol to Absolute Alcohol by Azeotropic Distillation with Benzene at 1 Atm

to column C, which produces the ternary azeotrope as the overhead product and benzene-free aqueous alcohol as the bottoms product. This latter product is fed to column D, which produces by ordinary distillation an overhead product that is 96 per cent alcohol and a bottoms product that is nearly pure water. The overhead from column D is recycled to column A for removal of the water. The benzene is recycled continuously in this system.

4.4 EXPLOITING SOLUBILITY DIFFERENCES

Gas absorption is the operation in which a soluble gas in a mixture of gases is absorbed into a liquid, and *stripping* is the inverse operation in which a volatile gas dissolved in a liquid is transferred to the gas phase. Absorber–strippers are used quite effectively in the enrichment of gas mixtures, and the effectiveness of this operation depends on the relative solubilities of the gases in the chosen liquid solvent. *Liquid–liquid* extraction is a similar operation effective in enriching solute–solvent mixtures. A second immiscible solvent is introduced in which the solute is also soluble, and an equilibrium is reached in the distribution of solute among the solvent phases. Here again the effectiveness of this liquid–liquid extraction is dependent on the relative solubility of the solute in the solvent phases.

Although these operations exploit solubility differences, the equipment used is quite different because of the different phase combinations handled: gas–liquid and liquid–liquid. In the next subsections we examine the data needed to evaluate the feasibility of separations based on solubility differences.

Absorbing–Stripping Operations

Figure 4.4–1 is the flow sheet for a typical absorber–stripper combination, which takes advantage of the decrease in the solubility with increase in temperature of a gas in a liquid. In the absorber the solute gas is absorbed by the cool solvent, the lean gas leaves the system, and the loaded solvent leaves the absorber for the stripper. In the stripper the loaded solvent is heated, driving off the absorbed gas, the gas leaves the system, and the stripped solvent is recycled to the absorber. Heating and cooling are required to change the stream temperatures, thereby adjusting the solubility of the gas. Normally, the absorber and stripper are merely towers filled with loose solid packing, even just broken bottles or stones, over which the liquids trickle and up through which the gases are forced, causing close gas–liquid contact.

The selection of the solvent liquid is critical to the economic operation of absorber–strippers. Preference is usually given to liquids in which the solute gas is

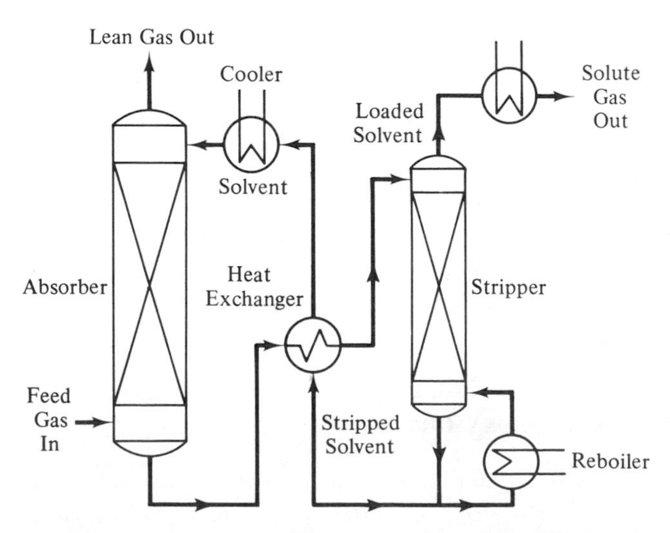

Figure 4.4–1. Typical Flow Sheet for Absorber–Stripper Combination

highly soluble; sometimes a reversible chemical reaction in the liquid phase can be used to advantage by increasing the solubility of a gas that enters into the reaction. Furthermore the solvent should be cheap, noncorrosive, stable, nonviscous, nonfoaming, and nonflammable, if possible. Since the exit gases are usually saturated with the solvent, losses may be high if a volatile solvent is used. Water is commonly used for gases soluble in water, oils for hydrocarbons, and special chemical solvents for gases such as hydrogen sulfide, carbon dioxide, and sulfur dioxide. Table 4.4–1 shows some of the vast information available on gas solubility, and Table 4.4–2 shows typical industrial solvent usage.

Example 4.4–1 Solvent Requirements for Carbon Dioxide–Air Separation *Carbon dioxide is to be recovered from a 70 per cent carbon dioxide–air stream, using an absorber–stripper. The solvent is a 15 per cent by weight aqueous solution of monoethanolamine. How much solvent must circulate through the absorber–stripper to reduce the carbon dioxide concentration in the exit air stream to 5 per cent?*

Certain postulates must be made more clearly to define the mode of operation of the absorber–stripper. First, temperatures must be assumed in the absorber and in the stripper, say, 40 and 100°C, respectively. These temperatures would be adjusted later to find the lowest cost of operation. Second, we postulate that the solvent streams leaving the absorber and the stripper are in equilibrium with the gases they contact. This would be true if sufficiently long contact were maintained, but in practice 25 to

TABLE 4.4–1 Smoothed Values for Solubility of Carbon Dioxide and Hydrogen Sulfide in 15.3 Weight Per Cent Monoethanolamine

Partial Pressure CO_2 (mmHg)	Moles Carbon Dioxide per Mole Amine		
	40°C	100°C	140°C
1	0.383	0.096	
10	0.471	0.194	
100	0.576	0.347	0.109
200	0.614	0.393	0.162
760	0.705	0.489	0.275
1000	0.727	0.509	0.300

Partial Pressure H_2S (mmHg)	Moles Hydrogen Sulfide per Mole Amine		
	40°C	100°C	140°C
1	0.128	0.029	
10	0.374	0.091	0.040
100	0.802	0.279	0.124
200	0.890	0.374	0.167
500	0.959	0.536	
700	0.980	0.607	

TABLE 4.4–2 Gas-Absorption Systems of Commercial Importance[a]

Solute	Solvent	Reagent Added to Solvent to Improve Solubility Properties	Degree of Commercial Importance	
			High	Moderate
CO_2, H_2S	Water		×	
CO_2, H_2S	Water	Monoethanolamine	×	
CO_2, H_2S	Water	Diethanolamine	×	
CO_2, H_2S	Water	K_2CO_3, Na_2CO_3	×	
CO_2, H_2S	Water	NH_2		×
CO_2, H_2S	Water	NaOH, KOH		×
CO_2, H_2S	Water	K_3PO_4		×
CO_2	Propylene carbonate			×
CO_2	Glycerol triacetate			×
CO_2	Butoxy diethylene glycol acetate			×
CO_2	Methoxy triethylene glycol acetate			×
HCl, HF	Water		×	
HCl, HF	Water	NaOH	×	
Cl_2	Water		×	
SO_2	Water			
SO_2	Water	NH_3		×
SO_2	Water	Xylidine		×
SO_2	Water	Dimethylaniline		×
SO_2	Water	Aluminum hydroxide sulfate		×
NH_3	Water		×	
NO_2	Water		×	
HCN	Water	NaOH	×	
CO	Water	Copper ammonium salts	×	

[a] A. L. Kohl and F. C. Riesenfeld, *Chemical Engineering*, **66,** No. 12, 127, 1959; T. K. Sherwood and R. L. Pigford, *Absorption and Extraction*, 2nd ed., McGraw-Hill Book Company, New York, 1952.

100 per cent more solvent is needed than is consistent with this equilibrium postulate. The equilibrium postulate enables an estimate of solvent requirements sufficient for these early stages in process discovery.

Using the equilibrium data in Table 4.4–1 for CO_2 in the aqueous monoethanolamine, we estimate that the solvent leaving the stripper at 100°C will contain a very small amount of dissolved CO_2, say, less than 0.10 mole CO_2/mole amine. The solvent leaving the absorber is postulated to be in equilibrium with the 70 per cent CO_2 stream, which has a partial pressure of CO_2 equal to $p_{CO_2} = (0.70)(760) = 530$ mmHg; the *dissolved CO_2 in that solvent stream amounts to about 0.68 mole CO_2/mole amine. The amount of CO_2 absorbed in the solvent is 0.68 − 0.10 = 0.58 mole CO_2/mole of amine circulated.*

Now examine the gas stream. Initially, it is 70 per cent CO_2, and finally it must be 5 per cent CO_2. Therefore, the amount of CO_2 that must be absorbed per mole of gas entering the absorber is 0.70 − 0.05 = 0.65 mole CO_2 absorbed/mole gas processed. The ratio of these gives an estimate of the amount of amine that must circulate per mole of gas processed.

$$\frac{0.65 \text{ mole } CO_2 \text{ absorbed/mole gas processed}}{0.58 \text{ mole } CO_2 \text{ absorbed/mole amine circulated}} = 1.1 \frac{\text{moles amine circulated}}{\text{moles gas processed}}$$

The preceding estimate is based on the assumption that equilibrium was reached. This would require a very large absorber. It is industrial practice to increase the solvent circulation rate by 25 to 100 per cent, thereby reducing the absorber size. Thus the solvent requirements are estimated to be 1.4 to 2.2 moles of amine circulated per mole of gas processed.

Liquid–Liquid Extraction

When immiscible liquids are brought into close contact, an equilibrium is reached in the distribution of solutes among the liquid phases. The mixing and then settling of the phases transfers the solute from the one solvent phase to a distribution between both phases. This transfer can be to the advantage of the engineer designing a process to clean dissolved material out of a liquid stream or to recover the solute.

For example, the recovery of benzoic acid from its dilute aqueous solution can, in theory, be accomplished merely by boiling off the water (heat of vaporization = 970 Btu/lb). However, if the benzoic acid is first extracted from the water by a benzene solvent, at least a tenfold increase in the benzoic acid concentration can be obtained, and also the benzene is much easier to vaporize (heat of vaporization = 170 Btu/lb). The benzoic acid recovery by benzene extraction and evaporation would be much easier than by the direct evaporation of the water.

The commercial uses of liquid–liquid extraction are enormous, a few being mentioned in Table 4.4–3. In the petroleum industry crude oil, which is a mixture of thousands of different hydrocarbons, is initially fractionated into a variety of products by exploiting boiling-point differences. Then these fractions are further fractionated by liquid–liquid extraction, using the greatest variety and largest extractors of any industry. As an example in another field, in the pharmaceutical industry penicillin is produced by fermentation in corn-steep liquor and lactose in 50,000-gallon tanks. The broth is acidified, which favors the transfer of penicillin to a butyl or amyl acetate solvent. A neutral solution of aqueous phosphate is used to strip the penicillin from the acetate solvent. Several extractions back and forth between the organic and aqueous solvents purify and concentrate the antibiotic before it is freeze dried.

TABLE 4.4–3 A Few of the Many Uses of Liquid–Liquid Extraction

Feed Solution	Extractive Solvent	Product and Remarks
Lubricating oil stocks	Furfural, phenol, liquid propane	Dewaxing, removal of varnishes and sludge, and adjustment of viscosity–temperature relationships of lube oils
Naphthas	Liquid sulfur dioxide, furfural, diethylene glycol	Extraction of aromatic from aliphatic hydrocarbons
Close boiling C_4 hydrocarbons	Aqueous cuprammonium acetate	Butadiene as raw material for synthetic rubber
Petroleum oils	Sodium hydroxide, potassium phosphate, and ethanolamine solutions	Removal of sulfur-containing compounds
Fermentation broth	Butyl and amyl acetate	Concentrations of penicillin, Aureomycin, Terramycin, and other antibiotics
Antibiotics in butyl and amyl acetate (from extraction above)	Aqueous phosphate buffer solution	Purification of antibiotics before freeze drying
Sulfite waste liquors from paper making	Toluene, xylene, methyl ethyl ketone	Vanillin, acetic, and formic acid
Animal and plant fats and oils	Sodium hydroxide or calcium carbonate solution	Fatty acid removal to form water-soluble soaps
	Propane, furfural	Separate oils by degree of unsaturation, vitamins A, D, and E concentration
	Hot deaerated water	Separate glycerol and fatty acids
Nuclear reactor fuel elements dissolved by acids	Kerosene–tributyl phosphate solution	Recover uranium and other nuclear reactants and products for reuse
Seawater	Methyl amyl or ethyl butylamine	Recovery of fresh water (not yet competitive with other desalination processes)

Figure 4.4–2 shows a simplified flow sheet for the separation of two materials, A and B, in solution by extraction with a solvent, S. In the extractor two phases are formed, the *extract* phase rich in component B and the *raffinate* phase rich in component A. By distillation the solvent S is recovered and recycled back to the extractor, leaving a purified raffinate enriched in A and a purified extract enriched in B.

The extractor usually is a series of mixers and settlers in which the feed and solvent streams are contacted and separated several times in some predetermined sequence to make the best use of the solvent extractive properties. Figure 4.4–3 shows several mixer–settler combinations: single contact, crosscurrent extraction, and countercurrent extraction. The staging of mixer–settlers into multicontact systems is

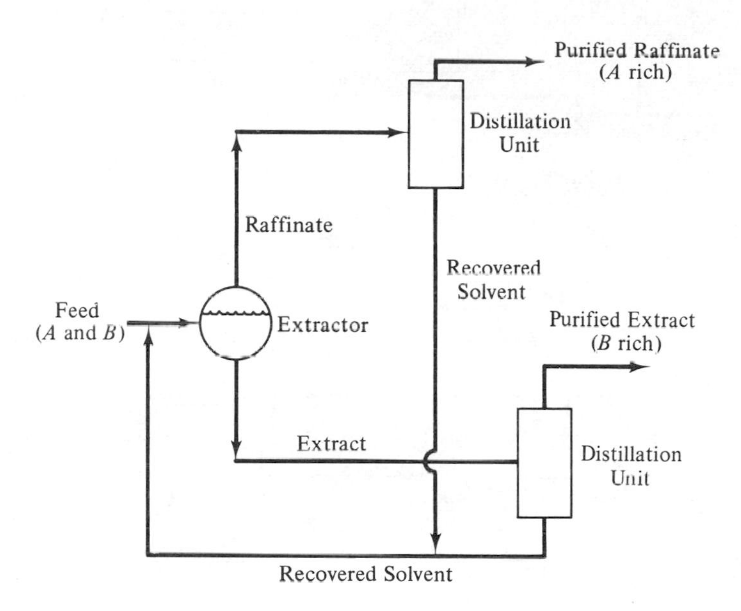

Figure 4.4–2. Liquid–Liquid Extraction with Solvent Recovery

an effective way of improving processing economy by reduction of solvent use.

Data on the solubility of three-phase systems (two solvents and the solute) are available from the standard sources cited at the end of the chapter. Frequently, the solvents themselves are partially mutually soluble, further complicating extraction technology. However, in this introduction to the subject we consider only immiscible solvents, in which case solubility data can be presented by the distribution equation

$$y = Kx$$

where y is the molar concentration of solute B in solvent phase A, and x is the molar concentration of solute B in the solvent phase C. Table 4.4–4 shows values of K for several aqueous systems.

Example 4.4–2 Benzoic Acid Extracted from Water *A waste-water stream from a process contains 0.05 mole benzoic acid/gallon, and issues from the process at 1000 gallons/h. How effective would benzene extraction be in recovering the benzoic acid? Suppose that 1000 gallons/h of benzene were used as the solvent phase.*

First we examine a general single-stage extractor, Figure 4.4–4, deriving the equations that describe its operation. Then we use these equations to assess the operation of the three configurations shown in Figure 4.4–3.

A material balance on benzoic acid gives

$$S_i y_{iF} + F_i x_{iF} = S_i y_i + F_i x_i$$

and since the two exit streams are in equilibrium

$$y_i = Kx_i$$

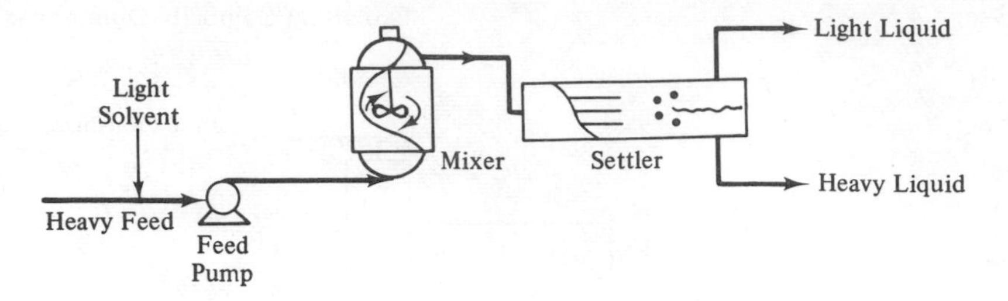

Single Contact

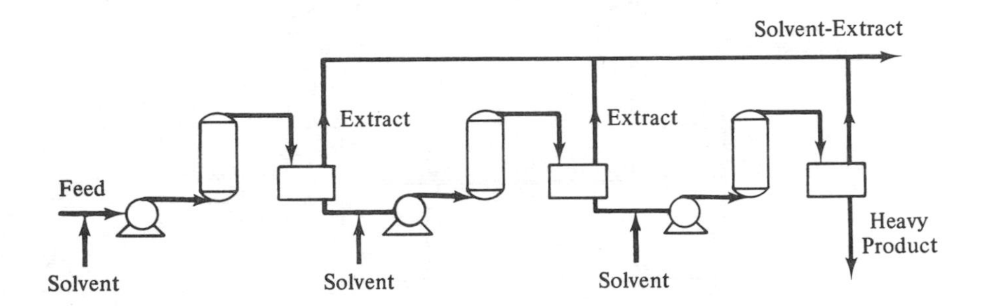

Cross-current Contact

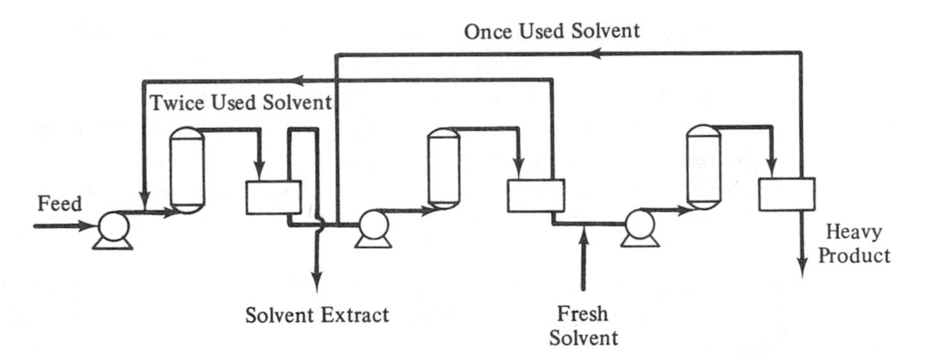

Counter-current Contact

Figure 4.4–3. Mixer–Settler Extractor and Several Configurations

Solving these two equations gives the following expressions for the concentration of benzoic acid in the effluent water stream.

$$x_i = \frac{x_{iF} + (S_i/F_i)y_{iF}}{1 + (S_i/F_i)K}$$

single-stage extraction

TABLE 4.4–4 Selected Distribution Coefficients for Aqueous Systems

Components		Temp,	Distribution
Soluts B[a]	A	°C	Coefficient K
Chlorine	Carbon tetrachloride	0	5.0
Bromine	Carbon tetrachloride	0	20
		23	27
		40	30
Iodine	Carbon tetrachloride	25	55
Ammonia	Carbon tetrachloride	25	0.0042
Chlorine dioxide	Carbon tetrachloride	0	0.85
		25	0.63
Ammonia	Chloroform	25	0.040
Sulfur dioxide	Chloroform	0	0.71
		20	0.71
Nitric acid	Ethyl ether	25	0.012
Ethanol	Benzene	25	1.1
Isopropanol	Benzene	25	0.50
Isopropanol	Toluene	25	0.21
Phenol	Carbon tetrachloride	25	0.36
Phenol	Chloroform	25	2.8
Phenol	Benzene	25	2.3
Phenol	Xylene	25	1.4
Acetone	Carbon tetrachloride	25	0.44
Acetone	Chloroform	25	5.5
Acetone	Toluene	0	0.48
		20	0.49
		30	0.51
Acetic acid	Carbon tetrachloride	25	0.059
Acetic acid	Bromoform	25	0.083
Acetic acid	Chloroform	20	0.028
Acetic acid	Ethyl ether	0	0.56
		15	0.48
		25	0.45
Acetic acid	Benzene	15	0.012
		25	0.023
Acetic acid	*o*-Xylene	25	0.042
Benzoic acid	Chloroform	10	4.2
		25	2.4
		40	3.2
Benzoic acid	Benzene	6	4.0
		20	4.3
		25	1.8
Diethylamine	Chloroform	25	2.2
Diethylamine	Benzene	25	0.63
Diethylamine	Toluene	18	0.48
		25	0.63
		32	0.90
Diethylamine	Xylene	25	0.20
Aniline	Toluene	25	7.7
Aniline	Xylene	25	3.0

[a] Solute B distributed between nearly immiscible components A and C; component C is water.

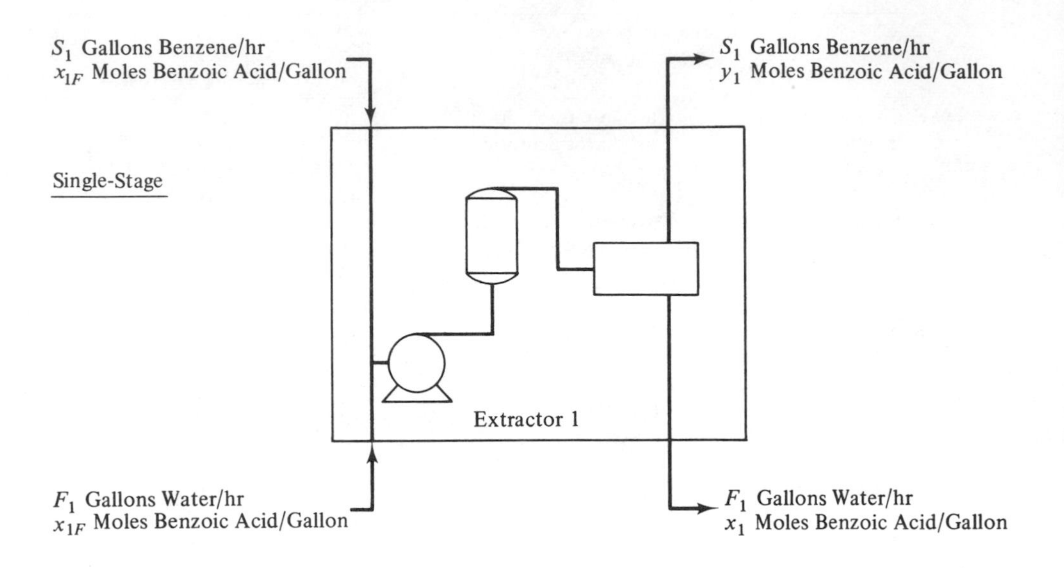

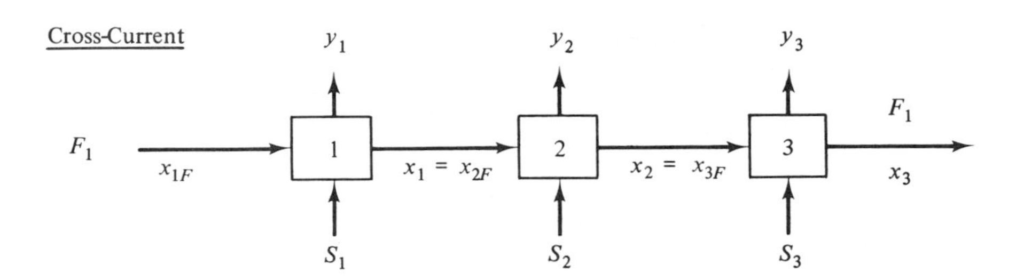

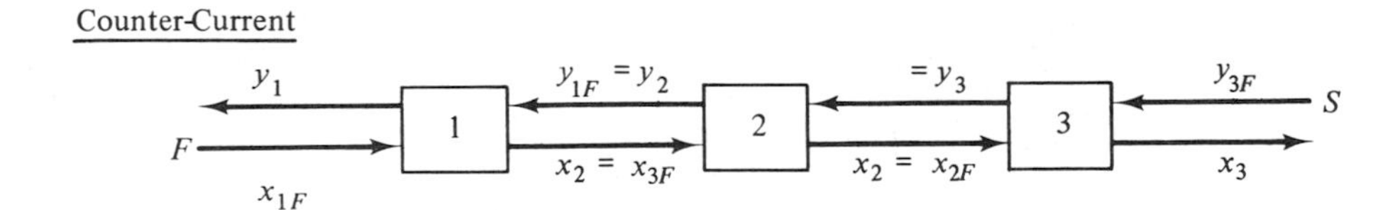

Figure 4.4–4. Extractor Notation

A single-stage extraction using benzene can be assessed directly:

$$S_i = 1000 \text{ gal/h}$$

$$F_i = 1000 \text{ gal/h}$$

$$y_{iF} = 0 \text{ mole benzoic acid/gal}$$

$$x_{iF} = 0.05 \text{ mole benzoic acid/gal}$$

$$K = 4, \qquad \text{from Table 4.4–3}$$

$$i = 1, \qquad \text{one stage}$$

Therefore,

$$x_1 = \frac{0.05 + 0}{1 + (1000/1000)4} = 0.01$$

The benzoic acid concentration in the water would be reduced by a factor of 5 in a single-stage extraction.

Now suppose that three-stage crosscurrent extraction were used with one third of the benzene used in each stage. Fresh solvent would be used in each stage, so $y_{iF} = 0$ *for* $i = 1, 2,$ *and 3. The water phase leaving stage 1 would be the feed to stage 2, and that leaving stage 2 the feed to stage 3. So*

$$x_1 = \frac{x_{1F}}{1 + (S_1/F_1)K}$$

$$x_{2F} = x_1$$

$$x_2 = \frac{x_{2F}}{1 + (S_2/F_1)K} = \frac{x_{1F}}{[1 + (S_1/F_1)K][1 + (S_2/F_1)K]}$$

$$x_{3F} = x_2$$

$$x_3 = \frac{x_{1F}}{[1 + (S_1/F_1)K][1 + (S_2/F_1)K][1 + (S_3/F_1)K]}$$

three-stage crosscurrent

Since

$$x_{1F} = 0.05 \text{ mole benzoic acid/gal}$$

$$F_1 = 1000 \text{ gal/h}$$

$$S_1 = S_2 = S_3 = 1000/3 \text{ gal/h}$$

$$x_3 = \frac{0.05}{\left[1 + \frac{(1000)(4)}{(1000)(3)}\right]^3} = 0.004 \text{ mole benzoic acid/gal}$$

Three-stage crosscurrent extraction will reduce the benzoic acid concentration in the water by a factor of 12.

Now consider the three-stage countercurrent extraction. Here the water phase effluent from the first stage is the feed to the second stage, but the solvent effluent from the third stage is the feed to the second stage. The fresh solvent enters the third stage and the water to be processed enters the first stage. The processed water leaves the third stage and the benzene solvent leaves the first stage.

For stage 1:

$$x_1 = \frac{x_{1F} + (S/F)y_2}{1 + (S/F)K}$$

$$y_1 = Kx_1$$

For stage 2:

$$x_2 = \frac{x_1 + (S/F)y_3}{1 + (S/F)K}$$

$$y_2 = Kx_2$$

For stage 3:

$$x_3 = \frac{x_2 + (S/F)y_{3F}}{1 + (S/F)K}$$

$$y_3 = Kx_3$$

$$y_{3F} = 0 \qquad \text{since pure benzene is used}$$

Solving these equations gives the following expression for the effectiveness of the countercurrent extraction system:

$$x_3 = \frac{x_{1F}}{[1 + (SK/F)][1 + (SK/F)^2]}$$

three-stage countercurrent extraction

When the numbers are put in,

$$x_3 = 0.0006$$

In summary, the single-stage contact reduces the benzoic acid concentration by a factor of 5, three-stage crosscurrent contact by a factor of 12, and three-stage countercurrent contact by a factor of 80. In industry one will see quite a bit of countercurrent contacting.

4.5 CONCLUSION

In this chapter we examined a few of the many physical phenomena that can be exploited to effect the separation of materials. The economy of the process depends largely on the phenomena selected to perform the separation tasks, and these separations must be tailored to the specific processing problem. Here we merely developed an understanding of the means of separation, and in the next chapter we will work on the problem of fitting separation methods into the processing problem.

The field of separations is so vast and challenging that there are those who spend a lifetime as separations specialists. We have mentioned here only the high points, and we must now move on in the development of the broad outlines of process synthesis.

REFERENCES

Excellent descriptions of a variety of separation devices are found in

Perry's Chemical Engineers Handbook, 5th ed., McGraw-Hill Book Company, New York, 1968.

A more advanced text on separations dealing with chemicals processing is

C. J. KING, *Separation Processes*, McGraw-Hill Book Company, New York, 1970.

A short summary of solid-waste separation is

Recovery and Utilization of Municipal Solids Wastes, U.S. Environmental Protection Agency, Government Printing Office, Washington, D.C., 1971.

An interesting summary of solids separation methods is

"Solids Separations," *Chemical Engineering*, Deskbook Issue, **78,** No. 4, Feb. 15, 1971.

See also

"Recent Advances in Separation Techniques," *American Institute of Chemical Engineers Symposium Series 120*, **68,** 1972.

PROBLEMS

The problems on separation technology are divided into two groups. The first group involves speculation on novel separation equipment. The solution should include a short essay on the physical and chemical property differences exploited and a sketch of the equipment and flow sheet. A discussion in class is often worthwhile on the variety of solutions obtained by the students. The remaining problems are quantitative extensions of the examples included in the chapter, and most often involve skills in material balancing.

1. **SOLID-WASTE RECYCLE.** Each day, as waste from our cities, thousands of tons of paper, glass, plastic, and metal are buried in sanitary landfill sites. These materials could find use as secondary raw materials if effective and cheap separation schemes could be devised. Suppose that the solid wastes from a city are milled to 1-in. size and that the composition is approximately

	% BY WEIGHT
Paper	60
Glass	10
Plastics	10
Metals	10
Vegetable and animal matter	10

Devise three different ways in which materials in these five classifications can be separated by exploiting differences in physical and chemical properties. Which separation scheme involves the least handling of material?

2. SOLID-WASTE RETRIEVAL.* The Bureau of Solid Waste Management reviews U.S. patents to provide information on possible schemes for the recovery of useful material from municipal solid wastes. Following are abstracts of patents on separation equipment. The original patents are to be examined and a brief report of the equipment is to be prepared, including a sketch of the working mechanism. The abstracts are from "Recovery and Utilization of Municipal Wastes," Public Health Service Publication No. 1908.

Patent No. 2,422,203—Specific Gravity Separation of Solids in Liquid Suspension —(June 17, 1947)

The invention relates to the recovery of solids from liquids by sink-float methods. Separation according to size or specific gravity is claimed. The invention claims an arrangement whereby the requirement to employ heavy media is eliminated. Specifically, a controlled centrifugal movement of a body of mixed solids in a liquid is claimed to permit clean and distinct separation between solids of different sizes and/or specific gravities in an aqueous pulp. The process was developed as an improved ore-beneficiation technique, although potential uses in coal-washery applications and the treatment of potash ores are claimed. Performance data on ores in the size range 6–200 mesh are provided. Performance attainable on various forms of solid waste cannot be established without experimental studies, except that it should be noted that potential applications are limited to relatively fine fractions.

Patent No. 2,474,695—Collecting and Separating Apparatus—(June 28, 1949)

This device was designed to separate particles from a fluid medium, to separate heavier fluids from light ores, and to remove dust, sand particles, and chips from castings and other items produced in foundries, mills, etc. A mixed fluid stream is accelerated in a duct connected to an elbow or bend. As the stream enters the elbow, heavier materials are separated from lighter ones because of differential centrifugal forces. A complex baffle arrangement is provided in the elbow to collect the separated material streams. No data indicating performance of the device are provided in the patent.

Patent No. 2,561,665—Continuous Classifier for Solids—(July 24, 1951)

This device was developed as an improved technique for separating solids of various specific gravities from a fluid medium, provided that all solids to be separated have a specific gravity greater than the fluid. A series of tanks are provided such that the cross-sectional area of flow path increases, tank-by-tank, in the direction of flow. A liquid suspension is introduced in a tank in a downward and upward path thereby inducing the settling of solids in each tank. Because the flow path of each subsequent tank is of larger cross-sectional area, the upward velocity of flow decreases, thus permitting the deposition of lighter solids in each stage. An adjustable baffle is provided in each tank which is claimed as a novel means for controlling flow and minimizing turbulence.

* Before this problem is assigned either the patents must be available in the library or copies ordered from the U.S. Patent Office.

Patent No. 2,689,646—Fluid Flotation Separator and Method for Separating Pulverized Materials—(September 21, 1954)

This device relates to a method for separating fines from pulverized materials using gas flotation. Problems with previous state-of-the-art are reported to be (1) difficulty in dislodging and rendering buoyant an efficient quantity of fines, which are distributed throughout the pulverized material; and (2) a tendency of fines to agglomerate and form heavy nodules which are not rendered buoyant and consequently not separated.

The new method involves conveying a pulverized material over an inclined porous surface at an appreciable speed. A stream of gas under pressure is passed through the porous surface from below and then through the pulverized material. As the pulverized material is conveyed over the porous surface, small portions of the moving material are agitated by the gas to break down the nodules. Simultaneously, small portions of the material adjacent to the upper surface of the material stream are elevated above the material and thereafter are permitted to fall in a cascading effect through the flotation gas flowing upwardly from the surface of the material stream. The fine particles of material in the cascading portion are rendered buoyant and become entrained in the flotation gas. The agitation and elevation of the pulverized material are produced by several deflector members positioned above the porous plate. After the flotation gas and the fine particles of material entrained therein have been conducted to a collector, the velocity of the flotation gas is reduced to permit the fine particles of material to settle and be collected.

Patent No. 2,689,648—Separation of Metallic from Nonmetallic Particles—(September 21, 1954)

This invention provides a method for separating metallic from nonmetallic particles. The method is based on the phenomena that nonmetallic particles may become electrostatically charged and thereby caused to react to forces created by an electrostatic field. The mixture to be separated is transported along a horizontal plane by means of a conveyor through an electrostatic field of alternating potential. Alternating forces created by this field impart sufficient vibration to the nonmetallic particles resulting in a stratification and concentration of metallic particles which do not so respond. Separation of the resultant layers of the two types of particles is accomplished mechanically. This device is an extension of long-established minerals-beneficiation technology whereby stratification and concentration of ores is accomplished by vibratory means alone. Performance data and feed requirements are not outlined in the patent, although it is expected that small particle size and low moisture content would be essential. Potential applications to solid waste recovery are judged to be minimal and would probably be limited to the removal of nonferrous metals from other nonmetallic components. The cleaning of compost may be one such potential application.

Patent No. 2,707,554—Grain Separators—(May 3, 1955)

This invention is claimed to be an improved machine for the removal of weed seeds from cereal grains (wheat, oats, barley, rye, etc.). The device is also claimed to segregate various grains from one another. The device provides a three-stage separation employing in combination a variable transverse diameter-dividing assembly, a weight-dividing (specific gravity) assembly, and a selective length-dividing assembly operating on a given mass of grain in the order listed above. Width separation is accomplished by one or more cylindrical screens with adjustable circumferential longitudinal openings which permit the division of the mixture into two fractions according to transverse diameter. Density separation is accomplished by classification in an air separator. Length separation is accomplished by means of a complex arrangement of rotating disks, the operation of which is difficult to describe in narrative form.

Patent No. 2,806,600—Separator for Separating Granular Poultry Feed from Poultry Bedding—(September 17, 1957)

This device is a very simple application of screening and consists essentially of a perforated cylinder mounted to rotate about a horizontal axis. Immediately below the cylindrical screen is a semicircular trough. A means is provided to deliver feed contaminated with bedding material to the rotating and tumbling screen; for collecting, in the trough, clean feed passing through the screen; and for collecting and discharging from the screen the bedding material separated from the mixture.

Patent No. 2,820,705—Method of Recovering Metals from Nonferrous Metallurgical Slags—(January 21, 1958)

Generally stated, the process is designed for recovering metals from nonferrous metallurgical slags containing reducible compounds of the metals and involves establishing a reduction zone in a furnace containing a matrix of carbon or other reducing materials for the reducible compounds. This zone is maintained at a temperature between the melting point of the slag and 1450°C. Also, an atmosphere of carbon monoxide and carbon dioxide is established in such proportions that log CO/CO_2 is, broadly, from -4 to $+2$.

The molten slag is contacted with the reducing material in the reduction zone by percolating it downwardly. This requires but a few seconds. However, during this short time under the conditions present in the reduction zone, the reducible compounds of the valuable metals to be recovered are converted to the free metals, while the iron content of the slag is substantially unaffected.

Accordingly, there is produced a treated slag product which contains the valuable metals in the reduced or free state. The treated slag containing these metals is then withdrawn from the reduction zone almost immediately, to prevent reduction of any iron compounds which might be present and concomitant contamination of the desired metal products. The treated slag then may be processed for separation of the metal fraction from the residual slag.

Potential applications in solid waste processing would relate to the recovery of high-temperature incinerator slags.

Patent No. 2,868,376—Heavy Media Separator—(January 13, 1959)

This device is an adaptation or modification of a spiral classifier. In general, classifiers of this type consist of a tank having an inclined bottom, an overflow for fines or "float," an elevated discharge for the "sink" or coarse rake products, and a spiral conveyor for agitating the pulp and raking the sink out of the tank. Three classifiers are used to separate a mass of grains of mixed sizes and/or different specific gravities into various grades or sizes.

This patent claims a modification of the spiral classifier to utilize heavy media, which in turn permits the segregation of mineral ores into a sink fraction, a middling fraction, and a float fraction. This device appears to be very similar to units manufactured by Denver Equipment Company.

Patent No. 2,986,277—Method and Means for Treating and Sorting Comminuted Substances—(May 30, 1961)

The invention relates to an apparatus for sorting comminuted particles on the basis of density, by means of a whirling motion. However, unlike a cyclone where particles move concurrently with air flow, this invention provides for a portion of the particles to move counter to the flowing gases.

The device is adaptable to particle size ranges on the order of 50 microns, and so would have extremely limited application for solid waste recovery.

Patent No. 3,064,806—Apparatus for Wet Sizing of Solid Materials—(November 20, 1962)

This invention relates to a means for wet sizing of finely ground (35–200 mesh) solid particles. The process combines hydraulic classification and wet screening and is claimed to be superior to either technique used alone. The invention consists of a stationary container equipped with an impeller to impart rotary motion to the mixed slurry, which is fed constantly to the unit; an inclined screen; and means for directing the upper portion of the slurry against the screen. Coarser materials are retained in the container and are removed after settling. Finer particles passing the screen are collected along with the overflow as a separate product. Because of the small particle sizes, direct application to solid waste processing will probably be extremely limited.

Patent No. 3,127,016—Sorting of Articles—(March 31, 1964)

This device employs the mechanics of partially elastic collision to effect separation. Particles or articles to be sorted are impinged one by one onto a stationary target element of desired elastic properties. Elastic properties of the colliding materials determine the amount of energy transfer and, hence the subsequent movement of the impinged particle. Provision is made to sort the particles into two fractions, depending on whether or not the resulting trajectory deflects the particles a certain degree.

Patent No. 3,327,857—Object Counting and Discriminating Device—(June 27, 1967)

This device is designed to automatically sort broken tablets from whole tablets, although extension of the principles involved to other applications is possible. A method is provided for comparing by photometric means every tablet to be sorted with a standard, i.e., integral tablet. Tablets which are smaller than the standard, because of being broken or other defects, are rejected. Those which compare favorably to the standard are retained. It is claimed that, with modification, the basic invention may be constructed to distinguish shape, color, or other physical characteristics. Operating principles are very similar to the Sortex optical separator.

3. **NUT MEAT SEPARATION.** How could a nut meat be separated automatically from its shell after crushing? These data will be useful.

	NUT MEAT	SHELL
Color	Brown	Red
Size after crushing	0.1–0.2 in.	0.1–0.2 in.
Density	0.850	0.600
Hardness	Soft	Hard
Smoothness	Smooth	Rough
Shape	Round	Half-shell
Surface property	Oily	Woody

The separation method should be reported as a patent abstract as in Problem 2.

4. **PEAS FROM PODS.** What property differences might be exploited to separate loose green peas from their pods at 1-ton/h rate? Prepare a patent abstract for your invention.

5. **BLACK SPECKS IN METHYL CELLULOSE.** During the drying of methyl cellulose, a white powder used as a thickener, black specks of charred material are formed. Although the black specks constitute only a very small fraction of the total product, they must be removed. How?

	METHYL CELLULOSE	BLACK SPECKS
Size	40–60 mesh	20–50 mesh
Color	White	Black
Density	0.70	0.80
Shape	Irregular	Flat
Solubility in water	Soluble	Insoluble

6. **ACTIVATED CARBON FROM SAWDUST.** Activated carbon is used in the purification of solutions, such as the cleanup of cane, beet, and corn sugar, and for the removal of tastes, colors, and odors from water, vegetable oils, alcoholic beverages, and pharmaceuticals. Recently, activated carbon has been successfully used to treat waste waters from chemical plants. It is estimated that 1 lb of activated carbon has more than 1 square mile of active surface area. In one method for producing activated carbon the raw material, sawdust or peat, is mixed with zinc chloride solution, dried, and calcined at 1500°F. The zinc chloride increases the surface area of the carbon. When the carbonization has been completed, the residual zinc chloride must be removed. What possible basis for separation might be used to separate this solid mixture?

Data on mixture:

	PURE CARBON	PURE ZINC CHLORIDE
Color	Black	White
Size	Zinc chloride coats carbon surface	
Density	2.0	2.91
Melting point	3500°C	262°C
Solubility	Insoluble in water	432 parts/100 parts H_2O at 25°C

7. **CRYSTAL MASS DRYING.** The solids discharge from a sodium bromide crystallizer is wet with mother liquor. A bone-dry solid product is desired. Invent at least three different and completely continuous processing schemes for removing the mother liquor from the solid. Specify the basis for each separation and explain why you chose it. Compare and contrast them with respect to estimated cost, ease of operation, reliability, and simplicity.

8. **DRYING OF SMOKELESS POWDER.** In the manufacture of smokeless gun powder, the water content of nitrated cotton is reduced by alcohol percolation combined with pressure dehydration.

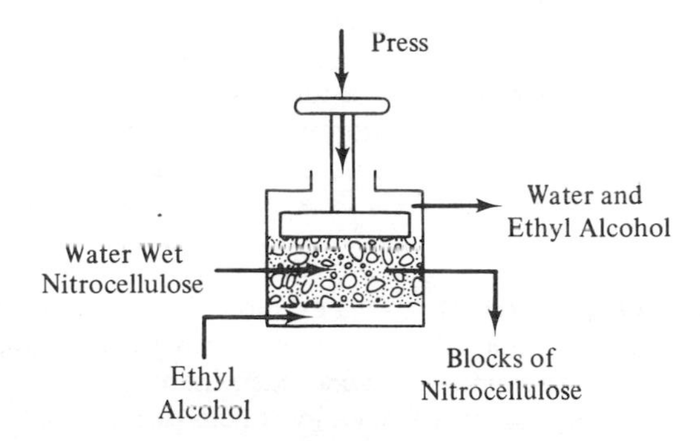

What is the basis of separation in this operation? Devise a piece of equipment which could carry out this operation continuously. What materials other than ethyl alcohol could be used in this operation? Explain your selections.

9. **LOW-TEMPERATURE DRYING OF SOLIDS.** In the production of food or pharmaceutical products, overheating must be avoided if product quality is to be maintained. Devise a technique that will remove water from a solid at a low temperature. If the temperature is below the freezing point of water, what basis can be used for separation? What pressure would be best for removal of solid water? Prepare a patent abstract on your invention.

10. **REDUCING OIL POLLUTION.** The separation of oil from water is an important and difficult problem facing the oil industry. One challenging problem relates to the oil–water mixtures discharged from large oil tankers. When an oil tanker discharges its cargo, it is common practice to refill several of the tanks on board with salt water. This

water is ballast, which improves the handling characteristics of the ship on the return trip. As the ship approaches land to pick up more oil, the water is pumped out of the tanks and into the ocean. Oil that clings to the walls of the tanks is mixed with this water and causes significant amounts of oil to be discharged directly into the ocean. The problem is to separate the oil–water emulsion that is pumped from the tank. Design a system to solve this problem.

Data

Density of oil	1.1 g/cm^3
Density of salt water	1.2 g/cm^3
Oil droplet size	$\sim 75\ \mu$
Oil concentration	0.5% by weight

A ferro fluid has been developed that consists of finely divided iron particles suspended in an organic solvent. Would this fluid be of any use in solving this problem? See "Magnetically Induced Separations of Stable Emulsions" AIChE *Symp Series* 124, vol. 68, 1972, p115.

11. **CRANBERRY DRAINING.** This problem illustrates how the basis of separation ought to change with the amount of material processed.

During the processing of cranberries the fruit is washed and placed on a screen to drain. A 1000-fold increase in production is planned for next season and the washing and liquid-removal sections of the process will require expansion. Two hundred square feet are presently required to drain the berries. Develop plans for the new liquid-removal section.

12. **COAL GASIFICATION PLANT.** This example illustrates the effect on the basis of separation of the concentration of the component being removed.

Two types of coal are supplied to a coal gasification plant. One type arrives from a coal field by railroad car, the other from a local field by a pipeline as a slurry. The first coal contains 10 per cent water and the second coal arrives in a slurry of 60 per cent coal and 40 per cent water. The coal fed to the plant has to be below 5 per cent water. Devise separation schemes for each source of coal.

13. **MULTISTAGE DISTILLATION.** If a mixture of benzene and toluene is partially vaporized or condensed, the compositions in the vapor and in the liquid are different, the more volatile component being enriched in the vapor phase. For the benzene–toluene mixture, the enrichment is expressed as

$$\frac{(\text{benzene/toluene})_{\text{vapor}}}{(\text{benzene/toluene})_{\text{liquid}}} = 2.5$$

If during processing it is arranged that half the feed entering a single unit leaves as vapor and half as liquid, how many stages must be used to produce a stream that is 75 per cent benzene from a feed that is 50 per cent benzene? What fraction of the benzene leaves as the enriched vapor, and what fraction leaves in the many liquid streams?

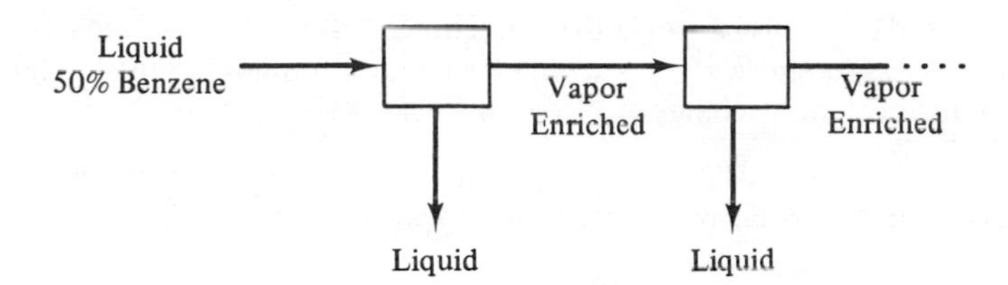

How does the structure of the process change if the liquid streams are fed back into the unit that has the enriched vapor feed with composition closest to the recycled liquid? In this way some of the benzene lost before in the liquid streams is recovered in the enriched vapor product. Such recycling operations are very important for increasing processing efficiency. See Example 4.3–1.

14. MILLING CIRCUIT. Small particles of a solid are to be reduced to powder in a ball mill. The ball mill is a long metal cylinder mounted horizontally and rotated. It is partially filled with hard metal or ceramic balls between which the particles are milled to a powder; the feed enters one end of the mill and the powder and unmilled feed leave the other end. Tests indicate that 70 per cent of the feed will not be sufficiently reduced in size during one pass through the mill. For this reason a screening operation is performed after milling to recycle the large particles back to the mill. If 10 tons/h of powder is to be produced, what must the mill capacity be in tons per hour?

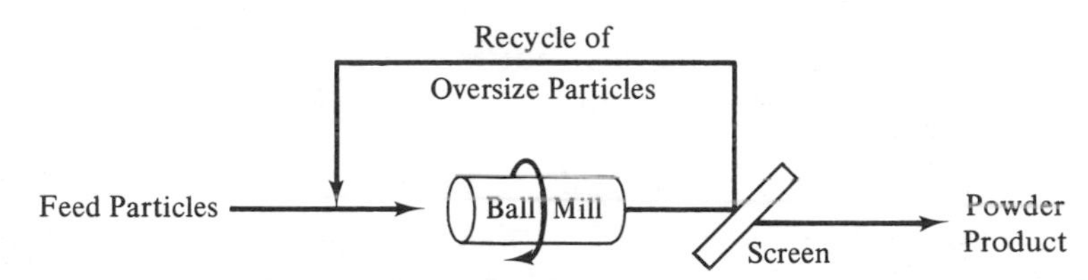

15. CRYSTALLIZER ANALYSIS. Solids can be crystallized out of a solvent if a sufficient amount of the solvent is evaporated. The equation describing the operation of such a crystallizer is

$$C = R\left[\frac{100\omega_o - S(H_o - E)}{100 - S(R - 1)}\right]$$

where C = weight of the crystals in the final magma

$$R = \frac{\text{molecular weight of the hydrated solute}}{\text{molecular weight of the anhydrous solute}}$$

S = solubility of the solute in the solvent at the final temperature (parts by weight of anhydrous solute per 100 parts by weight of total solvent)

ω_o = weight of anhydrous solute in original batch

H_o = total weight of solvent in batch at beginning of process

E = evaporation of solvent during crystallization

Derive this equation by use of material balances.

16. **ORE LEACHING.** Determine the percentage of the acid-soluble ore in Example 4.2–1 recovered in single-stage leaching, two-stage countercurrent leaching, and three-stage countercurrent leaching.

17. **ETHANOL ENRICHMENT.** If a 30 per cent ethanol in propanol solution is half vaporized, what fraction of the ethanol appears in the vapor phase and in the liquid phase?

18. **HYDROGEN SULFIDE SCRUBBING.** If in Example 4.4–1 hydrogen sulfide is to be scrubbed from the air stream, how much monoethanolamine must be circulated?

19. **COPPER SULFATE LEACHING.** A roasted ore contains 85 per cent insoluble gangue, 10 per cent soluble copper sulfate, and 5 per cent water. The ore is to be processed with water in a single-stage mixer–settler. In such a device the underflow sludge contains 0.8-lb solution/pound of gangue. If 1000 lb/h of ore is to be processed, determine the percentage of the copper sulfate leached out by 2000 lb/h of water; by 4000 lb/h of water. How much less water is required to reach, by two-stage countercurrent leaching, the same leaching as in the first case above?

20. **DESALTING SEA SAND.** How much fresh water is needed to wash a sea sand containing 85 per cent sand, 12 per cent salt, and 3 per cent water? The sand when dried is to contain less than 0.5 per cent salt. In a mixer–settler the sand underflow contains 0.5 lb solution/pound of sand. What is the salinity of the used wash water? How would two-stage countercurrent leaching improve processing efficiency?

21. **PHENOL EXTRACTION.** Suppose that the waste-water stream of Example 4.4–2 contains phenol in addition to the benzoic acid, both present in concentrations of 0.05 lb/gallon of water. What percentage of the phenol would be extracted in the single-, double-, and triple-stage extraction with the benzene solvent?

22. **LEAD–ZINC ORE PROCESSING.** A lead–zinc ore contains 24.6 weight per cent PbS (galena), 32.8 weight per cent ZnS (sphalerite), and 42.6 weight per cent inerts, chiefly $CaCO_3$ and SiO_2. The valuable minerals can be recovered from this ore by the following processes: The ore is ground and conditioned for 10 min at a pH of 8 to 9 by addition of Na_2CO_3 plus 0.3 lb NaCN/ton ore to suppress ZnS flotation. The PbS is floated in Denver-type cells by addition of ethyl xanthate plus a starvation quantity of frother (about 0.05 lb pine oil/ton ore). The PbS flotation tailings (waste) are conditioned by complexing the cyanide with $CuSO_4$ to activate the ZnS, and the ZnS is floated in Denver cells by addition of butyl xanthate as collector. Water is added in the ZnS conditioner to increase the liquid/solid ratio from 1.33 to 1.86. Results of the flotation processes are tabulated:

	Wt % Dry Solids		
	PbS	ZnS	L/S
Feed	24.6	32.8	1.63
PbS concentrate	90.0	3.4	1.94
Tailings 1	4.2	42.0	1.33
ZnS concentrate	4.3	91.1	2.03
Tailings 2	4.1	5.9	1.68

(a) Draw a flow diagram and make a complete material balance on the basis of 1 ton dry ore fed.

(b) Calculate the concentrating ratio and recovery of PbS and ZnS.

(c) Calculate the tons of liquid added in cell 2 per ton of dry ore fed to the process.

23. MERCURY POLLUTION OF THE ENVIRONMENT. The detection of mercury in fish in the late 1960s precipitated intensive engineering research to halt the escape of mercury to the environment. The Wisconsin River contained mercury from the paper industries located along its northern reaches. Mercury-based slimecides had been used to prevent the formation of slimes in paper making and mercury had escaped from the chlor-alkali processes, which provide the bleaches used in paper making.

T. W. Chapman and R. Caban performed basic experiments in the extraction of mercury from the salt solutions used in the chlor-alkali processes. The equilibrium data show that mercury has a strong tendency to remain in an organic phase (triisooctal amine in xylene) under acid conditions (pH 1 to 2), and has a strong tendency to remain in the aqueous phase (NaCl in water) under alkali conditions (pH 11 to 13).

This leads to the Chapman–Caban process* for trace mercury recovery. A waste aqueous stream containing salt and trace mercury is acidified and contacted with the organic phase in a mixer–settler. The mercury-loaded organic phase is then contacted with an alkaline aqueous phase in a second mixer–settler, thereby highly concentrating the mercury in the second aqueous phase.

(a) Determine the amount of organic that must be cycled per hour to scrub the mercury from the contaminated aqueous phase to the concentration shown.

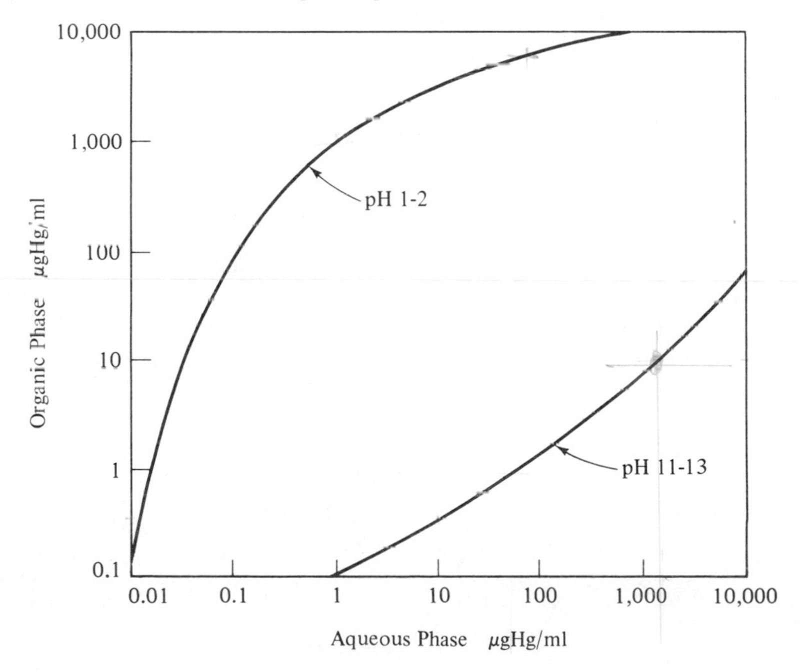

* See T. W. Chapman and R. Caban, *American Institute of Chemical Engineers Journal*, **18**, No. 5, 1972.

(b) If the clean organic may contain 10 μg of Hg/ml, how much fresh aqueous phase must be fed to the second mixer–settler?

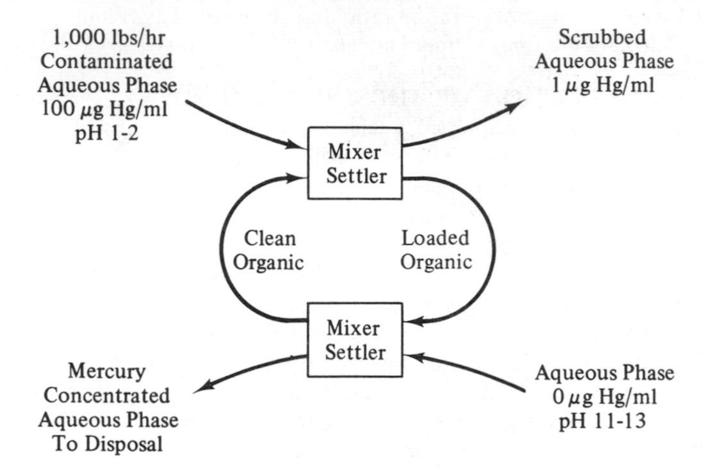

The Chapman-Caban Process

(c) What fraction of the mercury entering leaves in the mercury-concentrated aqueous phase? By what percentage has the volume been reduced of the mercury-contaminated aqueous phase to be disposed of?

5

Separation Task Selection

Man has three ways of acting wisely: Firstly, on meditation, this is the noblest; Secondly, on imitation, this is the easiest; and Thirdly, on experience; this is the bitterest.
Confucius

5.1 INTRODUCTION

At this point in process synthesis it is necessary to gamble on the sequence of physical and chemical tasks to be performed on each stream in the emerging process. Available are data on the chemistry to be exploited (Chapter 2), a mapping of the material flow through the process, which identifies the source and destination of each stream (Chapter 3), and experience in exploiting property differences for materials separation (Chapter 4).

Now we are faced with the problem of rectifying for each stream the differences that may exist between its source and its destination. For example, a source of bromine, needed to support the reaction chemistry, may be contaminated with chlorine and chloroform, and the reaction may require pure bromine. The presence of chlorine and chloroform in the source and their required absence in the destination is the difference that must be rectified now. What sequence of tasks *best* removes the chlorine and chloroform? By *best* we mean the sequence that leads to efficient and economic processing.

Differences in species composition demand our attention first, rather than differences in temperature, pressure, particle size, and other bulk properties. These auxiliary tasks are subordinate to the dominating problem of separating species from a mixture. This chapter is based on this heuristic:

> **Of the many differences that may exist between the source and destination of a stream, differences involving composition dominate. Select the separation tasks first.**

Webster defines:

> heuristic: (from Greek *heuriskein* to discover) serving to guide, discover or reveal: Valuable for empirical research but unproved or incapable of proof....

In this chapter a broad strategy emerges, which we might call the heuristics of task selection.

At each point in the process at which materials are to be separated, data on the chemical and physical properties must be accumulated. These data should include, for example, boiling point, melting point, volatility, solubility in a variety of solvents, density, size, phase, adsorptivity on surfaces, magnetic and electrostatic properties, and chemical reactivity. Differences in these properties in part point to the phenomena to be exploited; however, the selection of the basis of separation is difficult. The selection must be done heuristically by weighing the several attributes of each property difference and the unique features of the separations problem at hand.

The systematic collection, ordering, differencing, and weighing of the physical and chemical property information are the means by which the selection of separation schemes is approached. In particular, attention is focused on the position in the ordered property lists of the materials to be separated (those on the top or bottom of a property list are most easily pulled off a mixture), the gap existing between the properties of adjacent species (a big gap in property difference leads to clean separations), the extremes in temperature and pressure that must be reached to force the property difference (heating, cooling, compressing, and evacuating are costly operations), the conditions in which the materials enter and leave the separation stage (perhaps a high-temperature separation is worthwhile if the separated material must be hot for the next processing step), the current state of technology (we know a lot more about exploiting solubility differences than color differences), the amounts of the various species to be separated (the removal of trace amounts of mercury from water is a completely different problem from the removal of trace amounts of water from mercury), and other factors unique to the particular separation problem.

To ease the comprehension of this approach to task selection, we begin with an actual processing problem, the removal of trace impurities from liquid bromine.

Example 5.1–1 Removal of Trace Impurities from Bromine *In the production of fine chemicals, pharmaceuticals, and food products extraordinary care is taken to* *prevent impurities from entering the process with the raw materials, since it may be all but impossible to remove the impurities if they get built into products during processing. In such a situation, trace amounts of chloroform (300 ppm) and larger amounts of chlorine (2 per cent) are to be removed from liquid bromine, which is to be used as a raw material in fine-chemicals manufacture. How can this be done?*

First we determine the major chemical and physical properties of all the materials. In Table 5.1–1 the boiling points and solubilities in several obvious solvents are given

TABLE 5.1–1 Physical Property Data

			Solubility in					
	Mol wt	Bp (°C)	Br_2	Cl_2	$CHCl_3$	$CHBr_3$	Ether	H_2O (lb/100 lb)
Bromine Br_2	160	59	–	s	s	s	s	4
Chlorine Cl_2	71	−34	s	–	s	s	s	2
Chloroform $CHCl_3$	119	61	s	s		s	s	1
Bromoform $CHBr_3$	252	150	s	s	s	–	s	0.5

for bromine, chlorine, chloroform, and bromoform. Bromoform is a new chemical species with which we might be concerned, since it is the product of a chemical reaction among the species to be separated:

$$\underset{\text{(bromine)}}{\tfrac{3}{2}\,Br_2} + \underset{\text{(chloroform)}}{CHCl_3} \xrightarrow{250°C} \underset{\text{(bromoform)}}{CHBr_3} + \underset{\text{(chlorine)}}{\tfrac{3}{2}\,Cl_2}$$

If we take as our basis the production of 1000 lb/h of pure bromine, 20 lb/h of chlorine and 0.3 lb/h of chloroform must be removed somehow.

Table 5.1–1 shows the source of the separation problem; bromine, chlorine, and chloroform are mutually soluble.

Solubility in Water *In Figure 5.1–1 the materials are ordered according to solubility in water. Do these differences have processing significance? The bromine is more soluble than chlorine and chloroform, and hence there would be a tendency for the bromine to become enriched in the water phase, leaving behind a bromine liquid still more rich in chlorine and chloroform. Any washing or extraction operation using water would tend to work in the wrong direction in this case. This property difference might be useful in removing trace bromine from chloroform and chlorine, but not the reverse.*

Boiling-Point Differences *The removal of chlorine from bromine by distillation looks very attractive, since there is a 93°C difference in boiling points. However, the*

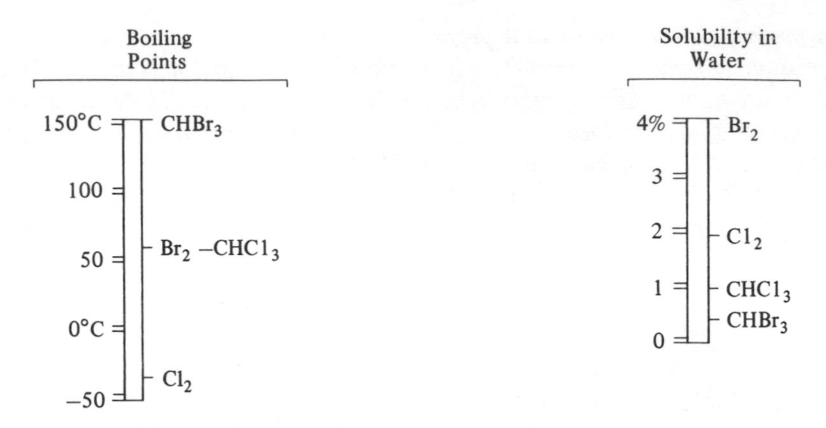

Figure 5.1–1. Ordering of Important Property Differences

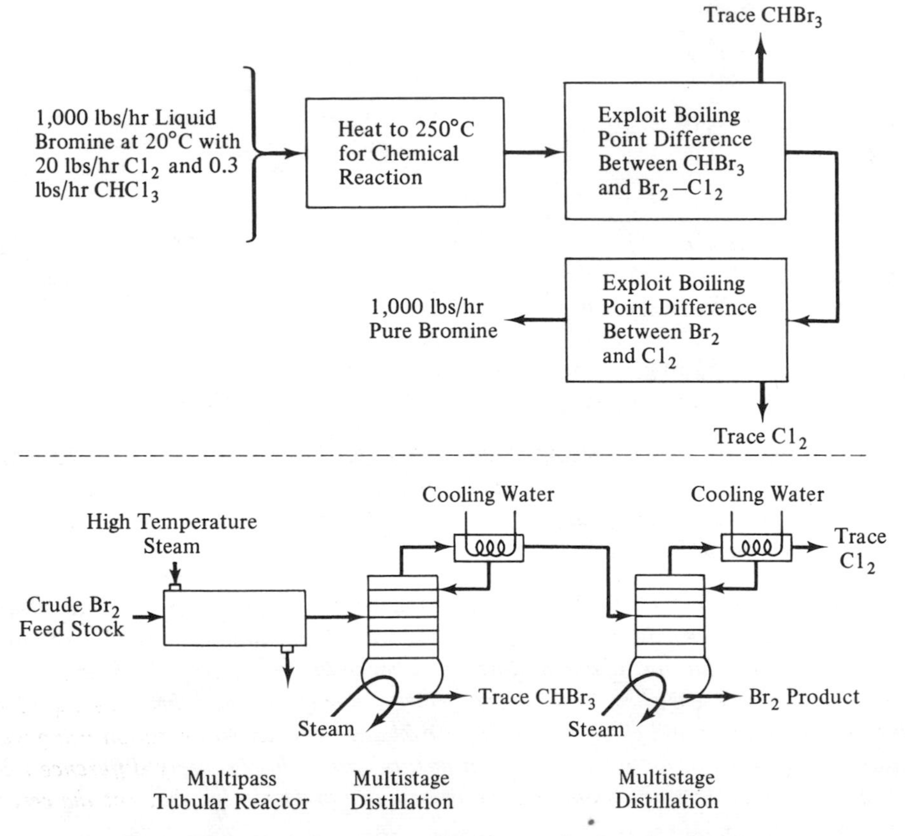

Figure 5.1–2. Task Sequence and Equipment for Bromine Purification

bromine and chloroform have almost identical boiling points, so distillation would be difficult for this separation. The difference in the boiling points of chloroform and bromoform immediately attracts attention, particularly since the one can be converted to the other by chemical reaction.

If the troublesome chloroform is destroyed by chemical reaction, the products are chlorine and bromoform, each of which is easily handled by distillation.

It looks like the processing plan in Figure 5.1–2 would be the sequence of tasks that leads to the efficient removal of the chlorine and chloroform. First, the entire stream is heated to about 250°C, at which temperature all the materials are vapor and the chloroform is converted into bromoform. Second, the vapors are cooled until the bromoform condenses as an enriched liquid; the enrichment might be performed in a multistage distillation tower. The vapor product is a bromine–chlorine mixture, and the liquid product is bromoform. Third, the trace chlorine is removed in a second multistage distillation tower.

This is what this chapter is all about.

5.2 REDUCTION OF SEPARATION LOAD

In many areas of processing the amount of material that must be processed for separation of species can be reduced by stream splitting and mixing. For example, it is quite common in the production of aviation and motor fuels to blend hydrocarbons from a variety of sources to meet octane and flash-point requirements. Such blending operations avoid the need for the species separation that might be required to upgrade the fuel source to meet specific destination requirements.

This section expands on the heuristic:

> **When possible reduce the separation load by stream splitting and blending.**

In a hypothetical separation problem, illustrated in Figure 5.2–1, material in sources *a* and *b* must be allocated to several destinations, *A*, *B*, and *C*. The amounts of the several species *i* that must appear in each destination have been determined by the species allocation. How can this allocation be accomplished best by a sequence of separations?

Two operations are exceedingly simple to perform, stream mixing and stream division. As long as there is no attempt to make streams of different species composition, a stream can be divided into as many parts as needed. And any several streams can be mixed together with ease to form a new stream of material. These two operations will be considered to cost nothing, and can be used whenever the need arises without concern for processing difficulty.

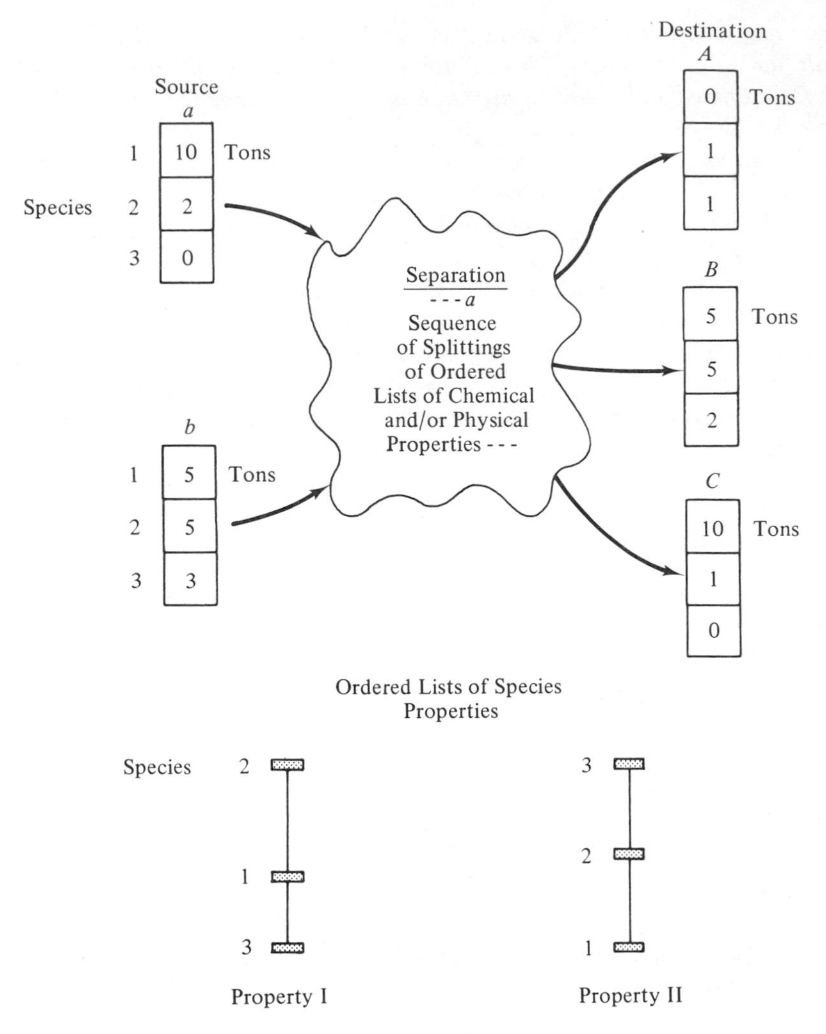

Figure 5.2–1

However, this is not the case if a stream is to be divided to form streams of different composition (a species separation). Separation is a costly operation, which must be done carefully and can only be done in certain ways. The certain ways are determined by the order in which the species are found in the list of physical and/or chemical properties upon which the separation is based.

Typically, in a single operation a mixture of species can be divided into two streams, one stream containing species above a breakpoint in the property list, and the other stream containing species below the breakpoint. For example, in the

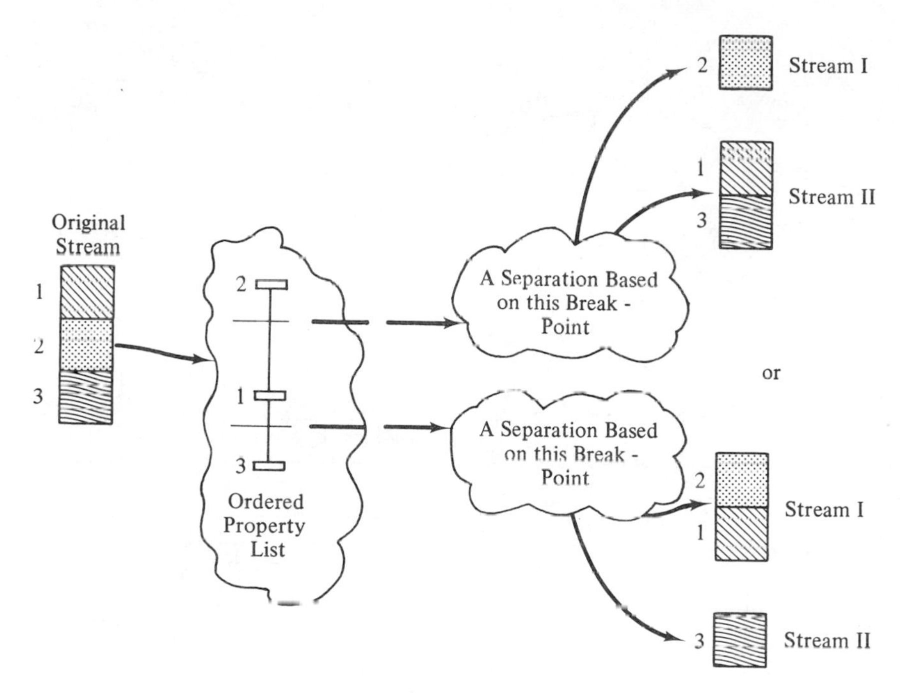

Figure 5.2–2. Limited Separations Possible by a Single Exploitation of Given Property Difference

property list shown in Figure 5.2–2, a stream containing species 1, 2, and 3 can be separated in two ways, depending on whether the property list is split between species 2 and 1 or between 1 and 3. In the first case are formed a stream consisting of species 2 and a stream consisting of species 1 and 3. In the second case are formed a stream consisting of species 2 and 1 and a stream consisting of species 3. These are the only separations that can be performed by a *single* exploitation of this particular property difference.

In practice, often the separation is not complete, and the streams are merely enriched with species above and below the breakpoint in the property list. However, for the moment we suppose that a complete separation is possible.

If a single operation does not do the job, multiple operations are necessary. As a first approximation, we may assess the cost or difficulty of an operation in proportion to the total amount of material processed. Thus the cost of a multiple-operation sequence is proportional to the total amount of material that enters the several separation operations. We seek the sequence of separations which reduces that total.

For example, the problem illustrated in Figure 5.2–1 using property I could be solved by the complete separation of all species, followed by the blending of the pure species to form the required products (Figure 5.2–3). The total difficulty of this

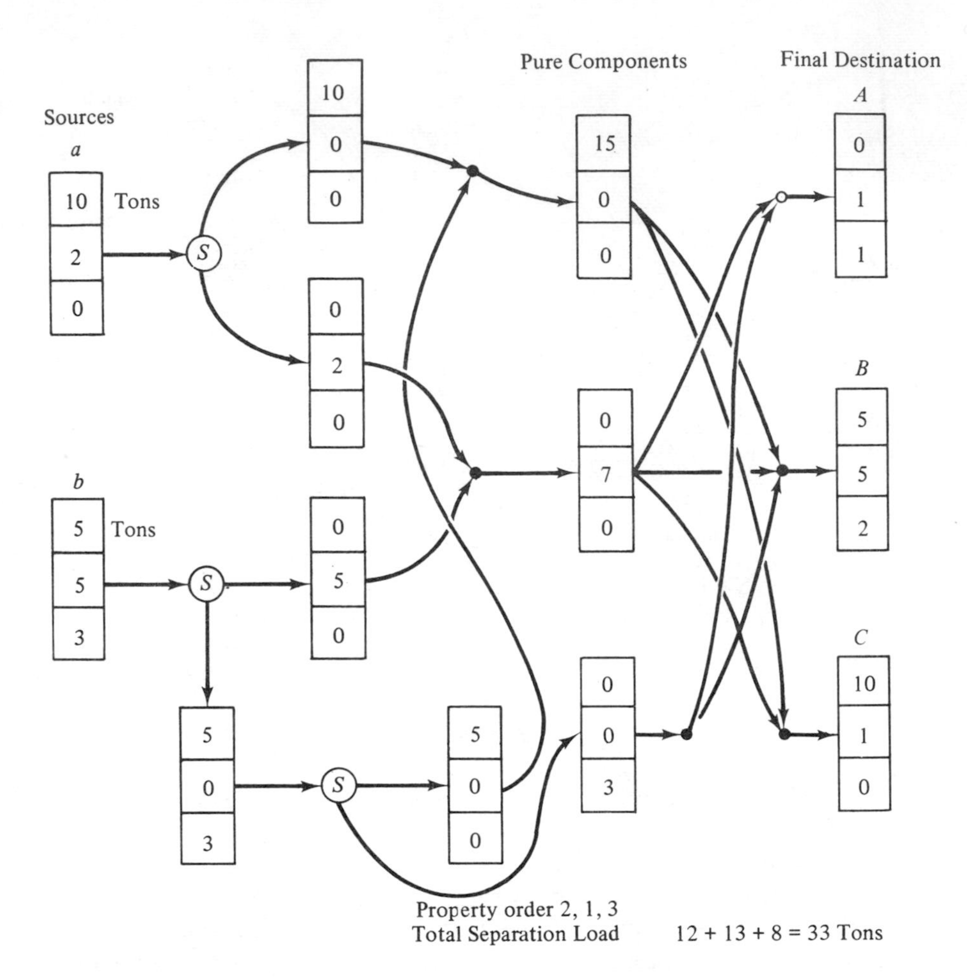

Figure 5.2–3. Complete Separation, a Poor Task Sequence

sequence of separations is estimated to be proportional to the total amount of material processed. This is an unnecessarily costly way of solving this problem, involving the handling of 33 tons.

We now show how to go about reducing the separation load. The plan is to divide and mix the source streams so as to get as close as possible to the destination streams without using any separations. Then we see how the separations can be used most effectively to complete the processing.

Figure 5.2–4 shows the beginning of the synthesis of a better separation plan. First the species are ordered 2, 1, and 3, corresponding to their order in the property list being exploited. This eases the visualization of the effects of separation. The percentage of each species in each stream is computed; this percentage does not change upon stream division and is useful in associating streams of like composition.

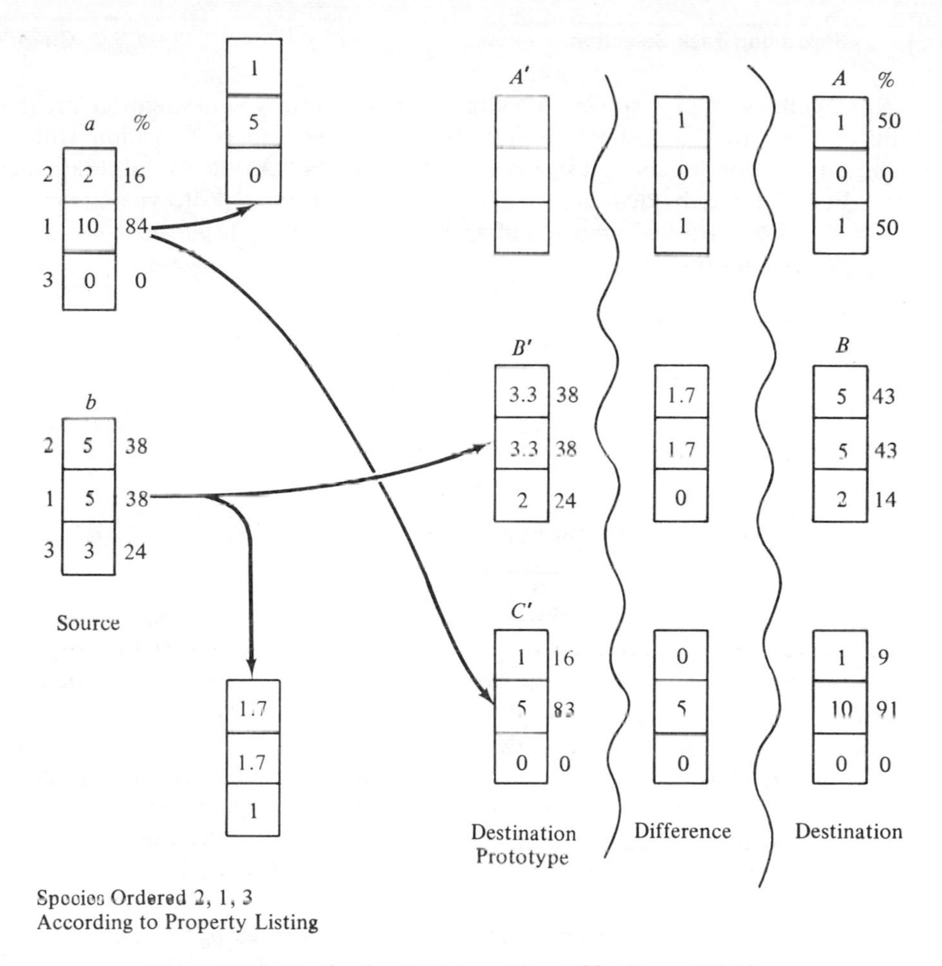

Figure 5.2–4. Destination Prototypes Formed by Stream Division

For example, these streams are similar in composition although different in amount.

	a	%			C	%
2	2	16	Composition similar to →	2	1	9
1	10	83		1	10	91
3	0	0		3	0	0

	b	%			B	%
2	5	38	Composition similar to →	2	5	43
1	1	38		1	5	43
3	3	24		3	2	14

Now, with the source streams we strive to build prototype destination streams, getting as close to the destination streams in amount and composition without forming a prototype stream that has more of any species than the actual destination stream. For example, by dividing stream *a* in half we form a prototype stream C', which contains the required amounts of species 2 and species 3, but does not match the requirements on species 1.

	a	%	C'	%	C	%	C–C'
2	2	16	1	16	1	9	0
1	10	83	5	83	10	91	5
3	0	0	0	0	0	0	0

prototype
of stream *C*

The difference between the destination stream and its prototype identifies the streams that must be found or created by separation to be mixed in with the prototype to meet the destination requirements. We see above that the stream C–C' identifies the need for 5 units of pure species 2.

Figure 5.2–4 shows this initial development of prototype streams. In Figure 5.2–5 the other streams formed by these initial divisions are separated to form the different streams required. The total amount of material undergoing separation is 10 tons. The system originally proposed in Figure 5.2–3 had 33 tons undergoing separation, and, for that reason, was on the order of three times more costly.

Example 5.2–1 Synthetic Skim Milk *Skim milk consists of 93 per cent water, 1.8 per cent casein, 0.7 per cent whey protein, 4 per cent lactose, and 0.5 per cent salts. During cheese making the casein is removed as cheese curd, and a whey results containing essentially all the other components of the skim milk. A synthetic skim milk can be formed by processing the whey so that the whey protein in the synthetic skim milk plays the role of the whey protein–casein component in the actual skim milk.*

Alcohol treatment will cause about 70 per cent of the whey protein to precipitate out of whey. How might we plan the sequence of tasks useful in producing a synthetic skim milk from cheese whey? Each year 25 billion lb of whey is produced in the United States, a typical cheese plant producing 300,000 lb/day of whey.

Let us take as a basis the whey formed by cheese making with 100 lb of skim milk, that is, a mixture of 93 lb of water, 0.7 lb of whey protein, 4 lb of lactose, and 0.5 lb of salts. This is to be transformed into a product consisting of 93 per cent water, 2.5 per cent whey protein (substituting for the casein and whey protein in skim milk), 4 per cent lactose, and 0.5 per cent salts. Figure 5.2–6a shows the source and destination data, where x lb of the whey protein is recovered ($0 \leq x \leq 0.7$ *lb and x is probably on the order of 70 per cent of 0.7 lb or 0.5 lb*).

Suppose that the whey is split into two parts, one part consisting of a fraction y of the whey, forming a prototype synthetic skim milk, and the other part undergoing

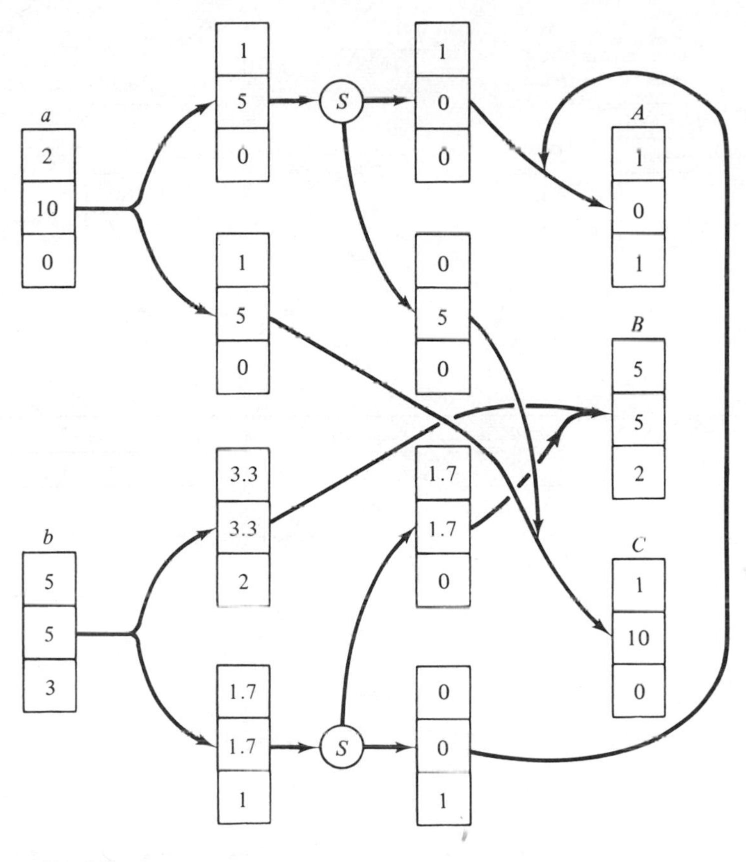

Figure 5.2–5. A Much Less Difficult Sequence of Separations (Separation load is only 10 tons rather than 33 tons.)

alcohol treatment to precipitate 70 per cent of its protein. The precipitated whey is then added to the skim-milk prototype to form the synthetic skim milk. This is shown in Figure 5.2–6b. Matching the components (the synthetic milk formed matched to the destination) gives

$$93y = 37x \qquad \text{water}$$

$$0.7[y + (1 - y)0.7] = x \qquad \text{whey protein}$$

$$4y = 1.6x \qquad \text{lactose}$$

$$0.5y = 1.25x \qquad \text{salts}$$

the solution of which is

$$x = 0.55$$

$$y = 0.22$$

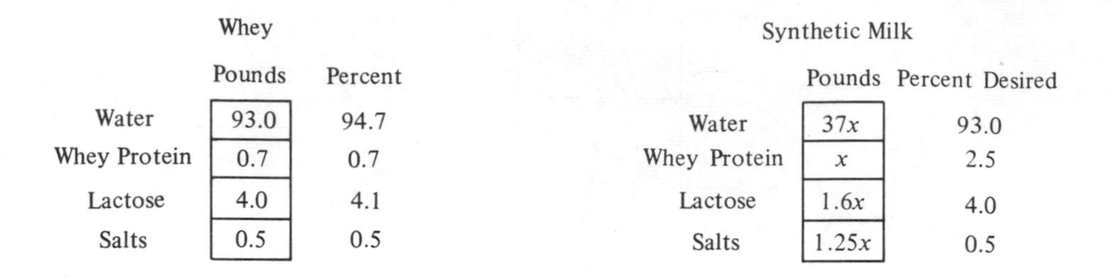

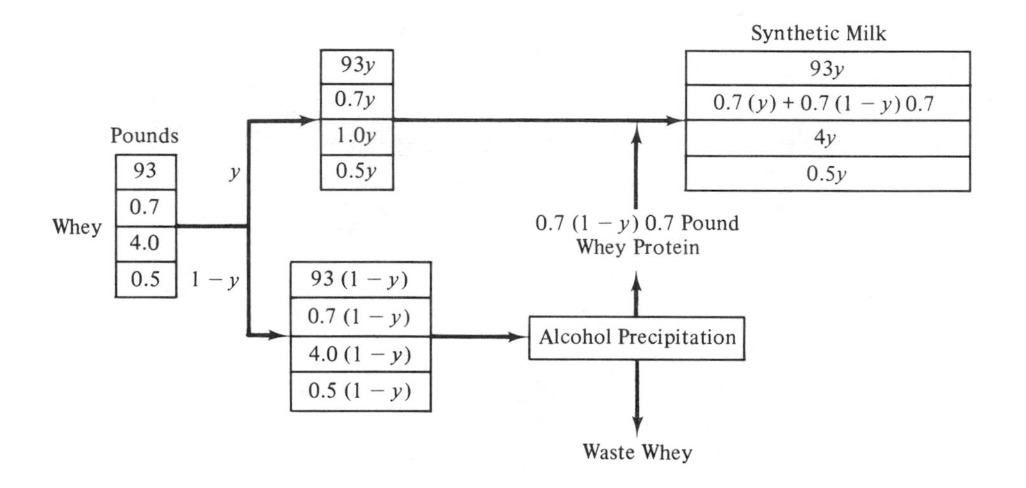

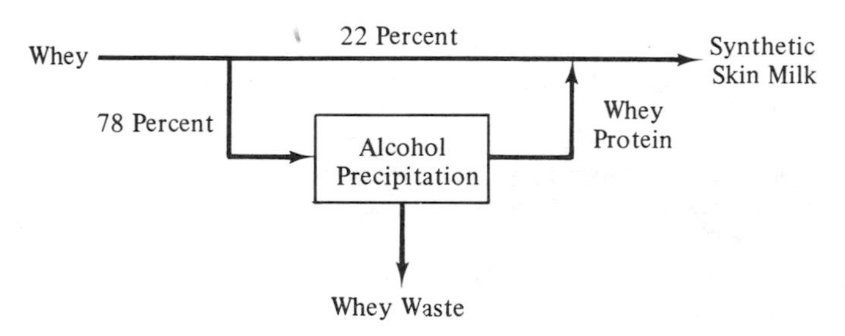

Figure 5.2–6. Reduction of Separation Load by 22 per Cent

Thus 22 per cent of the whey can go unprocessed, if 78 per cent undergoes alcohol precipitation, recovering sufficient whey protein to form the synthetic skim milk. The load on the costly alcohol precipitation protein separator has been reduced.

It should be noted that whey protein recovered by alcohol precipitation is not as soluble as the original protein, and the synthetic skim milk formed in this way is a slurry. This, however, is suitable as a skim-milk substitute for commercial baking purposes. Also, by proposing alcohol precipitation as a means of separation, we are forcing upon ourselves the problem of alcohol recovery from the waste stream. The addition of foreign species, such as the alcohol, often causes recovery problems that preclude the use of the species. There are better ways of doing this separation (see Section 5.7).

5.3 EARLY SEPARATIONS

Having avoided the separation problems as much as possible by stream splitting and blending, we now strive to reduce the repetitive processing of streams, which often occurs during multicomponent separation. The heuristic developed is:

> **All other things being equal, aim to separate the more plentiful components early.**

Suppose that now we have a stream of four components, 1, 2, 3, and 4, labeled in the order in which they appear in the property difference to be exploited; ordered, for example, by increasing boiling point, decreasing size, or solubility. Furthermore, suppose that these components are present in the stream in amounts D_1, D_2, D_3, and D_4, and the stream is to be separated into its pure components. Also, suppose that it is equally difficult to separate 1 lb of a mixture into two streams at any of the three possible breakpoints in the property list.

Within these restrictions the difficulty of performing the separation is proportional to the total amount of material that enters the separators. Since there are four components and since a single separator can form only two streams (one stream with components above the breakpoint and the other below), a total of three separators is required. We wish to sequence these three separations to minimize the total separation load. This leads to the heuristic of removing the most plentiful components early.

Figure 5.3-1 shows all possible ways of performing the separations and the total amount of material processed. For example, in separation sequence 1, component 1 is removed from the first separator, so its contribution to the total load is D_1; but component 2 passes through the separator in which pure 1 is removed and through the second separator in which pure 2 is removed, so its contribution to the total load is $2D_2$. In this way the total processing load for each separation sequence can be determined.

We seek the processing plan that minimizes the total separation load, those five sequences assumed to be equal in all other ways but processing load.

Notice that the components removed first have the smallest coefficient in the total load equations, and those removed last have the largest coefficient. This being the case, if one component is by far the most plentiful, we would aim to remove it first, so that it does not pass through several separators on its way to being removed. This is the origin of the heuristic of removing the most plentiful component first, all other things being equal.

In situations where one or several species are not present in greater amounts, the total loads can be computed and the discrimination made among the possible

Separation Sequence	Total Load
1	$D_1 + 2D_2 + 3D_3 + 3D_4$
2	$D_1 + 3D_2 + 3D_3 + 2D_4$
3	$2D_1 + 2D_2 + 2D_3 + 2D_4$
4	$3D_1 + 3D_2 + 2D_3 + D_4$
5	$2D_1 + 3D_2 + 3D_3 + D_4$

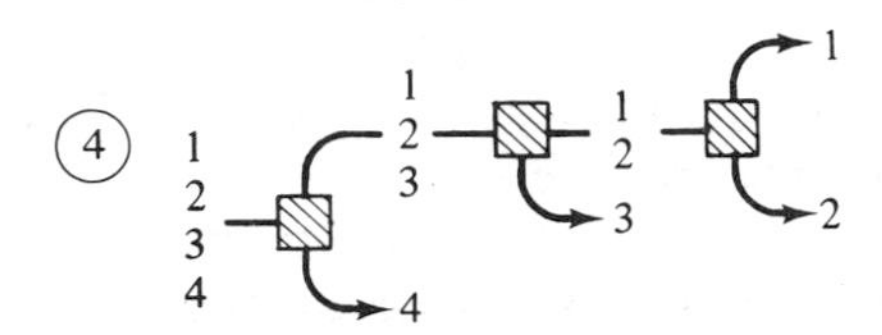

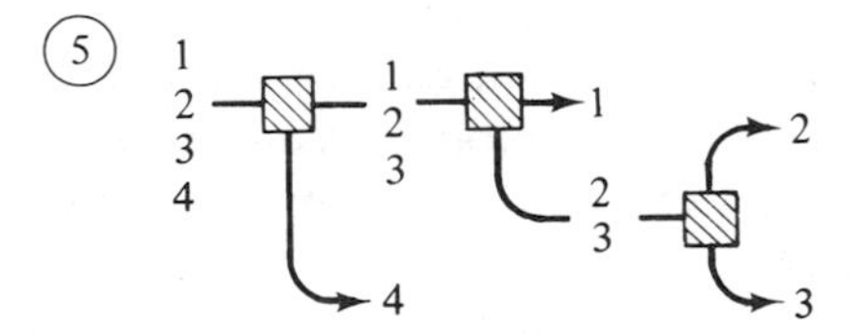

Figure 5.3–1. The Many Ways of Separating Four Components from a Mixture

separation sequences. For example, if the four components are present in equal amounts of D tons/h, the total loads for the respective separation sequences are

SEQUENCE	TOTAL LOAD
1	$9D$
2	$9D$
3	$8D$
4	$9D$
5	$9D$

We would process according to sequence 3, which involves a preliminary separation of the stream into two streams, each having two components, and the secondary separation of each of these two streams into pure components. We must remark, however, that this 11 per cent reduction in separation load is not exciting; it is the 100 or 200 per cent reduction that we seek in problems with greatly unequal amounts of the species. The small differences are apt to be altered by the other factors of separation examined in later sections. However, this heuristic commonly is used:

All other things being equal, aim to separate into equal-sized parts.

Example 5.3–1 Hematite Screening *Low-grade iron ore containing hematite is upgraded by roasting, magnetic separation, and pelletizing. The iron is present as* Fe_2O_3*, which is nonmagnetic. However, upon roasting with coal, magnetic* Fe_3O_4 *is formed:*

$$\underset{\text{(hematite)}}{6\ Fe_2O_3} + C \rightarrow \underset{\text{(magnetite)}}{4\ Fe_3O_4} + CO_2$$

The finely powdered Fe_3O_4 *is removed from the waste rock by magnetic separation and is pelletized for use in iron making.*

In an initial stage of ore processing a rough crushed ore must be screened into ore in the size ranges of 6 to 4 in., 4 to 2 in., and smaller than 2 in. If 8000 tons/h of the ore is to be processed, how should the screening be sequenced? The size distribution of the ore is 10 per cent in the 6- to 4-in. range, 10 per cent in the 4- to 2-in. range, and 80 per cent in the less than 2-in. range.

Most plentiful by far is the ore in the less than 2-in. size. The ore should be processed through a screen with 2-in. openings, and the material that does not pass through should be sent to a screen with 4-in. openings. This sequence, shown in Figure 5.3–2, has a total processing load of 9600 tons/h. The other possible sequence is also shown, and it involves a total load of 15,200 tons/h. The heuristic of removing the most plentiful first leads to a one third reduction of processing load.

Again we must ask the question, are all things equal other than the amounts of material in the various size ranges? For example, must we put the 4-in. screen first to protect the more delicate 2-in. screen from the big 6-in. rocks? If so, all other things are not equal, and the premise is not valid under which our minimum-load heuristic was developed. These value judgements are to be made by the engineer and cannot be reduced to rote.

Example 5.3–2 Separation of Benzene, Toluene, and Orthoxylene *Toluene and orthoxylene are products of the addition of methyl groups to benzene, and a reactor effluent contains all three species. How are these to be separated?*

The boiling points of benzene, toluene, and orthoxylene are 80, 110, and 144°C, respectively; these are sufficiently different to suggest the use of multistage distillation. Furthermore, the differences between the boiling points of successively more volatile components (40° and 34°) are close enough to suggest that it would be equally difficult

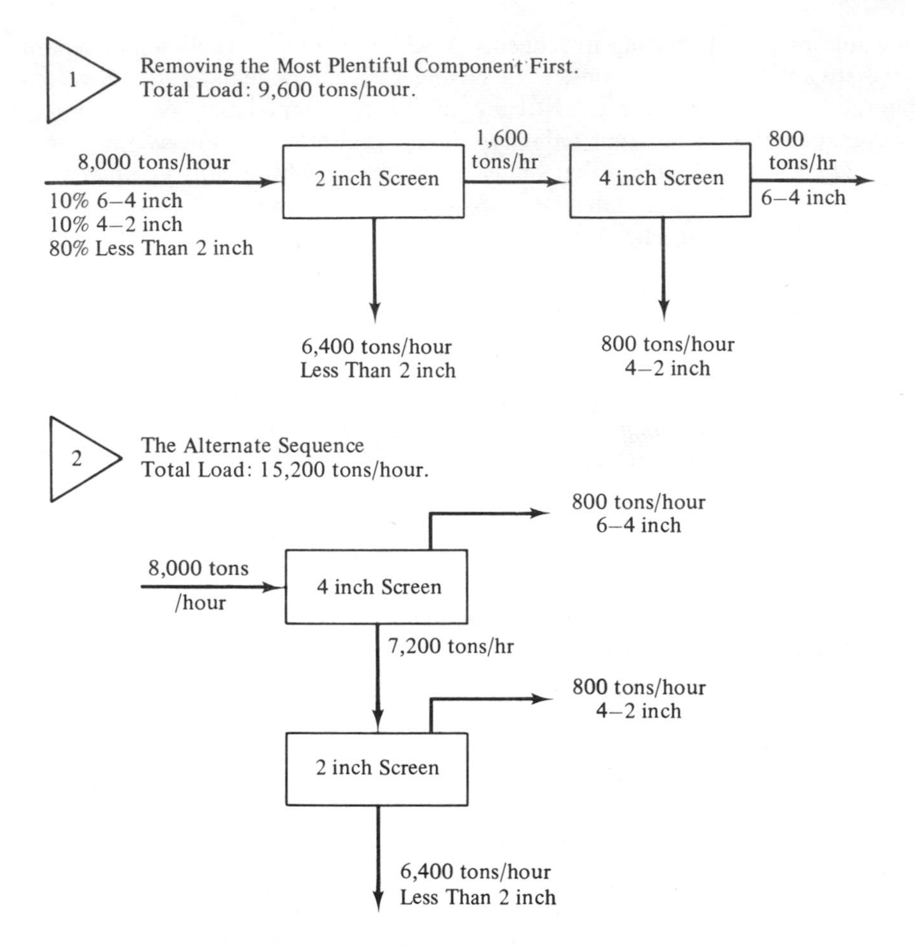

Figure 5.3–2. Ore Screening

to operate a distillation tower for splitting off either the benzene or the orthoxylene first. It is the separation load that might determine the order of separation, all other things appearing equal.

What separation sequence would be best if the hydrocarbon stream is 75 per cent benzene, 5 per cent toluene, and 20 per cent orthoxylene? Clearly that in Figure 5.3–3 results in the minimum total distillation load.

In addition to the removal of the most plentiful species early, we also strive to remove corrosive, toxic, and otherwise hazardous material early. For example, if a highly corrosive species is kept in the processing, all the equipment with which it comes in contact must be made of noncorrosive material. These expensive special considerations can be disregarded for the bulk of the process, if the corrosive material is removed early. All other things being equal, this is a valuable heuristic:

Remove corrosive and hazardous material early.

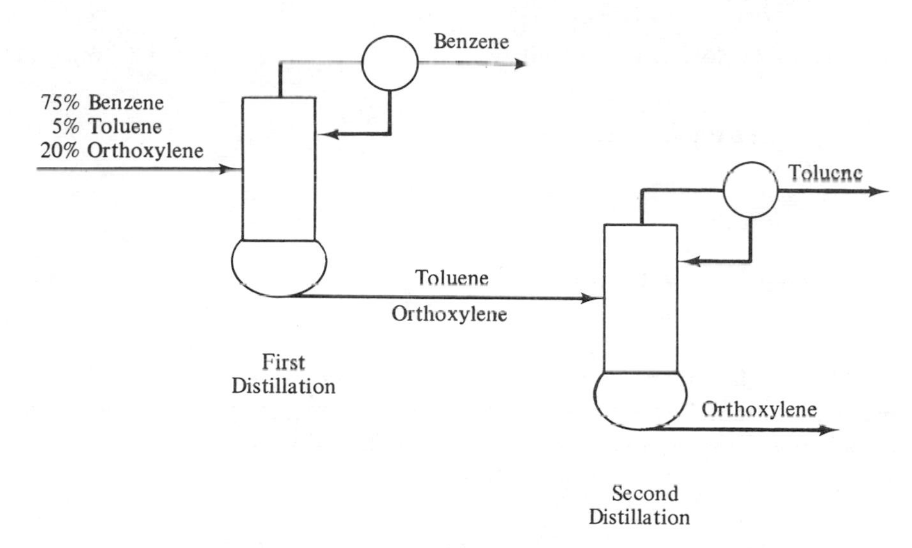

Figure 5.3–3. Benzene, Toluene, and Orthoxylene Separation

5.4 Difficult Separations Go Last

So far we have considered situations in which the difficulties of separations were equal for any and all breakpoints we might postulate in the ordered list of species property. This enabled us to focus on the development of heuristics dealing with minimum separation load. However, in many situations great differences exist in the difficulty of separation, and these differences must be accounted for accurately in the selection of the separation tasks.

In this section we develop and use the heuristic:

The difficult separations are best saved for last.

A difficult separation is one in which the properties are close for species on each side of the separation breakpoint. For example, the boiling point, size, solubility, or density of the separated species is close during distillation, screening, extraction, or settling separations.

In such a situation either the equipment has to be designed to greater tolerances or the separation phenomena must be applied again and again, as in multistage distillation, to obtain the required purities. For example, the number of countercurrent stages trays required in a distillation tower increases as the difference between the boiling points of the material decreases. However, the diameter of the tower and the heating and cooling loads are an increasing function of the amount of material being processed. To a *first approximation* we may estimate the relative cost of making a distillation separation as

$$\text{distillation separation cost} \propto \frac{\text{feed rate}}{\text{boiling-point difference}}$$

In general the cost of an individual separation will be approximated by

$$\text{separation cost} \propto \frac{\text{feed rate}}{\begin{array}{l}\text{property difference between}\\ \text{the two species on each side}\\ \text{of the separation breakpoint}\end{array}} = \frac{D}{\Delta}$$

This approximate estimate of separation cost is sufficiently accurate to use in planning the gross features of separation tasks that involve both different amounts of species to be separated and different separation difficulties.

In Figure 5.4–1 we examine the four-species separation problem considered in the previous section. For example, in sequence 1, where the species are separated in order of appearance, the first separator receives the entire feed stock and separates species 1 from its nearest neighbor species 2, the difficulty of separation being estimated at $(D_1 + D_2 + D_3 + D_4)/\Delta_{12}$, where the numerator is the total feed rate and the denominator is the difference in the properties of species 1 and 2. In this way, by following the sequence of separations, the total separation difficulty can be estimated.

What general trends can be observed? Suppose that the amounts of the species to be separated are equal ($D_1 = D_2 = D_3 = D_4 = D$) and that the separation between species 2 and 3 is a close one compared to the other separations [$\Delta_{12} = \Delta_{34} = \Delta$, but $\Delta_{23} = (\Delta/3)$]. The total difficulties of the processing of sequences are

SEQUENCE	TOTAL DIFFICULTY
1	$\frac{D}{\Delta}(4 + 9 + 2) = \frac{D}{\Delta}(15)$
2	$\frac{D}{\Delta}(4 + 6 + 3) = \frac{D}{\Delta}(13)$
3	$\frac{D}{\Delta}(2 + 12 + 2) = \frac{D}{\Delta}(16)$
4	$\frac{D}{\Delta}(2 + 9 + 4) = \frac{D}{\Delta}(15)$
5	$\frac{D}{\Delta}(3 + 6 + 4) = \frac{D}{\Delta}(13)$

The worst sequence is 3 in which the close (species 2–species 3) separation is attempted first. Sequence 2 and 5 are the best, and they both involve the most difficult separation last. We must keep in mind however that the equations developed here are to be used merely as guidelines, for they are not exact estimates of separation costs.

Total Difficulty

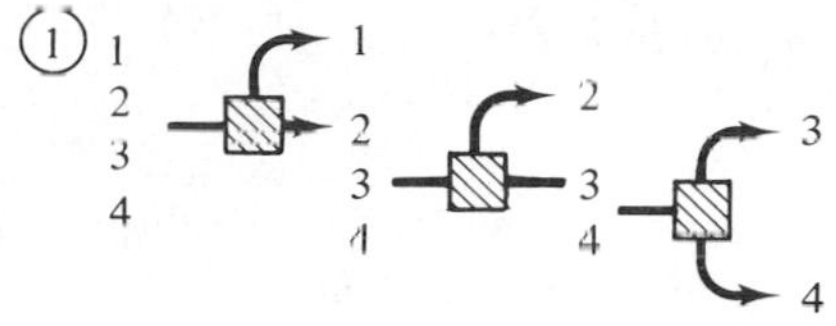

(1)
$$\frac{(D_1+D_2+D_3+D_4)}{\Delta_{12}}+\frac{(D_2+D_3+D_4)}{\Delta_{23}}+\frac{(D_3+D_4)}{\Delta_{34}}$$

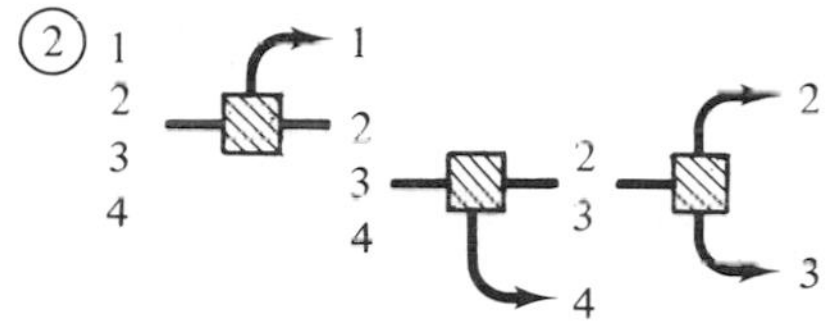

(2)
$$\frac{(D_1+D_2+D_3+D_4)}{\Delta_{12}}+\frac{(D_2+D_3)}{\Delta_{23}}+\frac{(D_2+D_3+D_4)}{\Delta_{34}}$$

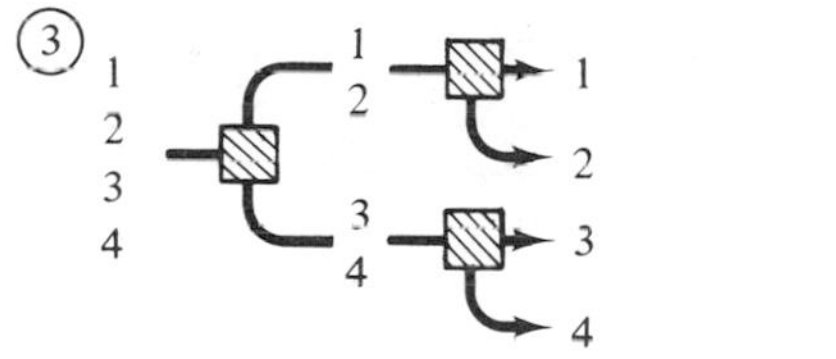

(3)
$$\frac{(D_1+D_2)}{\Delta_{12}}+\frac{(D_1+D_2+D_3+D_4)}{\Delta_{23}}+\frac{(D_3+D_4)}{\Delta_{34}}$$

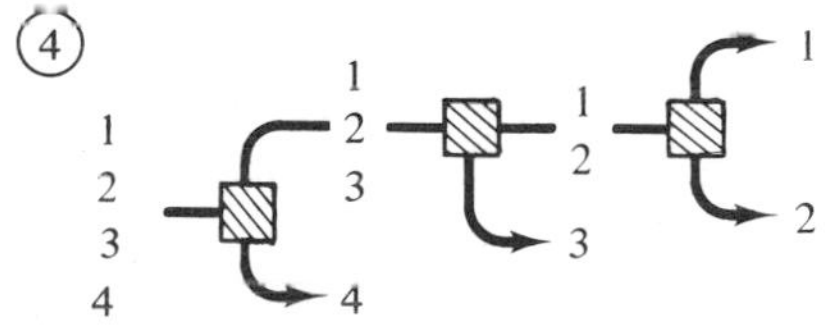

(4)
$$\frac{(D_1+D_2)}{\Delta_{12}}+\frac{(D_1+D_2+D_3)}{\Delta_{23}}+\frac{(D_1+D_2+D_3+D_4)}{\Delta_{34}}$$

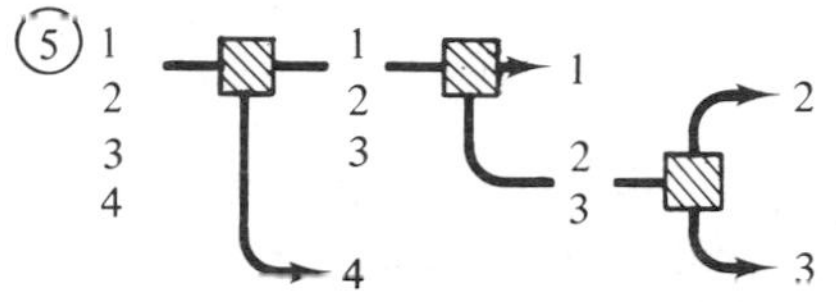

(5)
$$\frac{(D_1+D_2+D_3)}{\Delta_{12}}+\frac{(D_2+D_3)}{\Delta_{23}}+\frac{(D_1+D_2+D_3+D_4)}{\Delta_{34}}$$

Δ_{ij} = | difference in exploited property of species i and j |

D_i = amount of species i

Figure 5.4–1

From this kind of a study examining the form of the difficulty equations in Figure 5.4–1, evolves the heuristic, *the difficult separations are best saved for last*; and implied in this heuristic is *all other things being equal.* When the amounts to be separated are greatly different (all other things are not equal), consideration must be given to the heuristic of removing the most plentiful early.

Example 5.4–1 Separation of Methane, Benzene, Toluene, and Orthoxylene *A stream is to be separated of methane (bp − 161°C), benzene (bp 80°C), toluene* *(bp 110°C), and orthoxylene (bp 144°C) of a composition of, respectively, 50, 10, 10, and 30 per cent. What sequence of boiling-point exploitations will probably lead to the most economic separation?*

The most difficult separations involving the benzene, toluene, and orthoxylene should go last. Also, the methane is the most plentiful species, so following the most-plentiful-species rule, methane separation should go first. Thus on two counts the methane separation should be first.

Since the boiling-point differences between benzene and toluene (30°C) and toluene and orthoxylene (34°C) are essentially equal, the amounts to be processed might be the determining factor in establishing the separation sequence.

The plan of attack suggested is:

1. *Separate methane, an easy separation of plentiful species.*
2. *Separate orthoxylene, the next plentiful close separation.*
3. *Split benzene and orthoxylene, a close separation of the least plentiful species.*

5.5 SELECTION OF SEPARATION PHENOMENA

The differences in molecular and atomic structure that cause the property differences exploited for separation rarely cause differences in only one property. If species differ in boiling point, they may also differ in solubility, freezing point, diffusion rates, and other physical and chemical properties. Commonly, we must select one phenomenon from several phenomena that are superficially comparable.

In the previous section rules have been developed to guide in the sequencing of multispecies separations, and these rules also play a role in the selection of the separation phenomena. For example, we would tend toward a separation phenomenon that places the most plentiful species on the top or bottom of the property list so that species could be removed early, all other things being equal. However, in this section we go beyond these rules to develop the following heuristic:

> **All other things being equal, shy away from separations that require the use of species not normally present in the processing. However, if a foreign species is used to effect a separation, remove that species soon after it is used.**

Separation phenomena can be divided broadly into two categories, those which are *direct* and involve no foreign substance, and those which are *indirect* and involve the introduction of a foreign substance to drive the separation. Distillation, evaporation, pressure-driven membrane separations, screening, and crystallization are direct separations. Liquid extraction, solvent gas absorption, leaching, azeotropic distillation, and ion exchange are indirect separations, all requiring a foreign species that may not normally be present during processing. If the separation is performed by an indirect method, the next step in processing is usually the removal of the foreign species. This early removal occurred in 85 per cent of a large number of industrial processes surveyed.

We present the advice of R. E. Treybal in the use of liquid extraction as an indirect method of separation.* This experience applies to other indirect methods equally well.

> In the indirect methods, addition of a foreign substance such as an extraction solvent presents a number of problems. The added substance, because of the requirement that it be insoluble with the original mixture, is necessarily chemically different, and this complicates the choice of materials of construction to ensure resistance to corrosion. Sometimes large inventories of the added substance must be kept on hand, thus tying up relatively large amounts of capital. Such operations yield the separated product in the form of a new solution containing the added solvent, and usually some sort of solvent-recovery system must be devised, which in itself incurs an expense. There are inevitably losses of solvent, which must be replaced. The opportunity for contamination of the ultimate product by a material not normally expected to be associated with it is always present.
>
> On the other hand, because the added solvent is chemically different from the mixture to be separated, there are opportunities for unusual and sometimes even spectacular separations. In the case of distillation, the vapor and liquid phases between which the separating components distribute are chemically the same, and great differences in distribution cannot be expected. Indeed, in many cases a complete separation cannot be made at all by such a method. With liquid extraction, on the other hand, one may usually choose among many solvents until the best one is found. It is as if there were many kinds of heat available for distillation, each of which would provide for different separabilities. In fact, it is the ability to make separations according to chemical type, rather than according to physical characteristics such as vapor pressure, which so frequently makes liquid extraction so attractive.
>
> In any case, it is clear that liquid extraction will be useful either where direct methods fail or where, despite apparent disadvantages, it nevertheless provides a less expensive overall process than a competitive direct or chemical method. This can be said, of course, of any of the indirect mass-transfer operations. The following list indicates some typical fields of usefulness where liquid extraction has demonstrated its unique abilities as a separation method.
>
> 1. As a substitute for more direct methods where direct methods are more expensive.
>
> a. Separation of close-boiling liquids. A typical example is the separation of butadiene (bp, $-4.75°C$) from butylenes (bp, -5 to $-6°C$), which would be very expensive to do by distillation. Liquid extraction with aqueous cuprammonium acetate solution provides an easy separation.
>
> b. Separation of liquids of poor relative volatility. An example is the separation of acetic acid and water, which, despite their relatively large difference in boiling points, have poor relative volatility. Dilute acid solutions, particularly,

* R. E. Treybal, *Liquid Extraction*, McGraw-Hill Book Company, New York, 1963.

require vaporization of large amounts of water, and extraction of the acid with ethyl acetate, or with mixtures of ethyl acetate with benzene, provide a cheaper process.

c. As a substitute for high-vacuum, or molecular, distillation for mixtures whose boiling points are so high that they must be distilled in this manner. Long-chain fatty acids and vitamins may be extracted from naturally occurring oils by liquid propane as solvent instead of by distillation.

d. As a substitute for expensive evaporation. Thus, benzoic acid may be separated from dilute solution in water by evaporation of the water (latent heat of vaporization = 970 Btu/lb). It may also be extracted into benzene with a tenfold increase in concentration and the benzene removed by evaporation (latent heat of vaporization = 170 Btu/lb).

e. As a substitute for fractional crystallization. Tantalum and columbium may be separated by a tedious and very expensive series of fractional crystallizations of the potassium fluoride double salts, but the much easier separation by extraction of the hydrofluoric acid solution of the metals with methyl isobutyl ketone has now become standard practice.

2. As a separation means where direct methods fail.

a. Separation of heat-sensitive substances. Penicillin and most antibiotics, in dilute solution in fermentation broths, may not be concentrated by evaporation of the water at ordinary boiling temperatures because of chemical destruction of the product. Furthermore, they must be separated from the various other substances present in the fermentation broth. Consequently they are extracted, and thereby purified and concentrated, into an organic solvent.

b. Separation of mixtures that form azeotropes. These cannot be separated by direct distillation. Thus, methyl ethyl ketone and water may be separated by extraction of the water into concentrated, aqueous calcium chloride brine or of the ketone into trichloroethane.

c. Separations according to chemical type, where boiling points overlap. Aromatic hydrocarbons such as benzene, toluene, and xylene are separated from paraffin hydrocarbons of the same boiling range by extraction with liquid sulfur dioxide, furfural, or diethylene glycol.

3. As a substitute for expensive chemical methods. As an example we may cite the separation of uranium from vanadium in ore-leach liquors from each other and from undesired elements by extraction with kerosene solutions of dialkyl phosphoric acids at different pH's.

4. There are other uses of liquid-extraction techniques that have not been extensively exploited. Thus, carrying out a chemical reaction in the liquid phase may be greatly assisted by continuous extraction of one of the products, thereby improving the yield. Heat exchange between two insoluble liquids by direct contact, in the absence of an intervening metal wall, which may foul and which in any case retards the rate of heat transfer, is easily carried out. An industrial process that uses both of these is the continuous splitting of naturally occurring liquid fats into glycerol and fatty acids with water at high temperature.

Example 5.5–1 The Patented Process of W. G. Johnson* *The textile fiber Orlon is polymerized acrylonitrile. When copolymerized with butadiene, acrylonitrile becomes*

* See E. R. Hafslund, *Chemical Engineering Progress*, **65,** No. 9, 58–64, 1969.

the basis of synthetic rubbers of the GR-N type (Hycars, Chemigum, Perbunan). It is a highly reactive chemical used in a wide variety of chemical syntheses.

One reaction path to acrylonitrile is

$$\underset{\text{(acetylene)}}{C_2H_2} + \underset{\substack{\text{(hydrogen}\\\text{cyanide)}}}{HCN} \xrightarrow{\text{catalyst}} \underset{\text{(acrylonitrile)}}{C_3H_3N}$$

However, a recently patented reaction path (Sohio) uses cheaper raw materials:

$$\underset{\text{(propylene)}}{2\,C_3H_6} + \underset{\text{(oxygen)}}{3\,O_2} + \underset{\text{(ammonia)}}{2\,NH_3} \rightarrow \underset{\text{(acrylonitrile)}}{2\,C_3H_3N} + \underset{\text{(water)}}{6\,H_2O}$$

This reaction occurs in a fluidized bed reactor at 450°C at a pressure of 30 to 40 psi. The catalyst is bismuth phosphomolybdate on silica gel. We are concerned with a separation problem that must be solved to support this reaction commercially.

Table 5.5–1 shows the effluent from the reactor, consisting of inerts (mainly nitrogen from the air used to supply the oxygen), unreacted propylene, propane, acrylonitrile, water, and impurities. A separation system must be designed to:

TABLE 5.5–1 Reactor Effluent

Component	Weight %
Inerts	40
Propylene	39
Propane	8
Acrylonitrile	7
Water	5
By-product impurities	1

1. *Vent the inerts without loss of acrylonitrile or propylene.*
2. *Recycle the propylene to the reactor without recycling the acrylonitrile.*
3. *Purge the propane to prevent propane buildup in the reactor recycle.*
4. *Recover the acrylonitrile for final purification.*

Figure 5.5–1 shows this separation problem.

The most difficult separation is the removal of the propane from the propylene recycled to the reactor. The relative volatility of propylene to propane is 1.2 with the propylene being the more volatile. Since the propylene is five times more plentiful than the propane (see Table 5.5–1), the separation of the two would involve the taking of a large amount of propylene as the overhead in a distillation tower, and purging the smaller amount of propane as the bottoms. This requires high refluxes and high costs of compression and refrigeration. It is desirable to reverse the volatilities so that the purge propane is the less volatile compound and can be bled off the top of a distillation tower with the nitrogen inerts. The direct *separation of propylene–propane mixtures is*

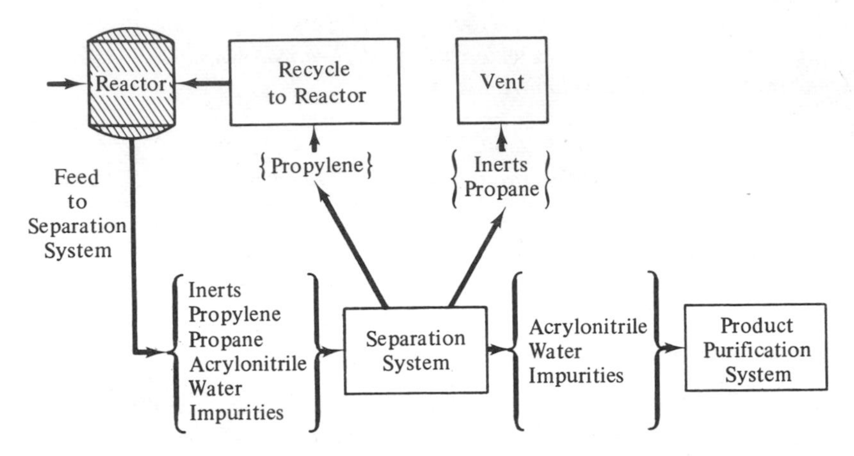

Figure 5.5–1. Source of Propane–Propylene Separation Problem

not economic in this case, since the more volatile component is by far the most plentiful and is not that much more volatile than the compound from which it must be separated.

The indirect *separation of extractive distillation is proposed in which a solvent is used to reverse the propane–propylene volatilities. If a proper solvent can be found, the indirect separation scheme shown in Figure 5.5–2 can be envisioned. The inerts and the propane can be taken off first, the propylene removed second, and the solvent recovered third. The reordering of the propylene–propane positions in the volatility list enables these breakpoints. To fit into this processing plan, the solvent must have these properties:*

1. *It should reverse the propylene–propane volatilities to make the propane more volatile.*
2. *The propylene should be relatively soluble in the solvent to ease the recovery.*
3. *The solvent should be less volatile than the propylene to ease the propylene–solvent separation.*
4. *The solvent should be compatible with and easy to remove from the acrylonitrile.*

Table 5.5–2 shows the ability of several solvents to reverse the propylene–propane relative volatility of 1.2 to something favoring propane as the more volatile. All of these would do an excellent job of reversing the propylene–propane volatilities.

However, if we keep in mind our heuristic—all things being equal, shy away from separation schemes that require the use of species not normally present in the processing—*acrylonitrile stands out. Acrylonitrile is the* only *solvent that both reverses the volatility and is not foreign to the processing. Using acrylonitrile as the solvent eliminates the solvent recovery step and results in the simpler scheme shown in Figure 5.5–3. In the separation of propane and propylene during the manufacture of acrylonitrile, acrylonitrile is not a foreign species. The use of this solvent is the basis of U.S. Patent 2,980,727 by W. G. Johnson (April 18, 1961) assigned to E. I. du Pont de Nemours & Company. Others who wish to use this processing innovation must license the know-how from the patent holder.*

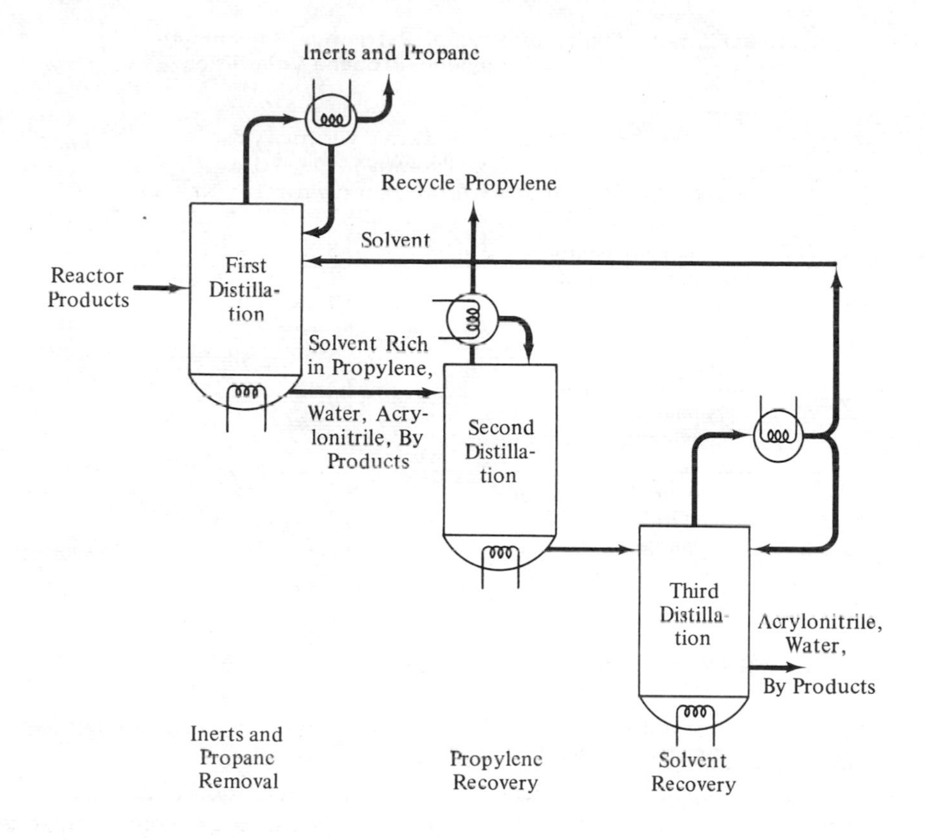

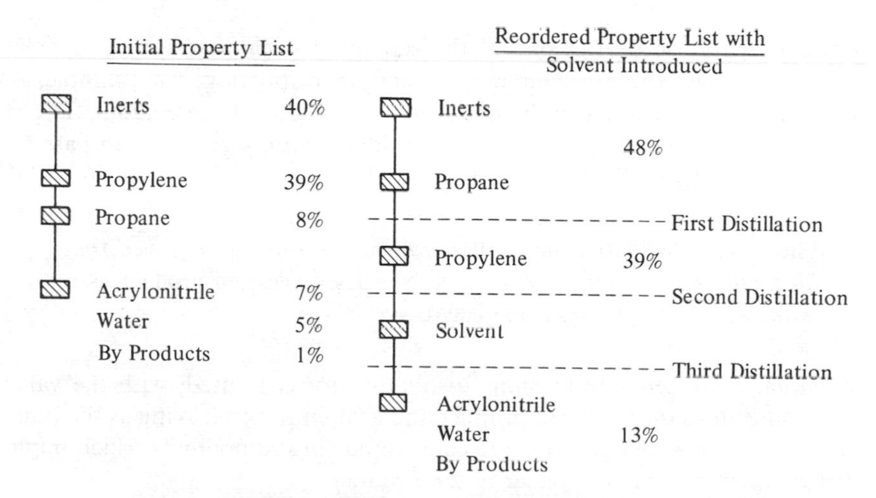

Figure 5.5–2. Indirect Separation Scheme Based on the Reversal of Propane–Propylene Volatility by a Proper Extractive Solvent

TABLE 5.5–2 Ability of Several Extractive Solvents to Reverse Propylene–Propane Volatilities

Extractive Solvent	Relative Volatility of Propane to Propylene When Flooded with Solvent
Trichlorobutyronitrile	1.8
{ Acetonitrile 65% Acrylonitrile 20% Water 15% }	1.7
Butyronitrile	1.6
Monochloroacetonitrile	1.5
Acrylonitrile	1.5
Isobutyl nitrile	1.5
Acetone	1.4
Propionitrile	1.3
Acetonitrile	1.3
Benzonitrile	1.3

5.6 EXCURSIONS FROM AMBIENT CONDITIONS

In the previous section we developed the practice that substances foreign to the processing should not be used to drive a separation unless there is a clear-cut advantage in using an indirect separation. We tend to favor direct separation phenomena such as evaporation, distillation, crystallization, dual-temperature exchange reactions, pressure-driven membrane processes, and other separations driven by the addition or withdrawal of energy. The driving energy may be the addition or removal of heat and/or pressure.

In some separation devices one of the separated streams is subjected to more harsh conditions than the other effluent stream. In distillation, for example, some species degradation may occur in the hotter lower portions of the column. Degradation products often have undesirable color and low volatility and tend to pass to the bottoms effluent. This leads to the heuristic:

> **When using distillation or similar schemes, choose a sequence that will *finally* remove the most valuable species or desired product as a *distillate*, all other things being equal.**

In many situations a separate finishing distillation tower is used, with the valuable product coming off as the overhead product and a slight trickle leaving as the bottoms product, merely to ensure that the large color-imparting compounds which might be generated by accident will not appear in the product.

Generally, a direct separation is driven by forcing the material away from the

Propylene recovery tower at the time of construction.

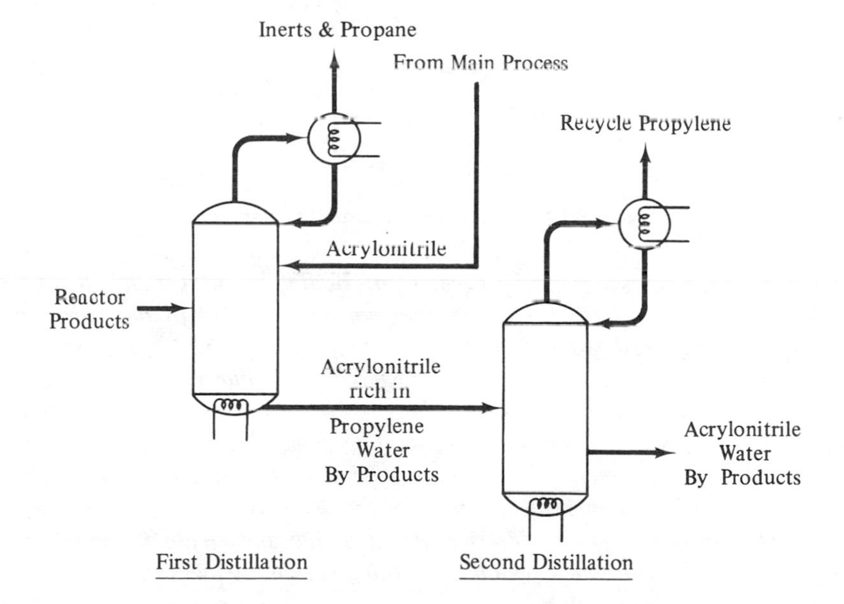

Figure 5.5–3. Simpler Extractive Distillation Unique to the Recovery of Propylene During Acrylonitrile Manufacture, an Indirect Separation Turned into a Direct Separation – U.S. Patent 2,980,727. (Photograph Courtesy of E.I. du Pont de Nemours and Co.)

normal atmospheric, or ambient, conditions, and the heuristic governing such excursions is:

> **All other things being equal, avoid excursions in temperature and pressure, but aim high rather than low.**

Figure 5.6–1 shows the relative costs of energy added or removed at different temperatures. The sharp increase in cost for excursions below room temperature and the relatively slow increase for excursions above room temperature is the origin of this heuristic. We tend to see many more high-temperature processes than low-temperature processes. Also, high-pressure equipment is less costly and easier to maintain than vacuum equipment. Thus one should tend to aim high rather than low in excursions away from ambient conditions, unless there is a sufficient reason for violating this rule.

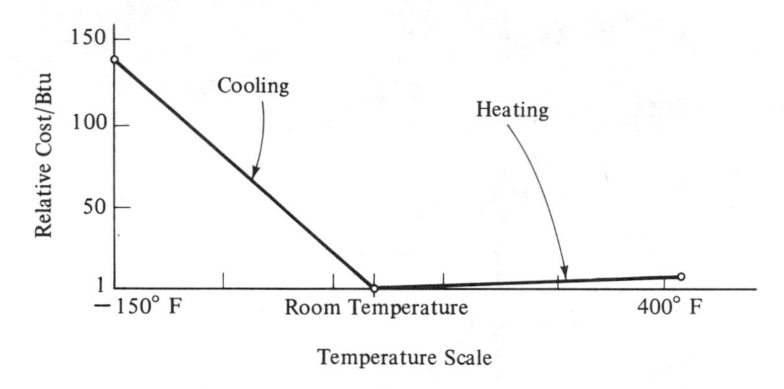

Figure 5.6–1. Cost of Temperature Excursions

Example 5.6–1 Fresh Water by Freezing or Boiling *Both ice and steam formed from salt water are salt free. Which phenomenon should lead to the more efficient process for recovering fresh water from the sea?*

If we operate at atmospheric pressure, freezing would require a temperature excursion to 30°F and boiling to 220°F, at energy costs on the order of $\$1.50/10^6$ Btu and $\$0.50/10^6$ Btu. However, the latent heat of freezing water is 144 Btu/lb and the latent heat of vaporizing water is 1000 Btu/lb. The heuristic reasoning must include not just the quality of energy requirements, but also the quantity.

The cost of energy consumption per pound of ice and steam formed is a more appropriate screening criterion, since all other things are not equal in these temperature excursions.

$$\text{freezing:}\quad (144\ \text{Btu/lb})(\$1.50/10^6\ \text{Btu}) = \$0.22/1000\ \text{lb}$$

$$\text{boiling:}\quad (1000\ \text{Btu/lb})(\$0.50/10^6\ \text{Btu}) = \$0.50/1000\ \text{lb}$$

On the basis of this criterion the freezing phenomenon is the more promising. In Chapter 7 we develop a process based on the freezing phenomenon, which produces water at

\$0.06/1000 lb, well below the energy costs estimated above. Task integrations enable the multiple use of the energy, resulting in the cost reduction.

The heuristic of aiming high in temperature and pressure excursions leads to interesting and very useful rules specific to particular separation phenomena. Such a rule for distillation system sequencing is typical of the special-purpose heuristics that can be developed:

> **During distillation, sequences that remove the components one by one in column overheads should be favored, all other things being equal.**

A distillation tower is driven partially by the addition of heat to the bottom of the tower, causing partial vaporization, and partially by the removal of heat from the top of the tower, causing partial condensation. Thus the heat must be added at the boiling point of the bottom stream (containing the least volatile components), and the heat must be removed at the condensation point of the overhead stream (containing the most volatile components). Figure 5.6–1 indicates the rapid increase in cooling cost with decrease in temperature. It is wise, therefore, to remove the most volatile components first in order to lower the tower-top temperature in subsequent distillations. Thence, the heuristic of removing the components one by one as column overhead products, other things being equal. Of course, this heuristic is valid if the tower top cooling must be done at temperatures below room temperature.

Example 5.6–2 Ethylene and Propylene Manufacture *Figure 5.6–2 is a panoramic view of the Sinclair-Koppers 500 million-lb/year ethylene plant in Houston, Texas, and Figure 5.6–3 is the schematic diagram of a distillation sequence used to separate the light hydrocarbon mixture. We see how this processing sequence follows the separation heuristics.*

1. Difficult Separations Last *The boiling points of propane (42°C) and propylene (48°C) are very close. The C_3 splitter is a last separation. The next most difficult separation is that of ethane (85°C) and ethylene (104°C). The C_2 splitter is also in a last position.*

2. Favor Overhead Removal *To get at these last separations, material more volatile and less volatile than the C_2 and C_3 fractions must be removed. We favor the removal of the most volatile first to lower the tower-top temperatures in later processing. The first distillation, the demethanizer, pulls off the most volatile components first, hydrogen (−253°C) and methane (−161°C). The depropanizer removes the least volatile C_8^+ heavies (> −1) at the last possible separation before the most difficult separation.*

3. Remove Valuable Products as Distillates *To ensure product purity and avoid inclusion of colored materials the C_2^{2-} and C_3^{2-} products should be tower-top products. We see that this is the case in Figure 5.6–3.*

Thus these three heuristics explain the gross features of this industrial separation scheme. As a point of interest, the ethylene and propylene are the valuable products

Figure 5.6–2. Panoramic View of the Sinclair-Koppers 500 Million-lb/Year Ethylene Plant Built by M. W. Kellogg Company at Houston, Texas. The pyrolysis furnaces are the rectangular units in the far left background. The tall tower at the right is the demethanizer. The tallest tower, painted in two colors, is the C_2 splitter. Just to the left of it are two large towers that compose the C_3 splitter, operating in series. (Courtesy The M. W. Kellogg Co., a division of Pullman, Inc., New York, N.Y.)
C. J. King, *Separation Processes*, McGraw-Hill Book Company, New York, 1970.

formed initially by the cracking catalytically of an ethane and propane mixture. The ethane and propane separated are to be recycled for additional cracking. The methane and hydrogen are used as fuel, and the heavies ultimately may appear as motor fuels.

5.7 WHEY FROM POLLUTANT TO PRODUCTS

It is important to demonstrate the usefulness and limitations of these heuristics for process discovery on real problems. The processing of cheese whey to transform an environmental pollutant into valuable food products is the real problem to which this section is devoted.*

* See R. I. Fenton-May, C. G. Hill, and C. H. Amundson, *Journal of Food Science*, **36**, No. 1, 14–21, 1971.

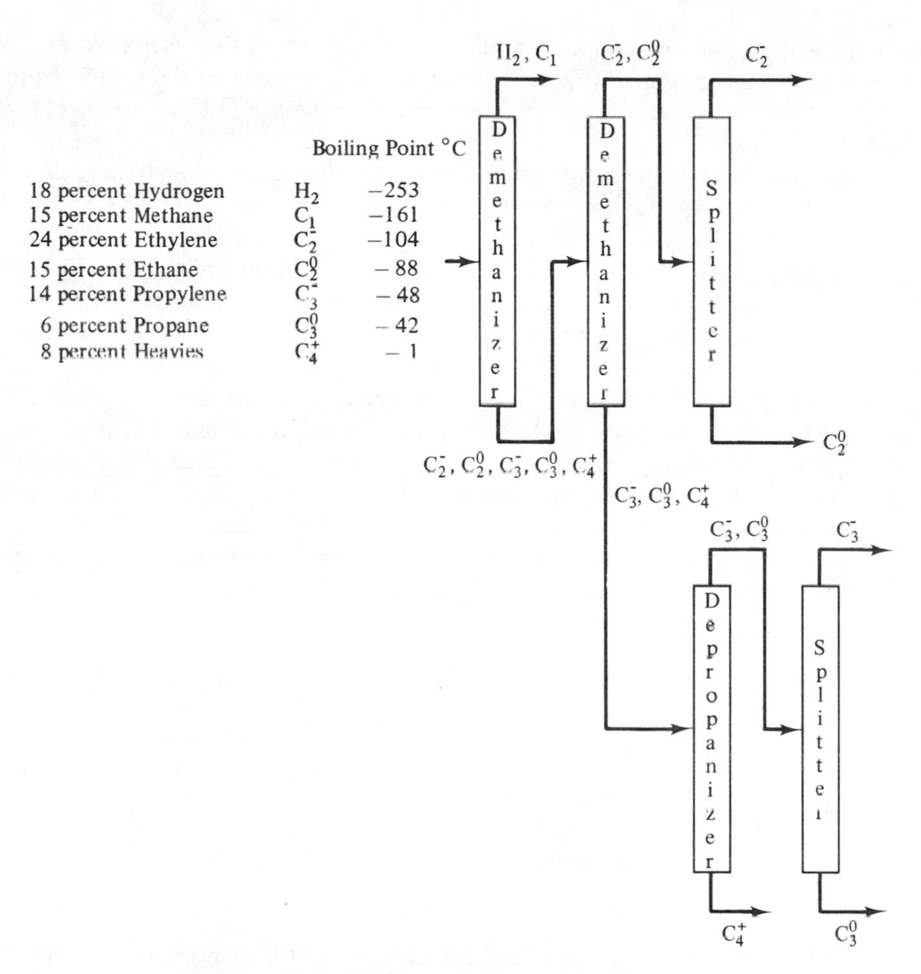

Figure 5.6–3. Typical Distillation Column Sequence for Separation of Products in Light Olefin Manufacture

Skim milk contains 3.0 to 3.5 per cent casein, 0.5 to 0.8 per cent whey protein, 0.1 to 0.2 per cent lower-molecular-weight polypeptides, amino acids, and urea (nonprotein nitrogen compounds), 4 to 5 per cent lactose, 0.5 to 1.0 per cent mineral salts, and various amounts of citric acid, lactic acid, and vitamins. During the production of cheese, the casein is coagulated and removed as a curd, leaving a cheese whey, which is nearly identical to skim milk with the casein removed. For every pound of cheese produced, 8 to 10 lb of whey is formed. Currently, the majority of the whey produced is disposed of on the land and in sewer systems, whey in its natural form not being a highly marketable product, even though it contains 12 to 20 per cent of the protein and most of the water-soluble constituents of milk.

Twenty-five *billion* pounds of whey are produced each year in the United States alone. The average cheese plant in Wisconsin produces 300,000 lb of whey per day.

The biological oxygen demand during the waste treatment of the whey effluent from such a cheese plant is comparable to that during the treatment of the sewage from a city of 60,000. The conversion of this whey pollutant to useful food products is an important problem.

In its raw form there is no extensive market for dried whey; in such high concentrations, the lactose acts as a laxative to mammals. The whey must be fractionated, somehow, into a predominately protein fraction and/or a predominately lactose fraction, which when dried can find a market in the food and chemical industry. The protein market is enormous. The lactose market is less extensive, even though new uses of lactose are being found, such as the use of lactose as a monomer in polyurethane plastics.

One way to avoid the lactose–whey protein separation problem is to grow in the whey the yeast *Saccharomyces fragilis.* This yeast feeds on the lactose, converting it to yeast protein. Desalting and drying the reacted whey leaves a whey protein–yeast protein product suitable for consumption as a protein food supplement. The protein–lactose separation problem is bypassed by this biochemical reaction.

However, let us focus attention on the problem of transforming the raw whey (Table 5.7–1) into a protein-rich low-salt dry solid and a lactose-rich low-salt dry solid. We begin by gathering data on the properties of the species, hoping that the means of separation will suggest itself.

TABLE 5.7–1 Whey Composition (w %)

Water	93.4
Protein	0.9
Lactose	5.0
Lactic acid	0.2
Inorganic mineral salts	0.5

In the literature contained in a modern technical library is to be found part of the data needed to begin process discovery; other data must be obtained by experiment. Let us suppose, however, that the data are at hand.

1. *Protein:* Whey protein consists predominately of α-lactalbumin, β-lactoglobulin, and the proteose–peptone fraction with molecular weights of 14,437, 36,000, and 200,000, respectively. These are relatively soluble in water, and precipitate out as a gel upon concentration of the whey by evaporation to about 50 to 60 per cent solids.
2. *Lactose:* Lactose, $C_{12}H_{22}O_{11}$, with a molecular weight of 342 is the main sugar in milk. It is soluble in water only up to 10 per cent solids. It has a laxative effect if used in high amounts in a mammalian diet.
3. *Lactic acid:* Lactic acid, $CH_3CHOHCOOH$, is formed by the action of *Bacillus acidi lactici* on lactose. It is a colorless viscous liquid of molecular weight 90, and is freely soluble in water, ethyl alcohol, and ether.

4. *Mineral salts:* The mineral salts are predominately inorganic sodium and calcium salts, with molecular weights in the 50 to 100 range. Moreover, in solution they exist as ions and for this reason are even more mobile. These salts limit the palatability of feed stuffs.
5. *Water:* This is the only volatile substance in whey. The vaporization of water, however, requires 970 Btu/lb. Vaporization is somewhat costly, steam (which is a good source of this energy) costing $0.50/10^6$ Btu.

Table 5.7–2 summarizes the major property differences that might be exploited. The protein and lactose differ greatly (of the properties mentioned here) only in molecular weight and solubility in water. These two property differences might be useful to separate protein and lactose. Upon evaporation the lactose would crystallize out first, and later the protein. The volatility difference could be used to separate water from the other constituents, but exploiting this difference alone, say, by evaporation, would leave the protein, lactose, lactic acid, and salts unseparated. The mobility in the electric field suggests the removal of salts and lactic acid by dialysis,

TABLE 5.7–2 Summary of Major Properties of Species in Whey

	Molecular Weight	Volatility	Electric Mobility in Solution	Solubility in Water
Water	18	Fairly volatile	Negligible	–
Protein	10,000–200,000	Nonvolatile	Negligible	Up to 50% solids
Lactose	342	Nonvolatile	Negligible	Up to 10% solids
Lactic acid	90	Nonvolatile	Fairly mobile	Very soluble
Salts	20–100	Nonvolatile	Highly mobile	Very soluble

Schematic Representation of Property Differences.

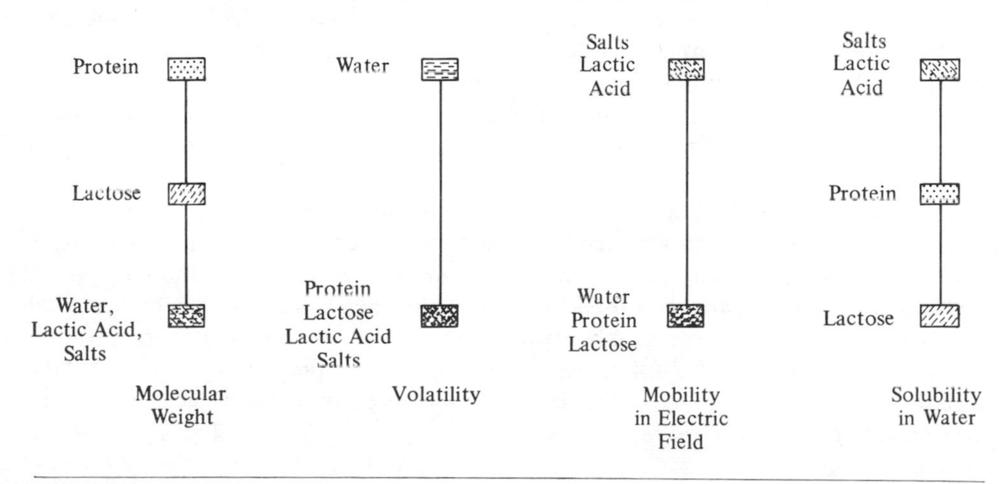

the migration of mobile ions through membranes across which an electric field is imposed.

However, most intriguing of all is the spread in the molecular-weight differences. In this property the protein and the lactose stand out separately from the water, lactic acid, and salts. This ordering in the property list attracts attention, since it is protein and lactose that we wish to recover. A technology exists, called *pressure-driven membrane separation*, by which molecules of greatly different size can be separated from their solution. Molecules differ in their ability to pass through plastic membranes, and by selecting the proper membrane, big molecules can be separated from smaller ones.

We now investigate the means for taking advantage of all these property differences.

Membrane Fractionation

The mere removal of water, say, by evaporation, does not separate the protein, lactose, and the salts. We must find property differences that allow a split of the species between the lactose and protein, between the salts and lactose, and so forth. The great differences in molecular weight suggest differences in the mobility of the molecules during the passage through the small openings in plastic membranes. Membranes are manufactured commercially that will allow the passage of molecules below a stated size, and these membranes are the basis of the pressure-driven membrane separation process of reverse osmosis and ultrafiltration.

Figure 5.7–1 shows the fabrication of a typical ultrafiltration cell. Two sheets of a specially prepared cellulose acetate based membrane are sandwiched over a porous backing material, the membranes are sealed at the edges, and sealed to a porous pipe at the end. The membrane is then covered with a wire mesh and rolled around the tube, producing a compact multilayer unit to fit in a larger tubular pressure vessel. In operation, the pressurized feed is forced through the mesh spacer, the water and small molecules pass through the membrane, and this *permeate* spirals through the porous material to the porous tube. The concentrated feed, which does not flow through the membrane, eventually flows out the end of the module through the mesh spacing. This fabrication is typical of a number of commercially available cell designs. By the selection of the proper membrane, these cells, which operate at several hundreds of pounds of fluid pressure, can be used to separate the whey according to molecular weight.

However, as the feed becomes more and more concentrated, the pressure needed to drive the fluids through the membrane may exceed the operating pressure of the cells. For this reason only part of the fluids can be forced through the membrane. Typically, these cells cannot be used to concentrate whey to beyond 25 per cent total solids (and with some wheys only 10 per cent solids). Beyond that concentration other property differences must be exploited. Let us see if pressure-driven membrane separators can be used to make the required adjustments in protein, lactose, and salt concentrations.

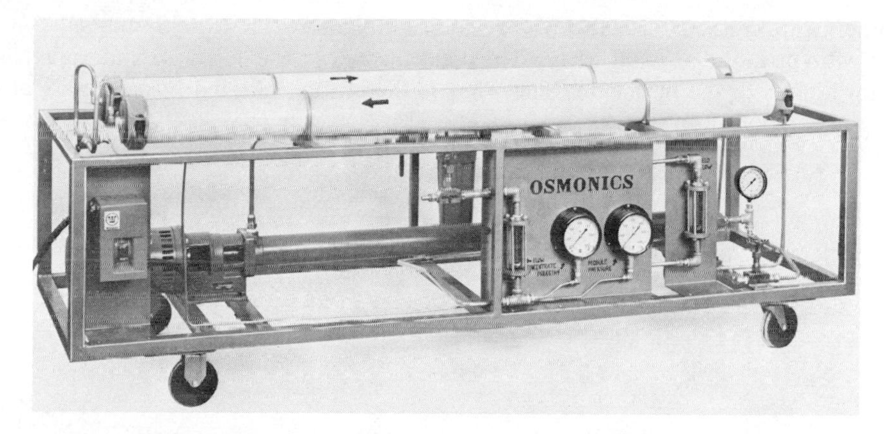

1. STAGED CENTRIFUGAL PUMP
2. REVERSE OSMOSIS VESSEL (100 sq. ft. of membrane per vessel)
3. VESSEL PRESSURE GAUGE
4. FEED FLOW INDICATOR
5. CONCENTRATE PRESSURE GAUGE
6. CONCENTRATE FLOW INDICATOR
7. FEED CONTROL VALVE
8. CONCENTRATE CONTROL VALVE
9. FEED INTO PUMP
10. CONCENTRATE
11. PERMEATE

APPROXIMATE DIMENSIONS—7′ LONG X 30″ HIGH X 24″ DEEP

Figure 5.7–1. A Pressure Driven Membrane Separation Process. (Courtesy of OSMONICS, INC., Hopkins, Minnesota.)

Consider now the ultrafiltration cell as a device that allows the passage of the species below a stated molecular weight, and that the cell operates up to the point where the remaining fluid is concentrated to 25 per cent solids. Suppose that a membrane is selected which holds back only the protein molecules. How effective would such a cell be in fractionating the whey?

As the water passes through the membrane, it carries with it lactose, lactic acid, and mineral salts in the same concentration as in the whey (Table 5.7–1). Only the protein is left behind completely. The feed material becomes more and more concentrated in protein as a larger and larger portion of the feed is passed through the membrane. Using a material balance, we can calculate the composition of the permeate and the concentrate as a function of the percentage of water removal.

Consider now 1 lb of whey of which a fraction X of the water is passed through the membrane. The original pound of whey (Table 5.7–1) has the composition

Species	Lb
Water	0.934
Protein	0.009
Lactose	0.050
Lactic acid	0.002
Mineral salts	0.005

The permeate consists of $0.934X$ lb of water with lactose, lactic acid, and salts in the same ratio as in the original whey. The concentrate consists of those materials which do not appear in the permeate. The whey is therefore separated into streams of the following composition:

	PERMEATE (LB)	CONCENTRATE (LB)
Water	$0.934X$	$0.934(1 - X)$
Protein	0	0.009
Lactose	$0.050X$	$0.050(1 - X)$
Lactic acid	$0.002X$	$0.002(1 - X)$
Salts	$0.005X$	$0.005(1 - X)$

Since the cell can concentrate only up to 25 per cent solids (75 per cent water), we can calculate the upper limit X^* on the fraction of water removable.

$$0.75 = \frac{0.934(1 - X^*)}{0.934(1 - X^*) + 0.009 + 0.050(1 - X^*) + 0.002(1 - X^*) + 0.005(1 - X^*)}$$

$$X^* \simeq 0.97$$

About 97 per cent of the water can be removed before the concentrate becomes 25 per cent solids.

Since the concentrate is to be dried eventually to a protein-rich solid, it is interesting to compute the composition of the solids as a function of the percentage of water passed through the cell. The results of this computation are shown in Table 5.7–3. This vividly shows how an ultrafiltration cell can be used to fractionate

TABLE 5.7–3 Composition of Whey Concentrate (on a Dry Basis) Obtained by Ultrafiltration[a]

Product	Dried Whey	Dried Ultrafiltration Concentrate			
Weight % \ PERCENTAGE WATER PERMEATED	0	80	90	95	97
Protein	13	39	57	72	83
Lactose	77	54	39	25	15
Lactic acid	3	2	2	1	1
Mineral salts	7	5	5	2	1
Percent solids in concentrate	7	10	12	17	25

[a]Calculated by material balancing.

the whey and also how material balancing can be used to assess the performance limitations of process equipment.

Earlier in this chapter we developed methods for the discovery of processes that had a low total separation difficulty. These methods were based on the assumption that property lists could be split cleanly into those species above and those species below a breakpoint in the property list. In the ultrafiltration of whey, Table 5.7–3, 83 per cent protein is the upper limit on the protein-rich stream (when dried). The clean splitting off of the protein is not possible, because the pressures required to drive the membrane process are excessive beyond a concentrate of 25 per cent solids. Those technical limits occur in every separation problem, and our earlier theoretical developments merely get us into the range of a feasible process.

Similar material-balance studies are to be performed on ultrafiltration membranes that restrict the passage of lactose and larger molecules, if the need for such a separation in processing arises.

Vaporization of Water

Were it not that pressure-driven membrane separators foul when solids concentrations reach 25 per cent in whey processing, we would be tempted to process the whey completely by taking advantage of molecular size differences. However, we are faced with the problem of water removal, even if partial processing is by ultrafiltration. To remove water, one merely supplies the energy necessary to vaporize it. The quality of the solids products left behind depends very much on the means by which the energy is applied and the water vaporized. Merely evaporating the water tends to leave a hard cake of solids, which must be removed from the heating surface and ground to the required solid form. The product formed by complete evaporation is often not of good quality for marketing.

However, if the water is vaporized from isolated drops of the solution with the energy supplied by warm air, granules tend to form that are easily handled. In *spray drying*, a solution is sprayed into a tower in which hot air is blown, and dried granules are collected as they fall to the bottom of the tower. Dried milk, coffee, and many other materials are manufactured in this way.

To spray dry effectively, the solutions must be as close as possible to being saturated with solids (about 50 per cent solids for the proteins and 10 per cent solids for the lactose). A spray dryer cannot be connected directly to the concentrate stream of the ultrafilter since the effluent is at the most 25 per cent solids and the spray dryer requires at the least 50 per cent solids. An evaporator must be used to bridge this gap in solids concentrations. Figure 5.7–2 is a schematic diagram of an evaporator and a spray dryer.

A Processing System

We are now in a position to synthesize a complete process, exploiting the property differences within the regions of available technology. Figure 5.7–3 shows a system that looks like a good candidate, deserving of some detailed analysis.

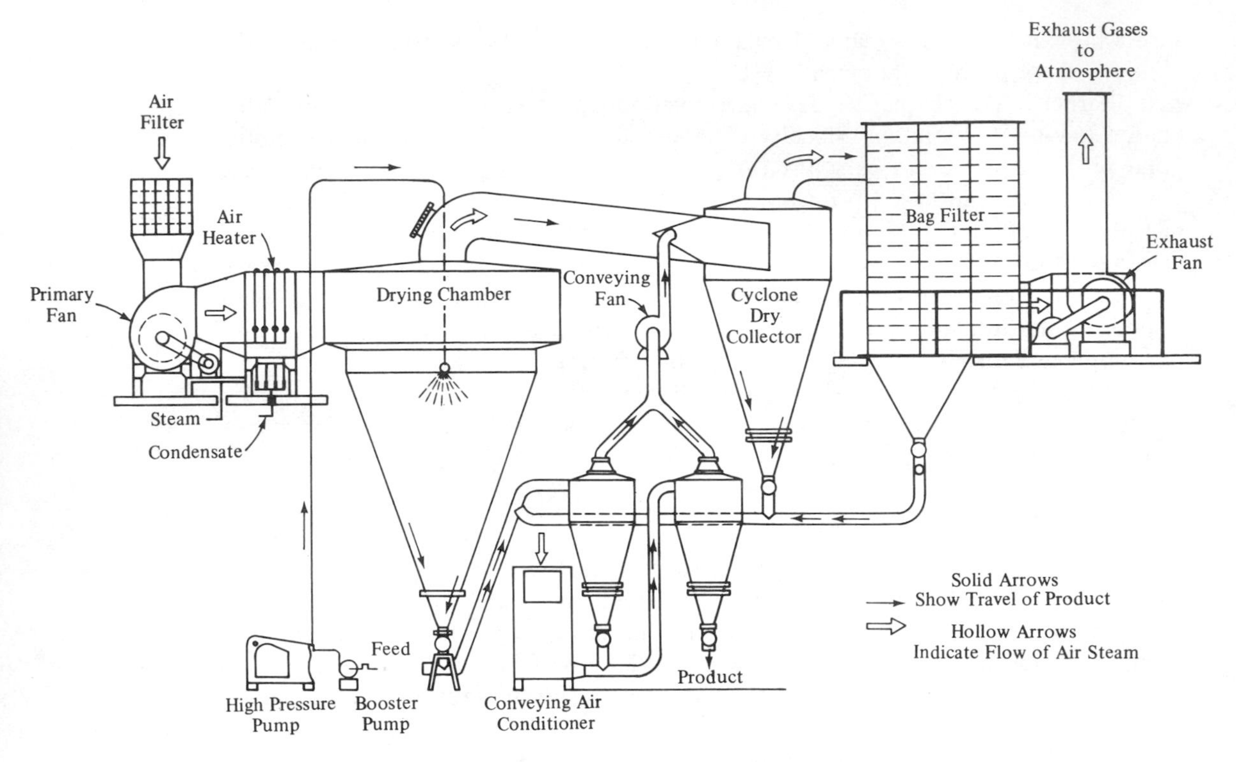

Figure 5.7–2. Steam-Heated Spray Dryer with a Dry Collection System. This dryer was designed for drying coffee solubles, but a similar arrangement can be used for drying pharmaceutical products, organic dyes, pigments, resins, and other high-value products that cannot be processed with wet collection equipment and that require lower drying temperatures. (Courtesy Swenson Evaporator Co., a Division of The Whiting Corp.)

First, we summarize the thinking that went into the synthesis of this plan. The separation of large amounts of water from the solids is a major problem to be faced; the vaporization of water is costly compared to pressurizing water for membrane separation. For this reason, separation by vaporization is reserved for those streams which are inaccessible to pressure-driven membrane processing. First, a lactose-permeable membrane is used to form a protein concentrate; then a lactose-impermeable membrane is used to form a lactose concentrate. Both concentrates are eventually spray dried to reach the desired product quality. In the problems the reader will have an opportunity to verify the material flows shown in Figure 5.7–3 and to innovate further.

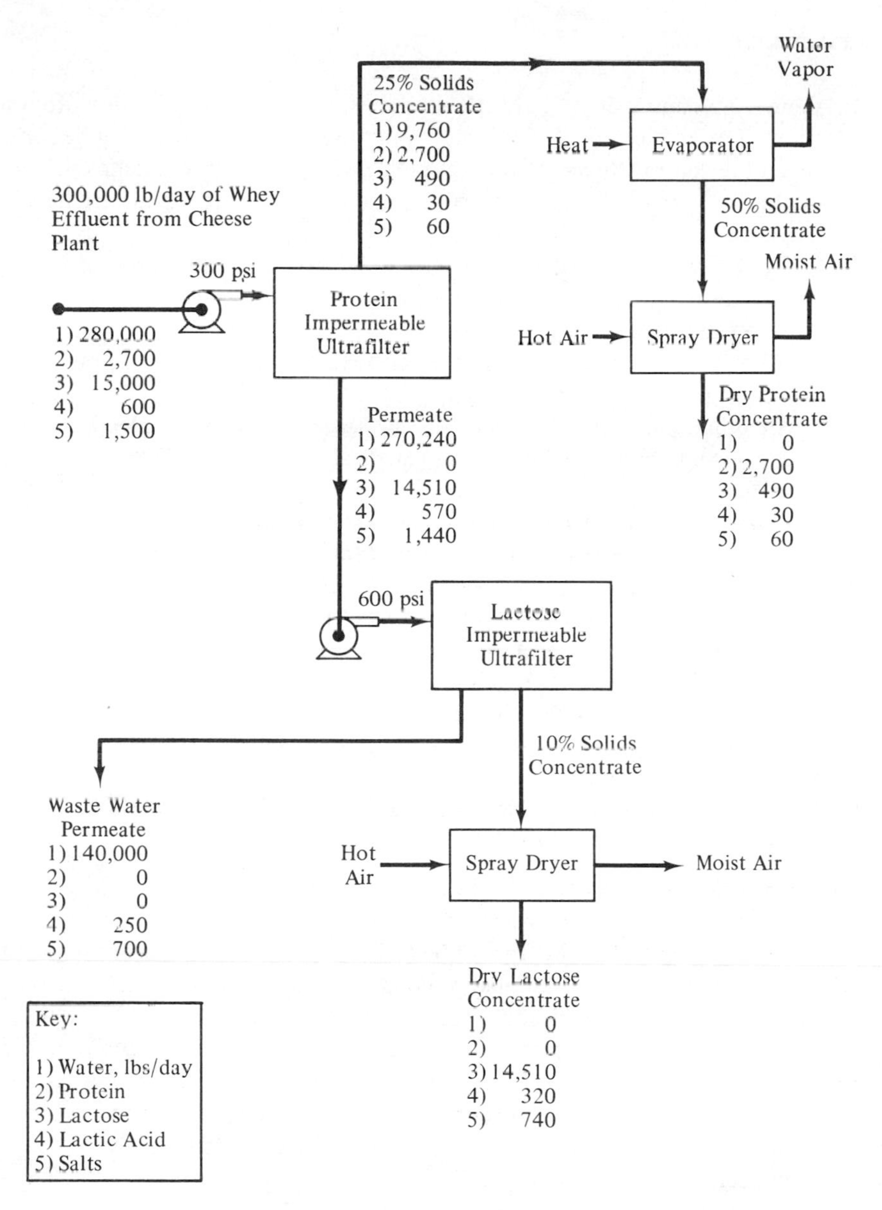

Figure 5.7–3. Plausible Whey Treatment Process

5.8 CONCLUSIONS

Differences in composition between stream sources and destination dominate task selection. The reduction of this dominant difference with specified separation tasks has been the central theme of this chapter, and the several heuristics developed lead toward the tasks that efficiently resolve composition differences. These heuristics are:

1. **Of the many differences that may exist between the source and destination of a stream, differences involving composition dominate. Select the separation tasks first.**
2. **When possible, reduce the separation load by stream splitting and blending.**
3. **All other things being equal, aim to separate the more plentiful components first. If amounts are equal, aim to separate into equal parts.**
4. **Remove the corrosive and hazardous material early.**
5. **The difficult separations are best saved for last.**
6. **All other things being equal, shy away from separations that require the use of species not normally present in the processing. However, if a foreign species is used to effect a separation, remove that species early.**
7. **Avoid excursions in temperature and pressure, but aim high rather than low.**
8. **In distillation, remove desired product species finally in a distillate.**
9. **In distillation favor the removal one by one of the more volatile components.**

These heuristics serve to weed out the processing plans that are definitely inferior, isolating the several promising process flow sheets. However, in many important situations the heuristics will be in conflict. For example, if the most plentiful component is also involved in the most difficult separation, one heuristic directs the early removal of that component and a second heuristic suggests that the component be removed last. The only completely accurate way to resolve these differences is by the complete design and cost analysis of the alternative processing plans. Frequently, these superficially competing separation plans must be carried through to completion to find the most economical plan. The heuristics make sure that we are not wasting effort on obviously noncompetitive plans.

REFERENCES

King presents one of the best summaries of task selection:

C. J. KING, *Separation Processes*, McGraw-Hill Book Company, New York, 1970, Chap. 14.

The systematic weighing of heuristics in conflict leads to techniques for the computer-aided synthesis of designs:

J. J. SIIROLA and D. F. RUDD, "Computer-Aided Synthesis of Chemical Process Designs," *Industrial and Engineering Chemistry Fundamentals*, **10,** 353, Aug. 1971.

G. J. POWERS, "Heuristic Synthesis in Process Development," *Chemical Engineering Progress*, **68,** no. 8, 1972.

J. HENDRY and R. R. HUGHES, "Generating Separation Process Flow Sheets," *Chemical Engineering Progress*, **68,** no. 6, 1972.

PROBLEMS

The first several problems deal with the direct application of the several heuristics presented in this chapter. Later we are concerned with the further development of heuristics suitable to resolve separation problems beyond the basic heuristics of this chapter.

1. **HUMAN KIDNEY AS A SEPARATOR.** The kidney is the organ that maintains the salt balance in the body and removes from the bloodstream the products of metabolism. What separations are performed by the kidney and how are they accomplished? Artificial kidneys have been engineered to support victims of renal failure. Outline the operation of such a system. (See E. N. Lightfoot, *Transport Phenomena in Living Systems*, John Wiley & Sons, Inc., New York, 1973.)

2. **CRUSHED ROCK SCREENING.** One thousand tons per day of crushed rock of the following size distribution is to be screened. Determine the screening sequence that involves the minimum rock handling. Why might one be attracted to the removal of the 5- to 6-in. rocks even though this increases the processing load? By what percentage is the load increased?

SIZE RANGE (IN.)	%
1–2	12
2–3	1
3–4	2
4–5	80
5–6	5

3. **AROMATICS SEPARATION.** Benzene, toluene, and xylenes are to be separated by distillation. The relative volatilities are 2.5, 1.0, and 0.4, respectively. The feed mixture contains the three components in equal amounts. It has been suggested to recover the xylenes as bottoms from a first distillation, and then separate the benzene and the toluene in a second distillation. What do you think of this plan?

4. **TASK SELECTION WITH GREATER EXPERIENCE.** A number of the illustrative examples in this chapter appear prior to the development of all the task-selection heuristics.

In this problem we go back to correct any errors resulting from our incomplete experience. Rework examples 5.1–1, 5.2–1, 5.3–1, 5.3–2, and 5.4–1.

5. **LACTOSE-IMPERMEABLE ULTRAFILTER OPERATION.** In Figure 5.7–3 the permeate from the protein-impermeable ultrafilter is the feed to the lactose ultrafilter. Prepare a table similar to Table 5.7–3 showing how the percentage of lactose, lactic acid, and salt in the dry lactose concentrate depends on the fraction of the material permeated. Does the material balance in Figure 5.7–3 match your results?

6. **WHEY PROCESSING MODIFICATION.** In Figure 5.7–3 cheese whey is fractionated into a high-protein solid and a high-lactose solid. Suppose that the lactose-impermeable membrane is used first (to 10 per cent solids) and the protein-impermeable membrane last (to 25 per cent solids). Determine the concentration of the dried lactose and protein solids from this process.

7. **WATER INJECTION IN PROTEIN FILTER.** In Problem 6 the placing of the protein-impermeable membrane last results in the incomplete fractionation of the cheese whey. It is proposed to inject water into the feed to the protein filter to increase the lactose permeation. How much water must be injected to reach the 83 per cent dried protein concentrate of Section 5.7?

8. **HIGH-PURITY BERYLLIUM.*** High-purity beryllium metal in sponge form can be produced from beryllium oxide by chlorination and sodium reduction:

$$C + BeO + Cl_2 \rightarrow BeCl_2 + CO$$

$$2\,Na + BeCl_2 \rightarrow Be + 2\,NaCl$$

The flow sheet outlines the reaction and purification steps. During chlorination the beryllium oxide is converted to the chloride. The residue is sent to a BeO recovery process and the $BeCl_2$ is sent to sublimation scrubbing. Once scrubbed free of impurities by molten salt, the chloride is reduced with metallic sodium to form crude sponge. The sodium chloride and sodium impurities in the crude sponge are distilled off to form the pure metal.

Physical Properties of Metals and Metal Chlorides

	MP (°C)	BP (°C)	VAPOR PRESSURE (MM HG) AT 950°C
Be	1283	2484	3.7×10^{-5}
$BeCl_2$	–	532	Very high
BeO	2530	Ca. 3900	Very low
Na	98	883	1310
NaCl	801	1465	4.4

* Beryllium metal is very dangerous, and extreme care must be taken to avoid breathing the metal dust.

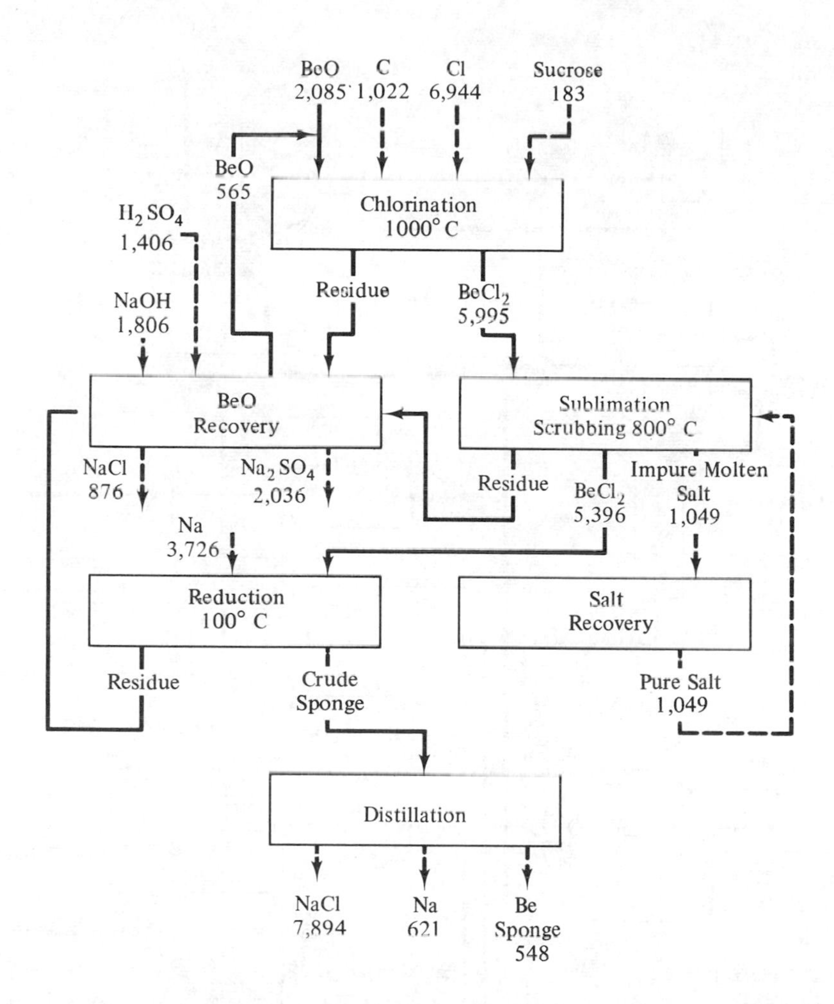

(a) Are the material balances on the chlorination process and the reduction process consistent with the reaction stoichiometry?

(b) What property differences are exploited in the sublimation scrubbing and distillation processes?

(c) What is the probable composition of the two residue streams?

(d) Speculate on the chemistry and separations occurring in the BeO recovery systems. Do the material balances check?

9. DETERGENT ACTIVE INGREDIENT. Shown is a process for the manufacture of alkyl aryl sulfonate from kerosene, chlorine, sulfuric acid, and benzene. This sulfonate is surface active and is used as the wetting agent in detergents to solubilize grease and oils. Examine the separation processes that support the four reactors, and determine which of the several heuristics explain the orders of separation.

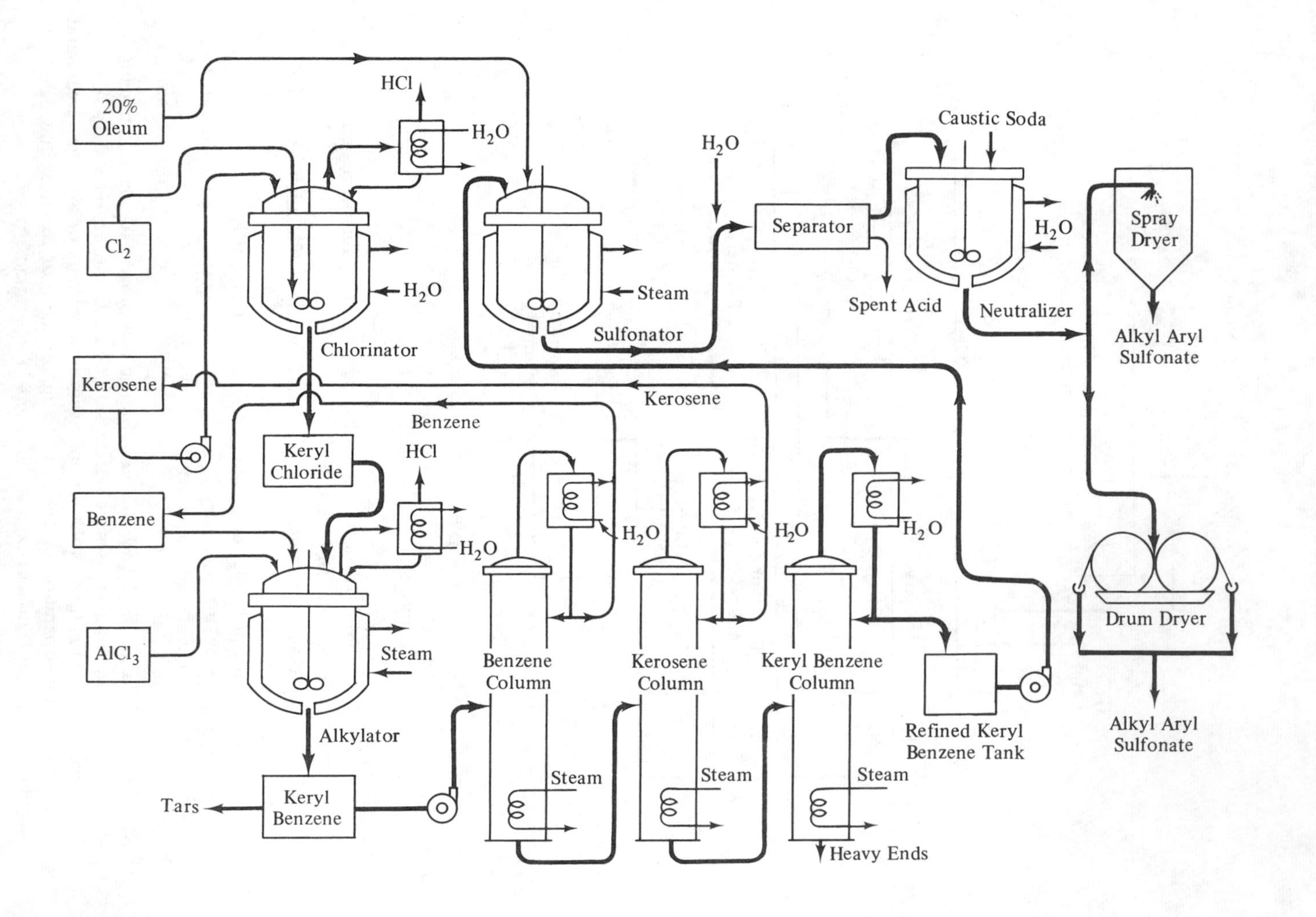

20% Oleum
HCl
H_2O
Cl_2
H_2O
Chlorinator
Steam
Sulfonator
H_2O
Separator
Spent Acid
Caustic Soda
H_2O
Neutralizer
Spray Dryer
Alkyl Aryl Sulfonate
Kerosene
Kerosene
Benzene
Keryl Chloride
HCl
Benzene
H_2O
H_2O
H_2O
H_2O
$AlCl_3$
Steam
Alkylator
Benzene Column
Kerosene Column
Keryl Benzene Column
Refined Keryl Benzene Tank
Drum Dryer
Alkyl Aryl Sulfonate
Tars
Keryl Benzene
Steam
Steam
Steam
Heavy Ends

The kerosene is chlorinated in the lead-lined, jacketed reactor at 60 to 70°C in the presence of an iodine catalyst:

$C_{12}H_{26}$	+	Cl_2	→	$C_{12}H_{25}Cl$	+	HCl
(kerosene)		(chlorine)		(keryl chloride)		(hydrogen chloride)

This reaction occurs with twice the stoichiometric amount of kerosene.

In the alkylator, benzene is added to the keryl chloride–kerosene mixture from the chlorinator. Six moles of benzene are added to each mole of keryl chloride to ensure the complete reaction of the chloride. The aluminum chloride is added as a catalyst in small amounts:

$C_{12}H_{25}Cl$	+	⬡	→	$C_{12}H_{25}$ – ⬡	+	HCl
(keryl chloride)		(benzene)		(keryl benzene)		(hydrogen chloride)

In the sulfonator, the purified keryl benzene is treated with sulfuric acid (1 : 1.25 mass ratio), and the completely sulfonated product is decanted from the spent acid. Then the sulfonated product is neutralized to form the final product.

(a) Determine the approximate composition of the stream leaving the alkylator, including the HCl.

(b) Using the compositions from above and the boiling-point data given, synthesize the sequence of distillations to separate the materials. How does this compare to the separations used in the commercial process?

Species	BP (°C)
Benzene	80
Kerosene	214
Keryl benzene	250
Hydrogen chloride	–85

10. **ALLYL CHLORIDE PRODUCTION.** Allyl chloride is a valuable chemical intermediate used in the manufacture of glycerol and other fine chemicals. A major source is the chlorination of propylene, the main reaction being

C_3H_6	+	Cl_2	→	C_3H_5Cl	+	HCl
(propylene)		(chlorine)		(allyl chloride)		(hydrogen chloride)

This reaction occurs at 300°C with a 100 per cent excess of propylene, and is accompanied by side reactions that produce a variety of chlorinated compounds. Shown are the reactor effluent composition, boiling points, and water solubilities. We wish to separate the pure allyl chloride, and recycle the propylene and chlorine reactor feeds.

Eight major separation schemes—some of which are reasonable, some of which are not—are shown on the following two pages.

A
B C D E F
B C D E F
B
C D E F
C
D E F
D
E F
E
F

1.

A B
A B C D E F
C
D E F
D
E F
E
F

2.

A
A B C
B C
B
C
A B C D E F
D
D E F
E F
E
F

3.

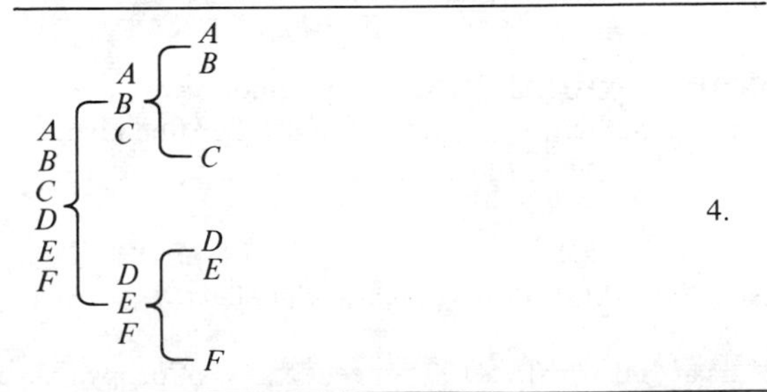

4.

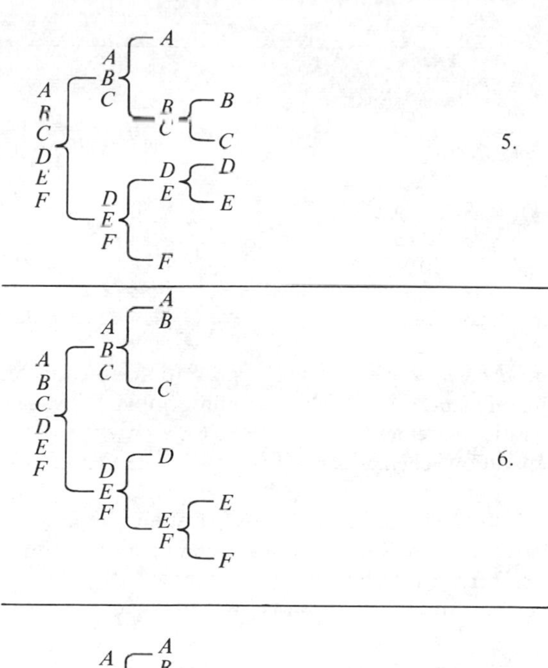
A B C D E F
A B C
A
B C
B
C
D E F
D E
D
E
F
5.
A B C D E F
A B C
A
B
C
D E F
D
E F
E
F
6.

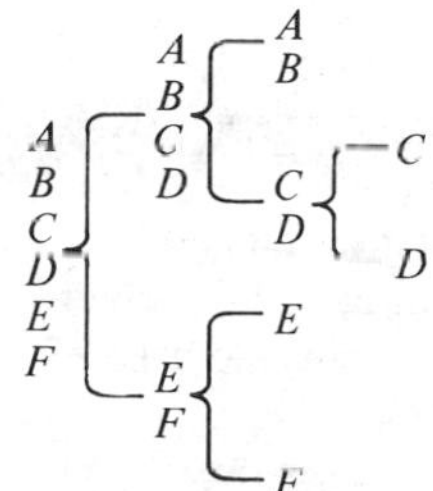
A B C D E F
A B C D
A
B
C D
C
D
E F
E
F

7.

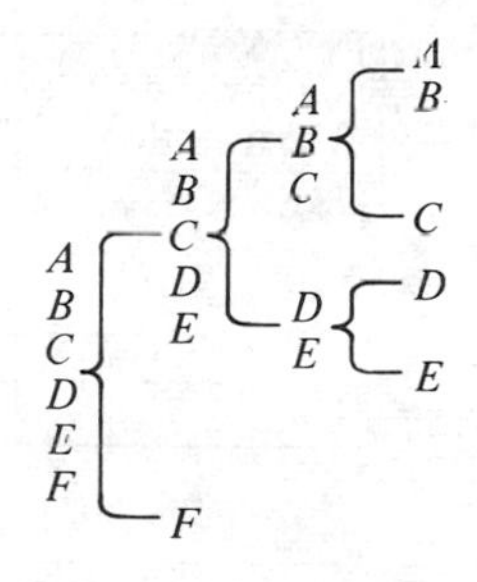
A B C D E F
A B C D E
A B C
A
B
C
D E
D
E
F

8.

Reactor Effluent Data

	Relative Amount (weight basis)	Bp (°C)	Solubility in H_2O (weight %)
A. 1,3-Dichloropropane	0.2	112	Insoluble
B. Acrolein dichloride	1.8	84	Insoluble
C. Allyl chloride	9.3	50	0.33
D. Chlorine	3	−34	1.46
E. Propylene	105	−48	0.89
F. Hydrogen chloride	93	−85	72

(a) Delete those schemes which involve separation not directed toward recycling the propylene, obtaining pure allyl chloride, and removing the reaction by-products. Suggest other separation schemes.

(b) Which separation schemes derive from the heuristic, remove the plentiful components early?

(c) Which derive from, save the difficult separations for last?

(d) Which derive from, remove the low-boiling materials early?

(e) Which derive from, remove corrosive material early?

(f) Based on the experience of parts a through e, which separation schemes dominate?

(g) Shown is the commercial process reported in the *Encyclopedia of Chemical Technology*, Vol. 5, John Wiley & Sons, Inc. (Interscience Division), New York, 1964, p. 211. Does this fit in with your analysis? Compare and contrast the commercial process with those which evolve by the heuristic reasoning of this problem. Why is the chlorine removed from the recycle propylene going to the heater?

(h) How might the differences in water solubility be used effectively?

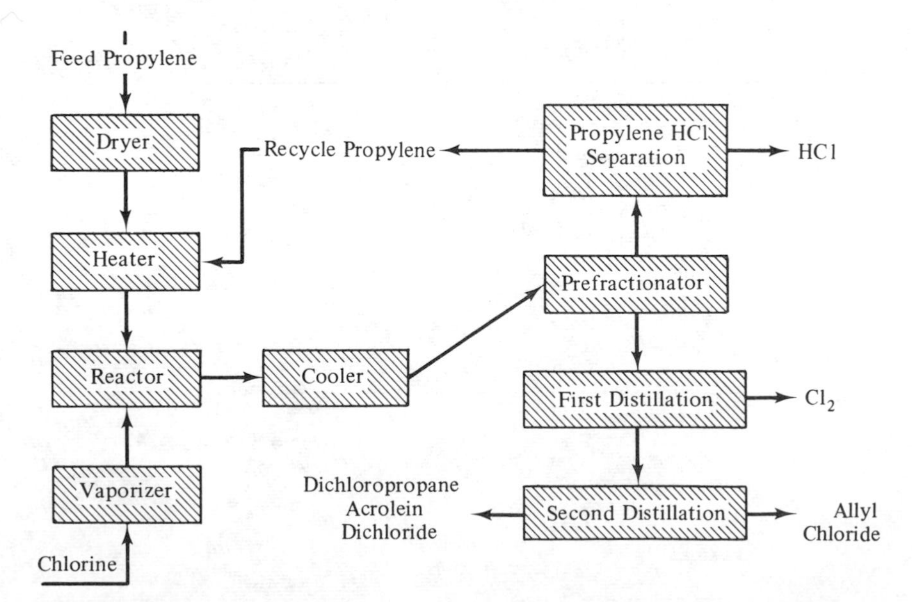

11. SEPARATION OF CHLOROMETHANES. Methane is reacted with chlorine to produce methyl chloride, methyl dichloride, chloroform, and carbon tetrachloride as the main products. Also contained in the reactor effluent is unreacted methane and the hydrogen chloride by-product. All the chlorine feed is consumed. The feed to the reactor is methane and chlorine in a 1 : 0.6 mole ratio. The chlorinated products are produced in the following mole ratios:

Methyl chloride	6
Methyl dichloride	3
Chloroform	1
Carbon tetrachloride	0.25

(a) Determine the mole per cent of each compound present in the reactor effluent.
(b) Devise the recommended distillation sequence to separate the reactor effluents.
(c) If the unreacted methane is recycled and more chlorine is added to make the required feed ratio, determine the net consumption and production of species for the process. What is the gross profit per pound of methane consumed?
(d) Would there be any economic advantages in using a second reactor, which when run with a surplus of chlorine in the feed converts any chloromethane to carbon tetrachloride?

SPECIES	BP (°C)	VALUE ($/LB)
Methane	−161	0.01
Hydrogen chloride	−85	0.07
Methyl chloride	24	0.16
Methyl dichloride	40	0.10
Chloroform	61	0.17
Carbon tetrachloride	76	0.12
Chlorine	−34	0.05

12. VINYL CHLORIDE PRODUCTION. E. F. Edwards and T. Weaver (*Chemical Engineering Progress*, **61** No. 1, 21, 1965) describe alternative processes for the production of vinyl chloride from ethylene. The following flow sheet exploits the reaction paths of Problem 5 of Chapter 2. Using these two sources of information and the separation heuristics of this chapter, rationalize the flow sheet by explaining why each operation leads toward efficient processing.

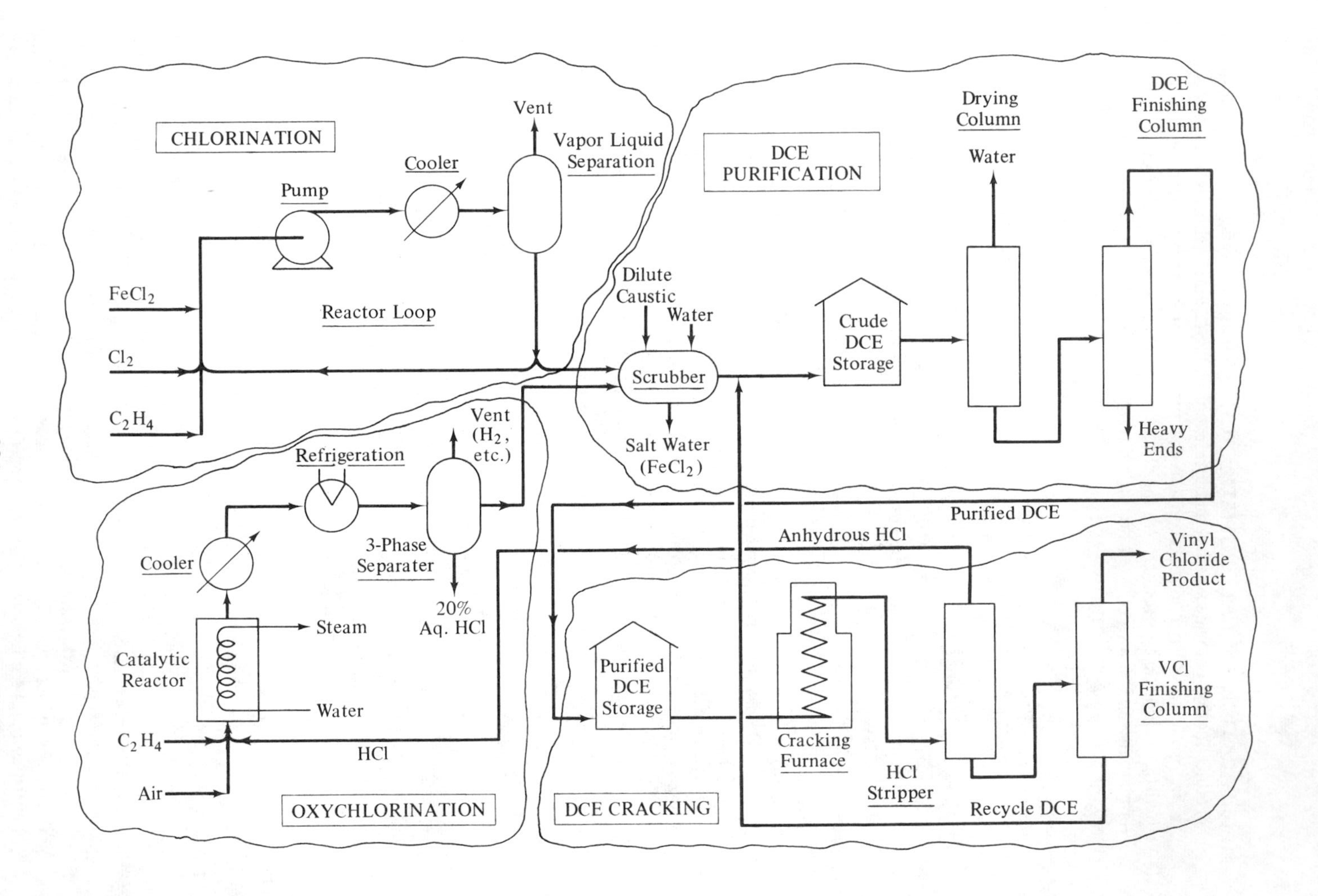
CHLORINATION
Pump
Cooler
Vent
Vapor Liquid Separation
$FeCl_2$
Cl_2
C_2H_4
Reactor Loop
DCE PURIFICATION
Dilute Caustic
Water
Scrubber
Salt Water ($FeCl_2$)
Crude DCE Storage
Drying Column
Water
DCE Finishing Column
Heavy Ends
Refrigeration
Vent (H_2, etc.)
Cooler
3-Phase Separater
20% Aq. HCl
Catalytic Reactor
Steam
Water
C_2H_4
Air
HCl
OXYCHLORINATION
Purified DCE
Anhydrous HCl
Vinyl Chloride Product
Purified DCE Storage
Cracking Furnace
HCl Stripper
VCl Finishing Column
Recycle DCE
DCE CRACKING

13. **LIMITS ON HEURISTICS.** A mixture of three species, in amounts D_1, D_2, and D_3 lb/h, is to be separated into pure components, and the property difference to be exploited has gaps of Δ_{12} and Δ_{23} between adjacent species. The difficulty of separation is estimated by $\sum D/\Delta$, where the sum is over all separations.

    (a) How many distinctly different separation schemes can be envisioned?

    (b) If $\Delta_{12} = \Delta_{23}$, the amounts of material processed determine the difficulty of the processing scheme. The heuristic *separate the more plentiful species first* is ambiguous in that *more plentiful* is not defined. What percentage of the total must the more plentiful be for the heuristic to be valid?

    (c) In cases where the more plentiful species is also involved in a difficult separation, the heuristic above contradicts the heuristic *save the difficult separations for last.* Under what conditions does the difficult-separation heuristic take precedence over the more-plentiful-species heuristic? Express your answer in terms of D's and Δ's.

14. **PRESENCE OF CORROSIVE MATERIAL.** Special care must be taken during the processing of a stream involving a corrosive material 1, present in amount D_1. Also present in the stream are species 2 and 3, in amounts D_2 and D_3. The special care can be estimated by a factor $f > 1$, which multiplies any separation difficulty estimate that involves the handling of species 1. For example, if species 3 is removed first, and then species 1 and 2 are separated, the corrosive species is present in both separations, and the processing difficulty would be estimated as $(D_1 + D_2 + D_3)f + (D_1 + D_2)f$.

    The heuristic *remove the more plentiful species first* contradicts the heuristic *remove the corrosive material early* if the corrosive material is present in small amounts. Under what conditions does the removal of small amounts of the corrosive material take precedence over the removal of larger amounts of noncorrosive material?

15. **CRUSHING AND SCREENING CIRCUITS.** In certain ore-processing situations the costs of screening ore according to size are much less than the costs of crushing the ore to reduce its size. In such a situation a careful placement of screening operations can reduce the cost of crushing, leading to elaborate crushing and screening circuits. The four major configurations from which more elaborate circuits are built are simple crusher, the normal-order circuit, the reverse-order circuit, and the reverse-order circuit without recycle. Shown are several of the elaborate circuits used in major ore-processing problems.

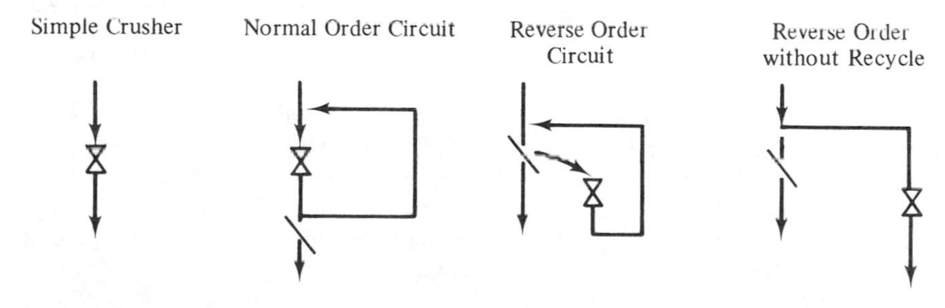

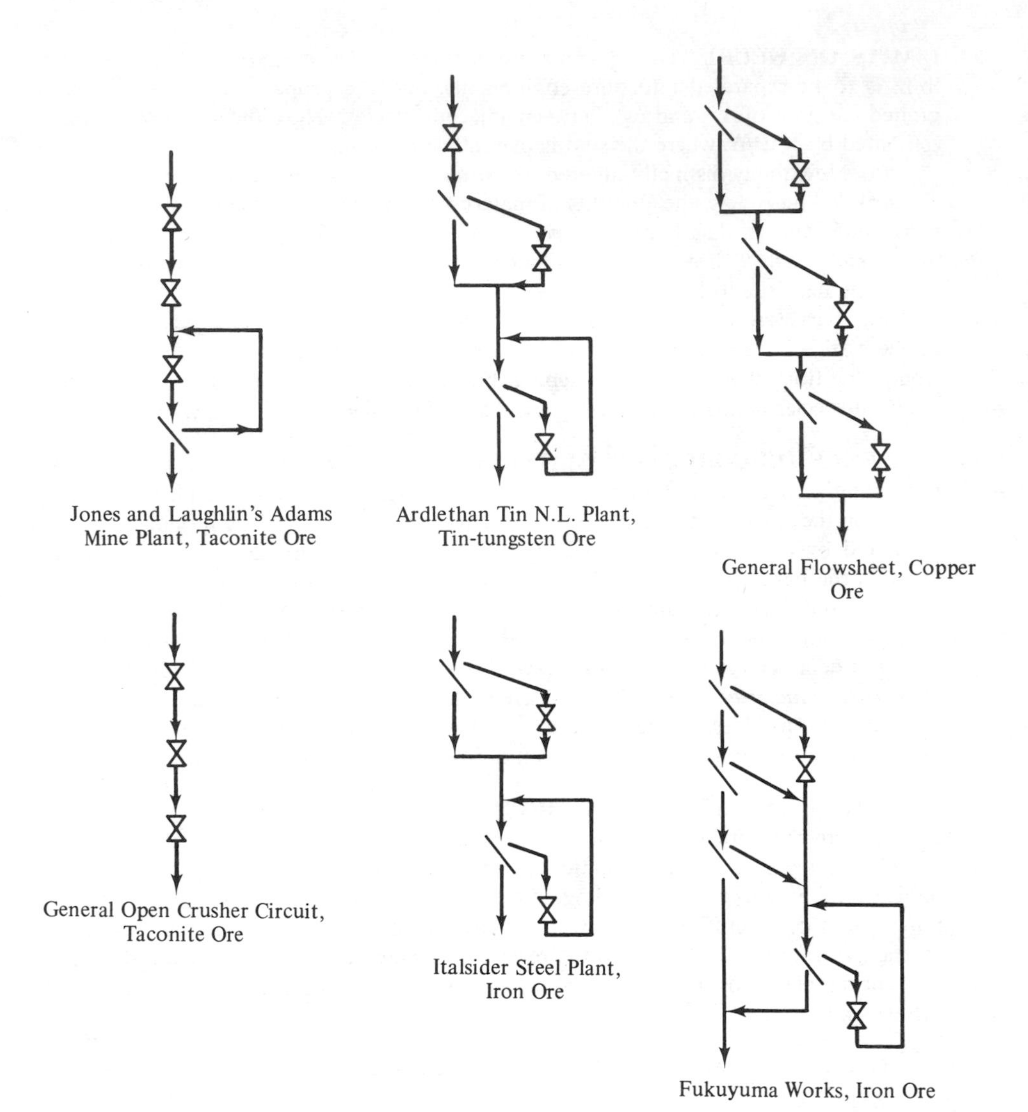

The heuristics of crushing and screening circuit design are special cases of the general heuristics developed in this chapter. The complete analysis of this engineering problem is beyond this introductory text (see E. Calanog, Ph.D. Thesis, University of Wisconsin, 1971). However, by simplifying the operation of screening and crushing operations, a useful problem in process development is formed, which is within the limits of the beginning student in engineering.

An ideal screen is one in which all the undersized material passes through the screen and all the oversized material does not pass. In reality some of the undersized material does not pass through and is entrained with the oversized. We presume that ideal screens are available and can be operated at negligible cost.

An ideal crusher receives material of diameter d_a and crushes it into material smaller than size d_b, forming a distribution of sizes linearly distributed from d_b to zero.

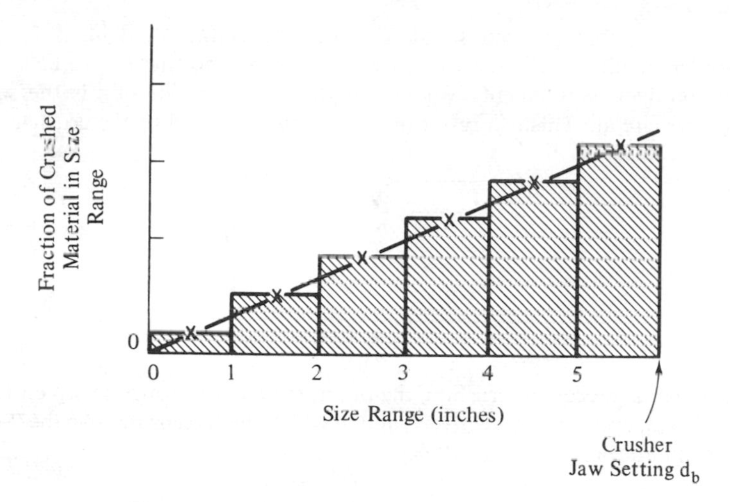

The cost of crushing is proportional to the amount of material, D tons/h, entering the crusher, and is proportional to the difference between the largest-sized material entering and the largest sized material allowed to leave the crusher:

$$\$/\text{hr cost} = KD(d_a - d_b), \qquad \text{where } K \text{ is a constant}$$

In reality the crushing distribution is more complicated, and the cost relationship involves more terms. However, we presume that ideal crushers are to be used.

Suppose that 100 tons/h of 6-in. hematite rock is to be crushed into ore of less than 2-in. size. How should this be done to reduce the total crushing costs?

(a) Determine the cost of processing the ore by simple crushing with no intermediate screening.

Answer: $K100(6 - 2) = 400K$ \$/h.

(b) Suppose that a first crusher was set at a 4-in. outlet size and that a normal-order circuit was used to process the ore. How might this change the processing cost of this idealized crushing and screening operation? This question should be answered in several parts.

(1) For each ton of ore entering the crusher set at 4 in., what fraction will pass through the 2-in. screen of the normal-order circuit? The linear distribution of crushed product is to be assumed.

Answer: $\frac{1}{4}$ tons/h.

(2) Show that if 100 tons/h of ore enters the circuit, the amount entering the crusher including recycle is 400 tons/h.

(3) Determine the cost of the normal-order circuit with the crusher set at 4 in.

Answer: $800K$ \$/h.

(c) We saw in part (b) that the normal-order circuit with a 4-in. crusher setting increased the cost of processing compared to simple crushing. Perhaps there is a crusher setting that lowers the cost. Determine the cost of processing the ore using a normal-order circuit with crusher setting of d_b. Is there a setting that gives a cost lower than the $400K$ \$/h of part (a)?

Answer: $(K100/4)d_b^2\,(6 - d_b)$ \$/h.

(d) Suppose that the first simple crusher with setting of 4 in. is followed by a reverse-order circuit without recycle. Clearly the second crusher has to be set at 2 in. to reach the product requirements. Show that the cost of processing is the same as the normal-order circuit. This is a result of our simplified model of the crusher economics.

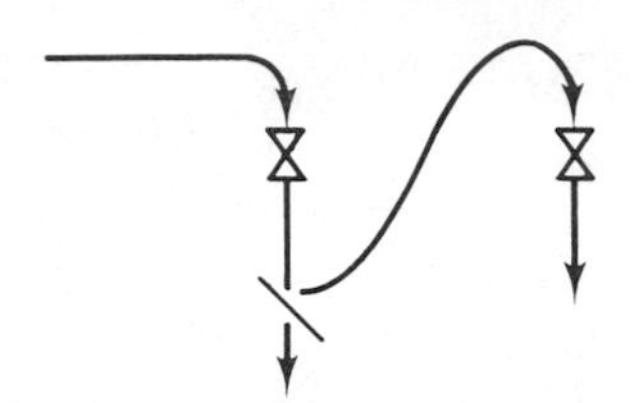

(e) What is the cost of crushing the ore in the circuit shown, in which the crushers are set at 5, 4, 3, and 2 in., respectively, and in which the screens remove the 2-in. product as undersized after each screening?

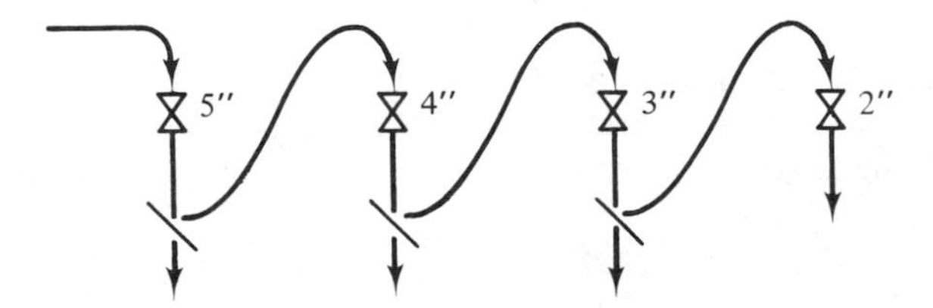

(f) Synthesize screening circuits that you anticipate to be of lower cost and attempt to make general rules on the synthesis of crushing and screening circuits.

(g) Comment on the several commercial screen–crusher circuits. Why might each be of value? Remember that the ore is generally of distributed sizes, and that our model of crushing and screening operations is overly simplified.

6

Task Integration

He who does not economize will agonize.
Confucius

6.1 Introduction

The central features of the process exist in our mind at this point in process synthesis as interconnected sequences of chemical-reaction tasks and separation tasks specially selected to lead toward economy of processing. This is the skelton of the process, to which we fasten the auxiliary processing operations, forming the completed process flow sheet.

The auxiliary operations are heating, cooling, compressing, expanding, crushing, pelletizing, and other operations needed to get the material into the special conditions required for the central tasks of chemical reaction and separation. These auxiliary tasks are necessary and are often quite expensive, the costs of which must be subtracted from the gross profit of the process. However, quite a bit can be done to reduce these costs by the careful integration of process operations, thus reducing the amount of energy that must be purchased to run the process.

To a large extent, many of the principles developed in earlier chapters lead directly to energy economy. For example, the reduction in solvent consumption accomplished by countercurrent extraction leads directly to a reduction in the energy consumed in solvent recovery. In a second example, the distillation heuristics, which recommend the early removal of low-boiling and plentiful species, lead to a lower consumption of less costly tower-top cooling and reboiler heating. Even the heuristic to avoid introducing foreign species to drive a separation is partially based on the energy saved by not having to remove the species after its purpose is served. So, by the time we reach the point where task integration can be examined in detail, the stage has already been carefully set and a large part of the economy of operation has been realized. Now we need to work on the final stages.

In Section 6.2 we outline the auxiliary process operations of major importance and show how the engineer seeks economy by the integration of their energy use, the waste energy from one task being used to drive other tasks. Then, in the bulk of the chapter, attention is focused on the principles of heat energy management. Central to the understanding of energy management is the principle of conservation of energy, a method of engineering analysis introduced in Section 6.3 and combined with the synthesis of economic operations in following sections.

6.2 INTRODUCTION OF AUXILIARY OPERATIONS

Chemical reactions and material separations often require unusual preparation of the material prior to the operation. We think of the chemistry and separations as the basic process operations and the other preparatory operations as auxiliary. The auxiliary operations of heating, cooling, compressing, expanding, crushing, agglomeration, and so forth serve the basic operations about which the process is conceived.

For example, in the preparation of taconite ore for use in steelmaking, the magnetite (Fe_3O_4) is separated from the predominate silica waste by magnetic separation. However, this separation is not possible until the mechanical bonds that hold the magnetite and silica crystal together are broken by crushing and grinding. Once freed of the silica, the magnetite is too fine for use in a blast furnace and must be pelletized to be suitable for marketing. In this example a size-reducing and a size-increasing operation serve to support the separation operation and to meet product-quality specifications. In every process flow sheet these auxiliary operations serve an important and often critical role.

Figure 6.2–1 shows a sequence of tasks one might synthesize to be the basis of a commercial process. We now illustrate how this plan might be modified as we seek economy by specifying the process equipment to be used. Stream *A* is to undergo a pressure change and be heated prior to mixing with a heated stream *B*. Two heaters would be required to implement this plan directly. However, suppose that the streams are mixed prior to the temperature change task; then one large and less expensive heater could be used rather than two small heaters. Enabling task consolidations of this type by changing the point of mixing is a common technique in task integration. Furthermore, the reacting, heating, and mixing operations may be consolidated into one piece of equipment, a steam-heated stirred tank reactor. Thus, in this example,

Original Task Sequence

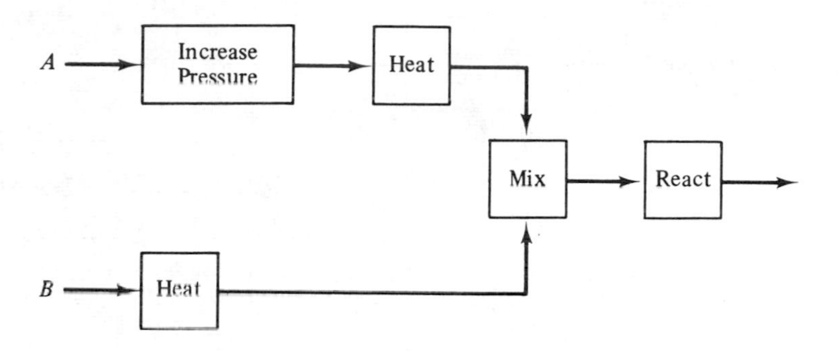

Simpler but Equivalent Task Sequence

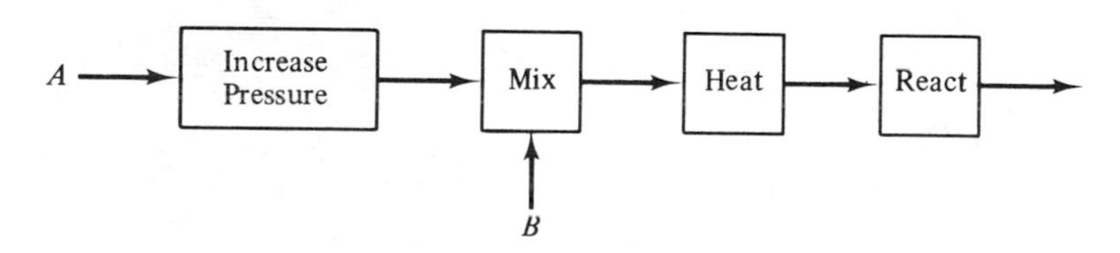

Equipment Selection

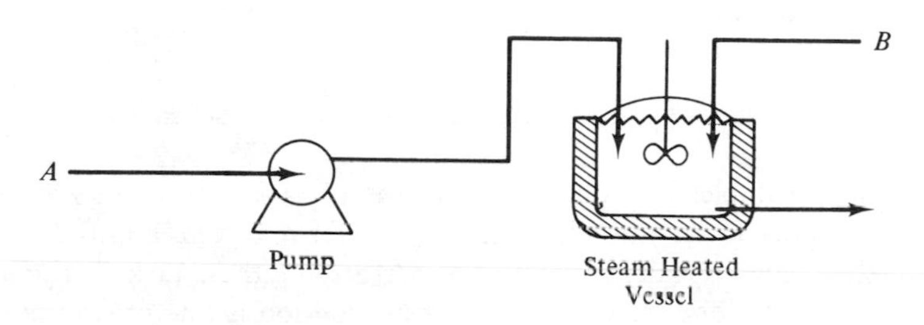

Figure 6.2–1. Parallel Flow Task Integration and Equipment Selection

the original sequence involving five tasks takes the form of two pieces of equipment, a pump and a reactor.

To illustrate the concepts of task combination and equipment selection, consider the sequence shown in Figure 6.2–2. Prior to mixing for reaction, vapor A is to be depressurized and vapor B pressurized. Perhaps the work obtained by the expansion of vapor A can be used to drive the pressurization of vapor B, thereby eliminating the need of an outside source of energy. An ejector is a pressurization device driven by a higher-pressure high-velocity fluid. In this case, a single ejector can perform four tasks: the energy integration, the depressurization of vapor A, the pressurization of vapor B, and the mixing of A and B.

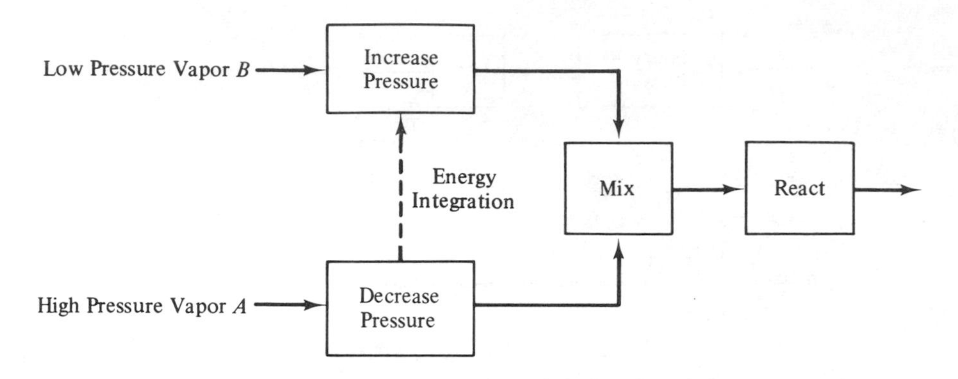

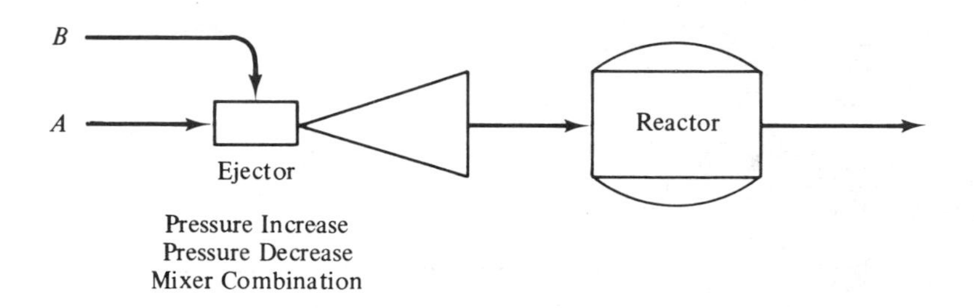

Figure 6.2–2. Task Integration and Equipment Selection

In another example, consider the sequences of tasks shown in Figure 6.2–3 in which wet solids are to be ground, heated, the water removed, and the solids settled out as a finely divided powder. One way of performing these tasks is by the use of a hot air mill and a cyclone separator. This consolidation is one of several that could be proposed.

Finally, the operational aspects of task integration can be critical, especially in situations involving energy integration. If integration is performed with only the steady operation of the process in mind, unnecessary problems may arise during start-up and shut-down. In Figure 6.2–4 is the task sequence for the manufacture of vinyl chloride by the exothermic reaction of acetylene and hydrogen chloride, a reaction that does not occur sufficiently below 180°C. Hence, the need to preheat the feed to the reactor. Once the reaction starts, heat is released and the effluents leave at 250°C and must be cooled to 50°C. These heating and cooling tasks might be integrated by a single heat exchanger. However, an additional outside heating source is needed during start-up to get the reactor feed above 180°C so that the hot reaction products will be generated to continue the feed heating. This spare heater would be used only during process start-up.

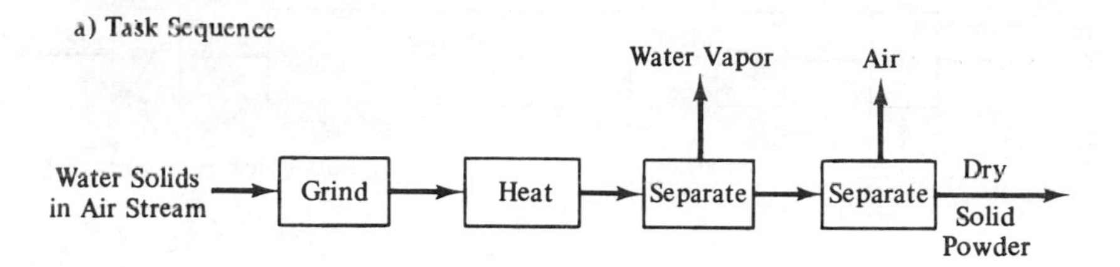

b) Equipment Selection

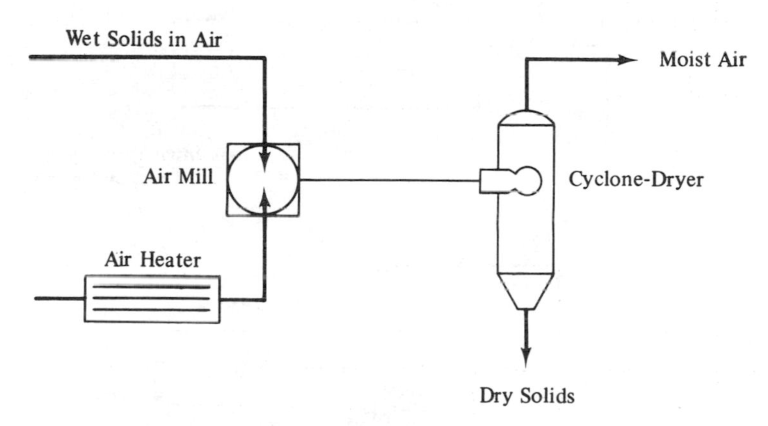

Figure 6.2–3. Task Consolidation in Powder Production

Example 6.2–1 Caustic Route to Vinyl Chloride *Figure 6.2–5 shows a task sequence for the performance of the reactions and separations to produce vinyl chloride from dichloroethane and sodium hydroxide:*

$$\underset{\text{(dichloroethane)}}{C_2H_4Cl_2} + \underset{\text{(caustic)}}{NaOH} \rightarrow \underset{\text{(vinyl chloride)}}{C_2H_3Cl} + \underset{\text{(water)}}{H_2O} + \underset{\text{(salt)}}{NaCl}$$

This is a reaction between two liquids, which produces a vapor and liquid product plus salts that dissolve in the liquid. The organic and the aqueous phases are only slightly miscible. The reaction is to be performed at a pressure of 4 atm and at 80°C.

The liquid–liquid reaction, the vapor–liquid separation, and the liquid–liquid separation are central to the task sequence. The reaction is best performed in a stirred tank, the vapor–liquid separation would occur satisfactorily in a stirred tank, and the liquid–liquid separation is handled best by a decanter. The neighboring mixing and heating tasks are capable of being performed in the stirred reaction vessel. The pressure change cannot be accomplished by the stirred reactor; hence we seek equipment suited for increasing the pressure of liquids, a pump. A pump also does a fairly good job of mixing liquids; hence this task may be consolidated into the pumping operation. However, for operational convenience, separate pumps for the caustic feed and the

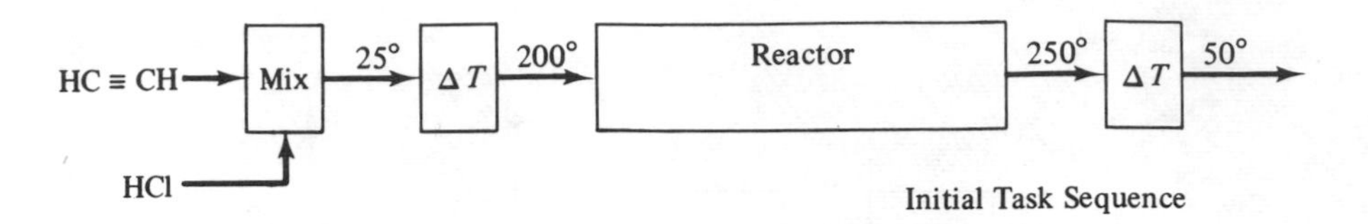

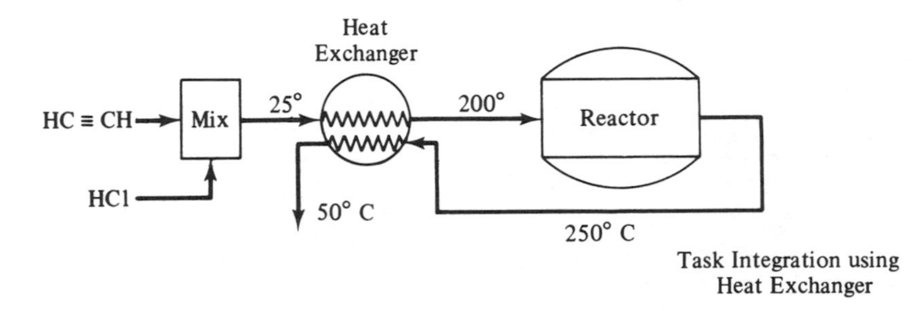

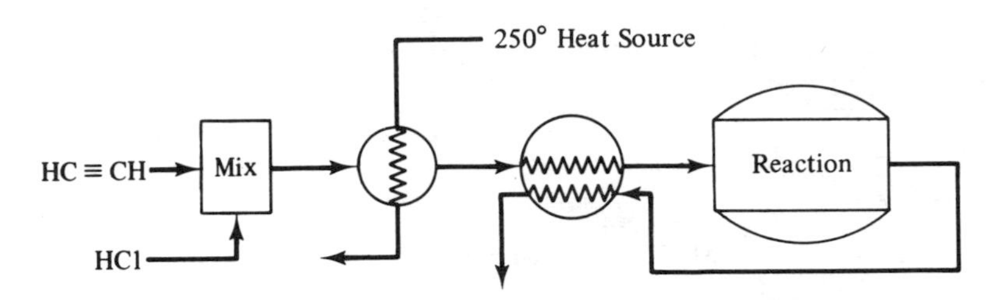

Figure 6.2–4. Vinyl Chloride from Acetylene

dichloroethane feed are selected, and the feed mixing task is consolidated into the stirred reactor.

The equipment flow sheet in Figure 6.2–6 shows the results of these consolidations. Two pumps feed the reactants into a single jacketed and baffled stirred tank reactor. A heating fluid is circulated through the jacket to perform the heating task. The vinyl chloride vapor is taken off the top of the reactor where it naturally gravitates. The liquid–liquid reaction takes place in the highly stirred region of the vessel, and in the baffled area the organic and liquid phases separate. The aqueous phase is decanted out of this baffled area. The recycle of the unreacted dichloroethane is accomplished by not allowing the organic phase to leave the reactor.

Such multipurpose reactors are quite common in industrial processing.

The developments of this section are not satisfactory, for they deal only with the qualitative aspects of task integration. To be of use, these qualitative aspects must

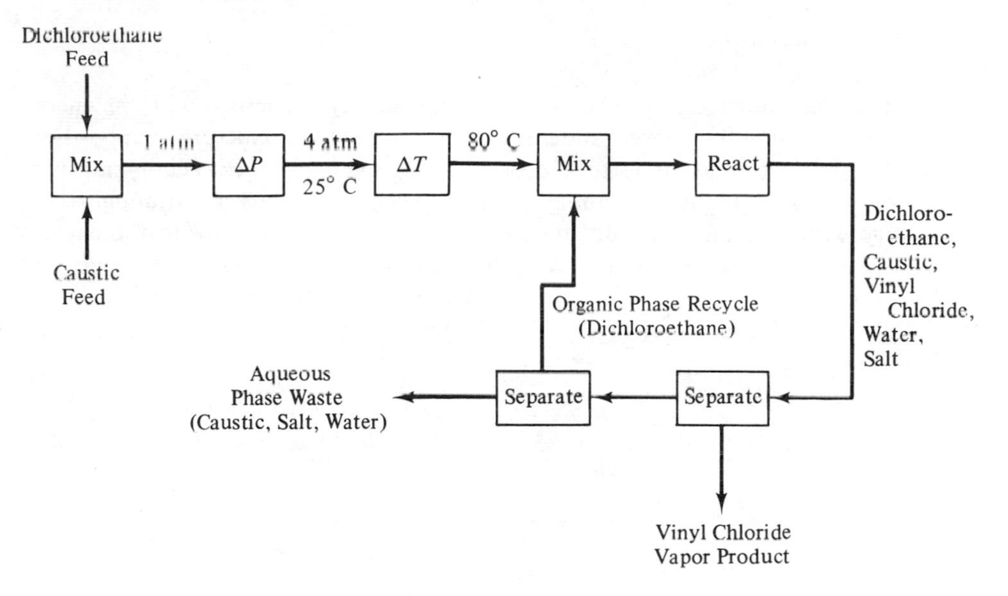

Figure 6.2–5. Task Sequence and Consolidation for Vinyl Chloride Production

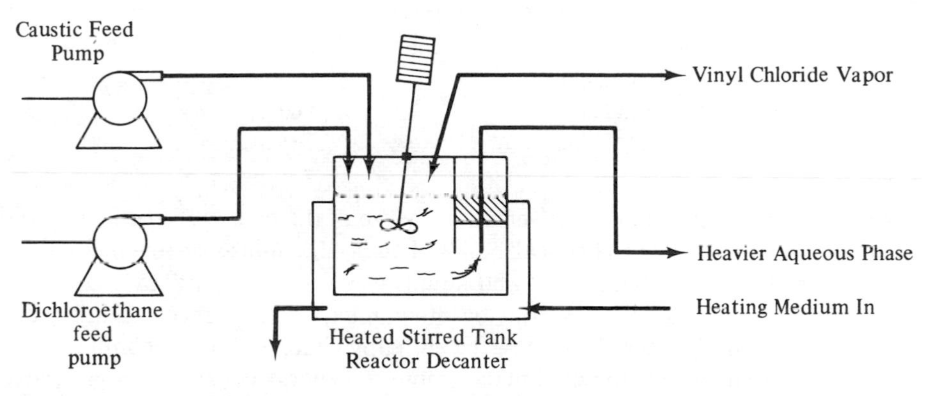

Figure 6.2–6. Equipment Flow Sheet for Figure 6.2–5

be combined with the quantitative techniques of energy balancing and with an understanding of the detailed performance characteristics of process equipment. In the remainder of this chapter we develop an understanding of energy balancing. However, we must defer to later study the vast field of the quantitative behavior of process machinery, often studied under the title of *unit operations*.

6.3 ENERGY BALANCE

In this stage of process engineering we are primarily concerned with the energy involved in changing the internal state of matter, its chemical structure, composition, phase, temperature, and so forth. To accomplish these changes, energy must be added or removed, at the substantial costs shown in Table 6.3–1. Although these costs vary with time and location, they serve to provide perspective to the energy-management problems faced in this chapter. Coal may cost \$4/ton at the mine, and \$10/ton far distant from the mine. Natural gas in the Southwest United States may cost only \$0.20/1000 ft^3, but in the New England states the cost may be as high as \$1.00/1000 ft^3.

TABLE 6.3–1 Typical Energy Costs for Heating and Cooling

	\$/MILLION BTU
Heating	
Natural gas	0.65
Coal and bunker C fuel oil	0.32
Electricity	2.70
Cooling	
Cooling water 50°F	1.50
Refrigeration 34°F	1.80
Refrigeration 0°F	3.00
Refrigeration −15°F	4.00

The *enthalpy*, H Btu/lb, is a convenient measure of the energy stored in material. This energy is stored by the chemical bonds that hold a molecule together, by the intermolecular forces that hold solids and liquids as a cohesive mass, by the random motion of the molecules of a gas, and by other internal phenomena. During processing, as the form of material changes, the enthalpy changes, and by computing the enthalpy change we are able to estimate the amount of energy consumed or generated.

In a simple batch operation in which M pounds of material is being changed from state 1 to state 2, the difference between the heat energy added to the material, Q Btu, and the mechanical work done by the material, W Btu, is given by the energy balance:

$$Q - W = M(H_2 - H_1) \qquad \text{Btu}$$

where H_2 and H_1 are the enthalpies of the material in Btu per pound in states 2 and 1.

In continuous operation in the steady state, with a variety of input streams M_i^{in} lb/h and output streams M_j^{out} lb/h, the energy balance for any region in the process takes the form

$$Q - W = \sum_j M_j^{out} H_j - \sum_i M_i^{in} H_i \qquad \text{Btu/h}$$

where

Q Btu/h = rate of heat energy addition to the region

W Btu/h = rate of work done by the region

$H_{i \text{ or } j}$ Btu/lb = enthalpy of material entering or leaving the region

M lb/h = flow rates of all streams entering or leaving the region

Of course, the steady state mass balance must also apply:

$$\sum_j M_j^{out} = \sum_i M_i^{in}$$

In these balances the heat energy Q is that energy transferred into the region by virtue of a temperature difference, and should the numerical value of Q be negative, the heat energy is being transferred out of the region of interest. The work term W is the mechanical work extracted from the material by the shaft of a turbine or the movement of a piston.

We see that an understanding of the energy effects in processing requires access to the enthalpy changes which occur as a result of the processing. Next, skill is developed in estimating the enthalpy changes involved in heating and cooling, in changes of phase, and in chemical reaction.

6.4 SENSIBLE HEAT EFFECTS

Changes in temperature can be sensed, say, by the touch of the hand, and for this reason the energy involved in temperature changes has come to be called *sensible heat.* The change in enthalpy with temperature is called the heat capacity at constant pressure, $C_p = \Delta H/\Delta T$. The amount of heat energy that must be added to change the temperature of material from T_1 to T_2 is

$$Q = M(H_2 - H_1) = MC_p(T_2 - T_1)$$

In this equation we presume that only the temperature of the material changes and not its pressure, phase, chemical composition, and so forth.

Figure 6.4–1 shows the heat capacities of some common gases and liquids, and Table 6.4–1 contains similar data for solids. Notice that the heat capacities are

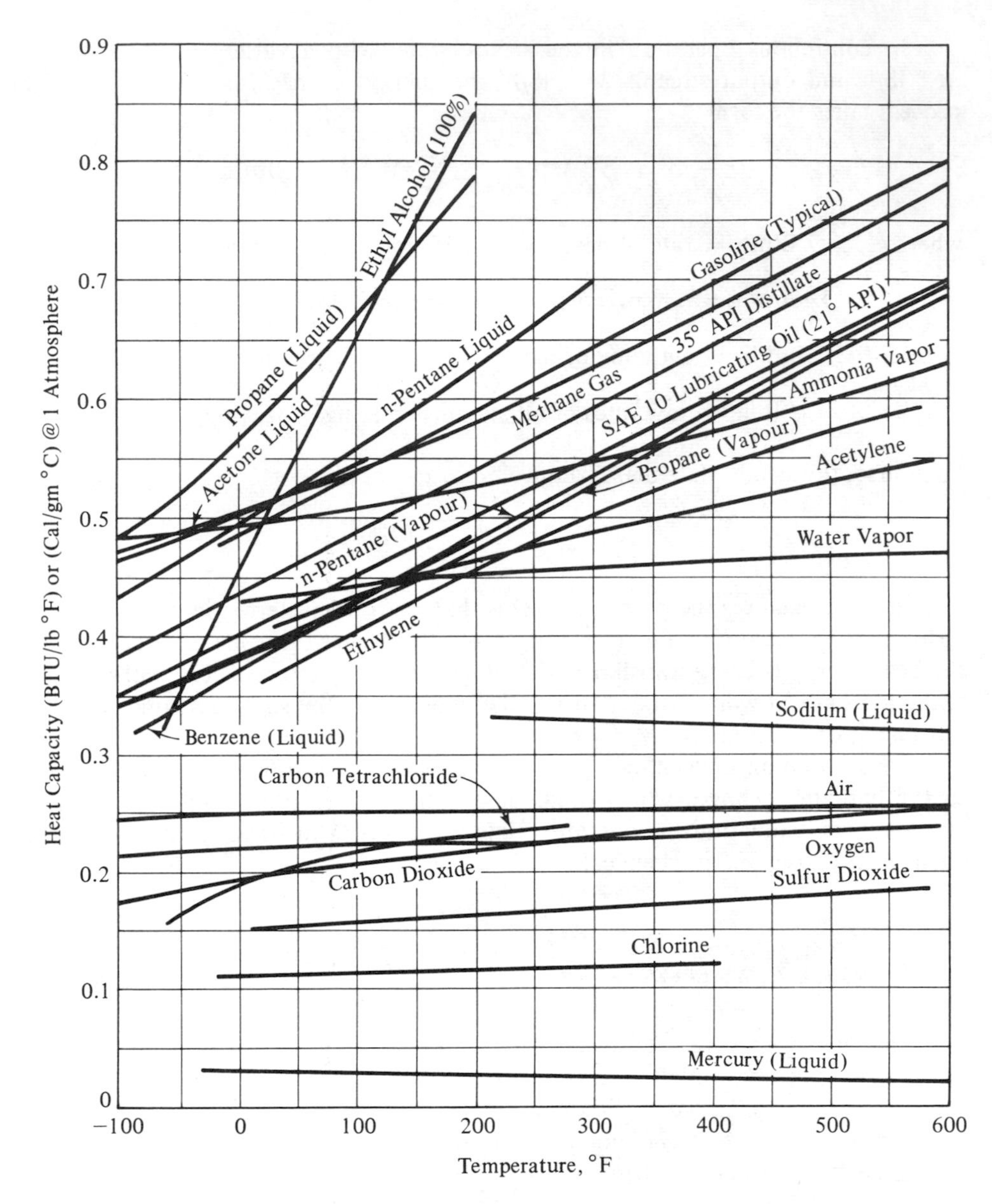

Figure 6.4–1. Heat Capacity of Various Gases and Liquids

TABLE 6.4–1 Heat Capacity of Various Solids

METALS	HEAT CAPACITY (BTU/LB °F or CAL/G °C)
Aluminum	0.374 (0°C), 0.405 (100°C)
Copper	0.164 (0°C), 0.169 (100°C)
Iron, cast	0.214 (20–100°C)
Lead	0.0535 (0°C), 0.0575 (100°C)
Nickel	0.0186 (0°C), 0.206 (100°C)
Steel	0.12
Tin	0.096 (0°C), 0.104 (100°C)
Zinc	0.164 (0°C), 0.172 (100°C)

MISCELLANEOUS	HEAT CAPACITY (BTU/LB °F OR CAL/G °C)
Alumina	0.2 (100°C), 0.274 (1500°C)
Asbestos	0.25
Brickwork	About 0.2
Carbon (mean values)	0.168 (26–76°C)
	0.314 (40–892°C)
	0.387 (56–1450°C)
Cellulose	0.32
Cement, portland clinker	0.186
Charcoal (wood)	0.242
Chrome brick	0.17
Clay	0.224
Coal	0.26 to 0.37
Coke (mean values)	0.265 (21–400°C)
	0.359 (21–800°C)
	0.403 (21–1300°C)
Concrete	0.156 (70–312°F), 0.219 (72–1472°F)
Fireclay brick	0.198 (100°C), 0.298 (1500°C)
Fluorspar	0.21 (30°C)
Glass (crown)	0.16 to 0.20
(flint)	0.117
(Pyrex)	0.20
(silicate)	0.188 to 0.204 (0–100°C)
	0.24 to 0.26 (0–700°C)
wool	0.157
Graphite	0.165 (26–76°C), 0.390 (56–1450°C)
Gypsum	0.259 (16–46°C)
Limestone	0.217
Magnesia	0.234 (100°C), 0.188 (1500°C)
Magnesite brick	0.222 (100°C), 0.195 (1500°C)
Marble	0.21 (18°C)
Quartz	0.17 (0°C), 0.28 (350°C)
Sand	0.191
Stone	About 0.2
Wood (oak)	0.570
Most woods vary between	0.45 and 0.65

temperature dependent, and that a suitable average value should be used when computing sensible heat effects for large temperature changes.

Example 6.4–1 Heating Iron to Its Melting Point *How costly would it be to heat 1 ton of cast iron from room temperature to its melting point of 1275°C? The heat capacity of cast iron, Table 6.4–1, is $C_p = 0.2$ Btu/lb °F. The temperature change envisioned is $\Delta T = (1275 - 25) = 1250°C = 2250°F$. The amount of iron $M =$ 1 ton = 2000 lb. Thus the total energy required is*

$$Q = M\,\Delta H = (2000 \text{ lb})(0.2 \text{ Btu/lb °F})(2250°\text{F}) = 9.0 \times 10^5 \text{ Btu}$$

The cheapest forms of energy cost about $0.30/1 million Btu. We calculate the cost of heating 1 ton of cast iron to its melting point as $[9 \times 10^{+5}(\text{Btu/ton})](\$0.30 \times 10^{-6}\text{Btu}) = \$0.27/\text{ton}$. *Of course, in such high-temperature operations much energy will be lost to the surroundings and the actual fuel consumption will be higher.*

Example 6.4–2 Sterilizing City Water *It has been proposed to decant the clear water from a river and to sterilize it by heating to near the boiling point. The water might then be useful to supplement a water supply. Municipal water costs less than $0.50/1000 gallons. Would the water formed from natural-gas-fired sterilizers be economically competitive?*

One thousand gallons of water weighs on the order of (1000 gallons)(8 lb/gallon) = 8000 lb, the temperature change envisioned is $(212 - 60) \cong 150°F$, *the heat capacity of water is 1 Btu/lb °F, and the cost of natural-gas heating is $0.65/1 million Btu. Thus the sterilizing cost in a simple heater is, per 1000 gallons of water,*

$$\left(\frac{8000 \text{ lb}}{1000 \text{ gal}}\right)(1 \text{ Btu/lb °F})(150°\text{F})(\$0.65/10^6 \text{ Btu}) = \$0.78/1000 \text{ gal}$$

This primitive form of sterilization is too expensive to be competitive with the municipal water available at less than $0.50/1000 gal. However, proper energy management can bring this cost down. In practice, chemical treatment of drinking water, say, by chlorine addition, is less costly and has a valuable residual action not present in heat treatment.

6.5 ENERGY OF PHASE CHANGE

The sensible heat effects include only those caused by the change in temperature of material, and do not account for changes in phase or chemical composition. Table 6.5–1 includes data on the heat effects involved in the melting and vaporization of a variety of materials. By definition,

$$\Delta H_{\text{vaporization}} = H_{\text{vapor}} - H_{\text{liquid}} \quad \text{at boiling temperature}$$

$$\Delta H_{\text{fusion}} = H_{\text{liquid}} - H_{\text{solid}} \quad \text{at melting temperature}$$

These are specific heats of phase change, in the units of kilocalories per gram-mole.

TABLE 6.5–1 Heats of Fusion Vaporization and Formation: Several Common Materials (Multiply by 1800 to Get Btu/lb-mole)

Compound	Formula	Formula Weight	Melting Temp (°K)	ΔH Fusion (kcal/g-mole)	Normal Bp (°K)	ΔH Vap at Bp (kcal/g-mole)	$-\Delta H_{formation}$ at 25°C (kcal/g-mole)	State
Acetic acid	CH_3CHO	60.05	328.9	2.89	391.4	5.83	97.8	l
Acetylene	C_2H_2	26.04	191.7	0.9	191.7	4.2	−54.2	g
Ammonia	NH_3	17.03	195.40	1.351	239.73	5.581	16.1	l
Ammonium nitrate	NH_4NO_3	80.05	442.8	1.3	(decomposes at 483°K)		87.3	s
Benzene	C_6H_6	78.11	278.693	2.351	353.26	7.353	−11.6	l
Boron oxide	B_2O_3	69.64	723	5.27	–	–	302.0	s
Butane	nC_4H_{10}	58.12	134.83	1.114	272.66	5.331	35.3	l
iso-Butane	iso-C_4H_{10}	58.12	113.56	1.085	261.43	5.089	37.9	l
i-Butene	C_4H_8	56.10	87.81	0.9197	266.91	5.238	−0.3	g
Calcium carbonate	$CaCO_3$	100	(decomposes at 1100°K)				288.5	s
Calcium chloride	$CaCl_2$	110.99	1055	6.78	–	–	190.0	s
Calcium oxide	CaO	56.08	2873	12	3123		151.9	s
Calcium phosphate	$Ca_3(PO_4)_2$	310.19	1943	–	–	–	988.9	s
Calcium silicate	$CaSiO_2$	116.17	1803	11.62	–	–	378.6	s
Carbon	C	12.010	3873	11.0	4473	–	0	s
Carbon dioxide	CO_2	44.01	$217.0^{0.2\ atm}$	1.99	(sublimes at 195°K)		94.0	g
Carbon disulfide	CS_2	76.14	161.1	1.05	319.41	6.40	−21.0	l
Carbon monoxide	CO	28.01	68.10	0.200	81.66	1.444	26.4	g
Carbon tetrachloride	CCl_4	153.84	250.3	0.60	349.9	7.17	33.3	l
Chlorine	Cl_2	70.91	172.16	1.531	239.10	4.878	0	g
Copper	Cu	63.54	1356.2	3.11	2855	72.8	0	s
Cumene	C_9H_{12}	120.19	177.125	1.7	425.56	8.97	9.8	l
Cyclohexane	C_6H_{12}	84.16	279.83	0.6398	353.90	7.19	34.3	l
Cyclopentane	C_5H_{10}	70.13	179.71	0.1455	322.42	6.524	25.3	l
Diethyl ether	$(C_2H_5)_2O$	74.12	156.86	1.745	307.76	6.226	–	–
Ethane	C_2H_6	30.07	89.89	0.6834	184.53	3.517	20.2	g
Ethanol	C_2H_6O	46.07	158.6	1.200	351.7	9.22	66.4	l
Ethylbenzene	C_8H_{10}	106.16	178.185	2.190	409.35	8.60	3.0	l
Ethyl chloride	CH_3CH_2Cl	64.52	134.83	1.064	285.43	5.9	−25.1	g
Ethylene	C_2H_4	28.05	103.97	0.8008	169.45	3.237	−12.5	g
Ethylene glycol	$C_2H_6O_2$	62.07	260	2.685	470.4	13.6	–	–
Ferrous sulfide	FeS	87.92	1466		(decomposes)		22.5	s
Formic acid	CH_2O_2	46.03	281.46	3.03	373.7	5.32	–	–
Glycerol	$C_3H_8O_3$	92.09	291.36	4.373	563.2	–	–	–
Helium	He	4.00	3.5	0.005	4.216	0.020	0	g
Heptane	C_7H_{16}	100.20	182.57	3.354	371.59	7.575	53.6	l
Hexane	C_6H_{14}	86.17	177.84	3.114	341.90	6.896	47.5	l
Hydrogen	H_2	2.016	13.96	0.028	20.39	0.216	0	–
Hydrogen chloride	HCl	36.47	158.94	0.476	188.11	3.86	22.0	g
Hydrogen sulfide	H_2S	34.08	187.63	0.568	212.82	4.463	−1.8	g
Iron	Fe	55.85	1808	3.6	3073	84.6	0	s

TABLE 6.5–1 (*Cont.*)

Compound	Formula	Formula Weight	Melting Temp (°K)	ΔH Fusion (kcal/g-mole)	Normal Bp (°K)	ΔH Vap at Bp (kcal/g-mole)	$-\Delta H_{formation}$ at 25°C (kcal/g-mole)	State
Iron oxide	Fe_3O_4	231.55	1867	33.0	(decomposes at 1867°K)		196.5	s
Lead	Pb	207.21	600.6	1.22	2023	43.0	0	s
Lead oxide	PbO	223.21	1159	2.8	1745	51	52.4	s
Magnesium	Mg	24.32	923	2.2	1393	31.5	0	s
Magnesium chloride	$MgCl_2$	95.23	987	10.3	1691	32.7	153.4	s
Magnesium oxide	MgO	40.32	3173	18.5	3873	–	143.8	s
Methane	CH_4	16.04	90.68	0.225	111.67	1.955	−17.9	g
Methanol	CH_3OH	32.04	175.26	0.757	337.9	8.43	57.0	l
Methyl cyclohexane	C_7H_{14}	98.18	146.58	1.6134	374.10	7.58	45.5	l
Naphthalene	$C_{10}H_8$	128.16	353.2	–	491.0	–	–	–
Nitric acid	HNO_3	63.02	231.56	2.503	359	7.241	41.4	l
Nitrogen	N_2	28.02	63.15	0.172	77.34	1.333	0	g
Nitrogen dioxide	NO_2	46.01	263.86	1.753	294.46	3.520	−8.1	g
Nitrogen oxide	NO	30.01	109.51	0.550	121.39	3.293	−21.6	g
Oxygen	O_2	32.00	54.40	0.106	90.19	1.630	0	g
n-Pentane	C_5H_{12}	72.15	143.49	2.006	309.23	6.160	41.3	l
l-Pentane	C_5H_{10}	70.13	107.96	1.180	303.13	–	–	–
Phenol	C_6H_5OH	94.11	315.66	2.732	454.56	–	–	–
Phenylhydrazine	$C_6H_8N_2$	108.14	292.76	3.927	516.66	–	–	–
Phosphoric acid	H_3PO_4	98.00	315.51	2.52	($-\frac{1}{2}$ H_2O at 486°K)		306.2	s
Phosphorus (red)	P_4	123.90	863	19.40	863	10.00	0	s
Phosphorus (white)	P_4	123.90	317.4	0.60	553	11.88	–	–
Propane	C_3H_8	44.09	85.47	0.8422	231.09	4.487	28.6	l
Propene	C_3H_6	42.08	87.91	0.7176	225.46	4.402	−4.9	g
n-Propylbenzene	C_9H_{12}	120.19	173.660	2.04	432.38	9.14	9.2	l
Silicon dioxide	SiO_2	60.09	1883	2.04	2503	–	203.9	s
Sodium chloride	$NaCl$	58.45	1081	6.8	1738	40.8	98.2	s
Sodium cyanide	$NaCN$	49.01	835	4.0	1770	37	21.4	s
Sodium nitrate	$NaNO_3$	85.00	583	3.8	(decomposes at 653°K)		111.5	s
Sodium sulfate	Na_2SO_4	142.05	1163	5.8	–	–	330.9	s
Sodium sulfide	Na_2S	78.05	1223	1.6	–	–	89.2	s
Sulfur (rhombic)	S_8	256.53	386	2.40	717.76	20.0	0	s
Sulfur dioxide	SO_2	64.07	197.68	1.769	263.14	5.955	71.0	g
Sulfur trioxide	SO_3	80.07	290.0	6.09	316.5	9.99	94.4	g
Sulfuric acid	H_2SO_4	98.08	283.51	2.36	(decomposes)		193.9	l
Toluene	$C_6H_5CH_3$	92.13	178.169	1.582	383.78	8.00	−2.9	l
Water	H_2O	18.016	273.16	1.4363	373.16	9.7171	68.3	l
m-Xylene	C_8H_{10}	106.16	225.288	2.765	412.26	8.70	6.1	l
o-Xylene	C_8H_{10}	106.16	247.978	3.250	417.58	8.80	5.8	l
p-Xylene	C_8H_{10}	106.16	286.423	4.090	411.51	8.62	5.8	l
Zinc	Zn	65.38	692.7	1.595	1180	27.43	0	s
Zinc sulfate	$ZnSO_4$	161.44	(decomposes at 1013°K)			–	233.9	s

To compute enthalpy changes that involve both sensible heat effects and phase change effects, we merely trace the thermal path the material might experience. For example, the change in enthalpy of a *solid* at temperature T_1 as it is changed to a *vapor* at temperature T_2 is computed as

$$H_2 - H_1 = \underbrace{C_{p_{\text{solid}}}(T_{\text{melting}} - T_1)}_{\text{heating solid}} + \underbrace{\Delta H_{\text{fusion}}}_{\substack{\text{melting}\\ \text{solid}}} + \underbrace{C_{p_{\text{liquid}}}(T_{\text{boiling}} - T_{\text{melting}})}_{\text{heating liquid}}$$

$$+ \underbrace{\Delta H_{\text{vaporization}}}_{\text{vaporizing liquid}} + \underbrace{C_{p_{\text{vapor}}}(T_2 - T_{\text{boiling}})}_{\text{heating vapor}}$$

Example 6.5–1 Melting Cast Iron *In Example 6.4–1 we calculated the cost of \$0.27 to provide the nearly 1 million Btu required to heat 1 ton of cast iron from room temperature to its melting point. Now we estimate the cost of melting the iron.*

Table 6.5–1 gives a heat of fusion for iron as $\Delta H_{\text{fusion}} = 3.6$ *kcal/g-mole. Converting this to Btu per pound gives*

$$(3.6 \text{ kcal/g-mole})\left(1800 \frac{\text{Btu/lb-mole}}{\text{kcal/g-mole}}\right)\left(\frac{\text{lb-mole}}{56 \text{ lb iron}}\right) = 116 \text{ Btu/lb}$$

The heat energy that must be added to melt 1 ton of iron, already at its melting point, is

$$Q = M \, \Delta H_{\text{fusion}} = (2000 \text{ lb/ton})(116 \text{ Btu/lb}) = 2.32 \times 10^5 \text{ Btu/ton}$$

at a cost of

$$(2.32 \times 10^5 \text{ Btu/ton})(\$0.32/10^6 \text{ Btu}) = \$0.07/\text{ton}$$

Example 6.5–2 Fresh Water from the Sea *How costly is it to obtain fresh water merely by boiling seawater and collecting the salt-free vapors? The energy required to heat water from room temperature, say,* $T_1 = 60°F$*, to its boiling point,* $T_2 = 212°F$*, and to form the vapor is*

$$Q = \Delta H_{\text{vaporization}} + C_p(T_2 - T_1)$$

$$\cong 970 \text{ Btu/lb} + 1 \text{ Btu/lb °F } (212 - 60)$$

$$= 970 + 152 = 1122 \text{ Btu/lb}$$

at a cost of

$$(1152 \text{ Btu/lb})(\$0.32/10^6 \text{ Btu}) = \$0.0036/\text{lb}$$

Of course, we also ought to include the cost of heating the seawater, which leaves as concentrated brine. A reasonable cost for fresh water is \$0.50/1000 gallons. The cost of fresh water by simple boiling is

$$(\$0.0036/\text{lb})\left(\frac{8000 \text{ lb}}{1000 \text{ gal}}\right) = \$30/1000 \text{ gal}$$

This is 60 times too expensive. However, proper energy management, such as that illustrated later in this chapter and in Chapter 7, can bring this cost down to a competitive value.

Example 6.5–3 Ice Melting *During the production of fresh water from the sea by a freezing process, 1000 tons/h of a slush at 32°F, consisting of 60 per cent ice and 40 per cent water, is to be processed (see Fig. 6.5–1). This stream is mixed with 200 tons/h of steam at 1 atm pressure at 300°F. What is the condition of the final mixture?*

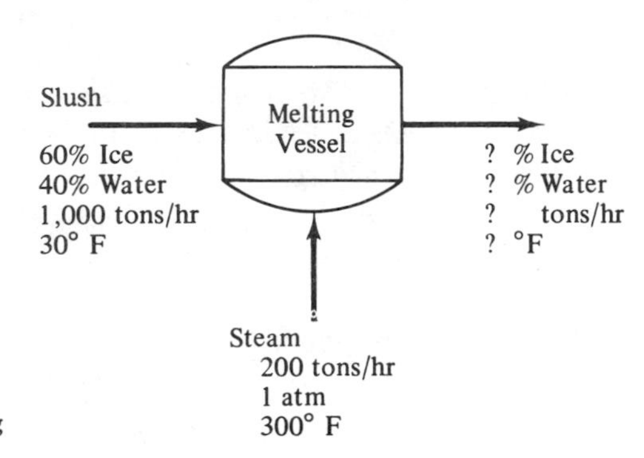

Figure 6.5–1. Slush Melting

Since $Q = W = 0$, the energy balance over the melter takes the form

$$m_{\text{slush}}H_{\text{slush}} + m_{\text{steam}}H_{\text{steam}} = m_{\text{out}}H_{\text{out}}$$

A mass balance gives

$$m_{\text{slush}} + m_{\text{steam}} = m_{\text{out}}$$

$$1000 + 200 = 1200 \text{ tons/h}$$

If we take water at 32°F as the enthalpy datum point (i.e., an arbitrary base where $H = 0$), the enthalpy of the slush is easily estimated:

$$H_{\text{slush}} = 0.60H_{\text{ice}} + 0.40H_{\text{water}}$$

The two right-hand terms are easily computed as

$$H_{\text{water}} = 0 \text{ at } 32°\text{F} \qquad \text{by definition of datum point}$$

$$H_{\text{ice}} = -\Delta H_{\text{freezing}} = -144 \text{ Btu/lb} = -288{,}000 \text{ Btu/ton}$$

Thus $$H_{\text{slush}} = -172{,}800 \text{ Btu/ton}$$

To estimate the enthalpy of the steam, we visualize it being heated from water at 32°F to steam at 300°F, and merely accumulate the energy input:

$$H_{steam} = \underbrace{C_{p_{water}}(212 - 32)}_{\text{heating of water}} + \underbrace{\Delta H_{vaporization}}_{\substack{\text{latent heat of}\\ \text{vaporization}}} + \underbrace{(300 - 212)C_{p_{vapor}}}_{\text{superheat of steam}}$$

$$= 1 \times (180) + 970 + (0.4) \times (88)$$

$$= 1200 \text{ Btu/lb} = 2{,}400{,}000 \text{ Btu/ton}$$

An energy balance over the melter, with $Q = W = 0$, is

$$m_{out}H_{out} = m_{slush}H_{slush} + m_{steam}H_{steam}$$

We can now estimate the enthalpy of the outlet stream as

$$H_{out} = \frac{m_{slush}H_{slush} + m_{steam}H_{steam}}{m_{out}}$$

$$= \frac{(1000)(-172{,}000) + (200)(2{,}400{,}000)}{1200}$$

$$= 250{,}000 \text{ Btu/ton}$$

The specific enthalpy is greater than zero (which is the enthalpy of water at 32°F). Hence the effluent is not a mixture of ice and water; it is water. Since it is water, the temperature is easily determined by use of the heat capacity data:

$$H_{out} = C_p(T_{out} - 32) = 2000(T - 32) \text{ Btu/ton}$$

From our previous calculation,

$$H_{out} - 250{,}000$$

Hence

$$T_{out} = 32 + \frac{165{,}000}{2000} = 32 + 125$$

$$= 157°\text{F}$$

By this enthalpy balance we estimate the output from the melter to be 1200 tons/h of water at 157°F.

Example 6.5–4 *Analysis of Washer–Melter Operation* *In Chapter 7 processes are developed to produce fresh water by freezing and melting seawater. Figure 6.5–2 is a possible design for the ice washing and melting operation, which we now analyze.*

An ice–brine slush from a freezing operation is to be washed with fresh water and drained. This cleaner slush is then melted to produce the freshwater product. We wish

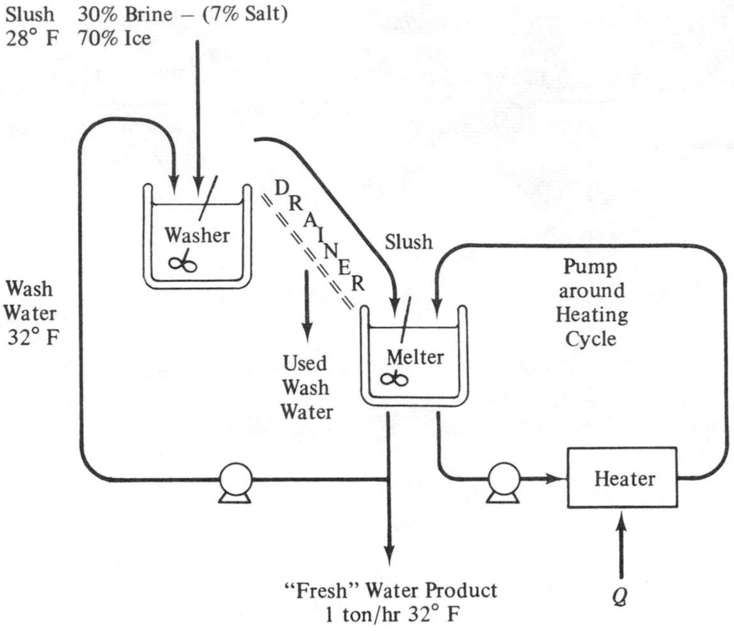

Figure 6.5–2. Slush Washing and Melting Portion of a Desalination Process

to determine the salinity of the freshwater product and the energy consumption as a function of the fraction f of the product water that is recycled. Suppose that 1 ton/h of fresh water is to leave the process.

The slush both entering the washer and entering the melter contains 30 per cent water and 70 per cent ice. The ice is salt free, but the water in the slush entering the washer is 7 per cent salt and entering the melter has a salt content determined by the amount of water recycled for the washing.

- ***Recycle Material Balance*** *If 1 ton/h of water is to leave the process, then* $f/(1-f)$ *tons/h must be recycled for washing, and* $1/(1-f)$ *tons/h must leave the melter.*
- ***Melter Material Balance*** *If* $1/(1-f)$ *tons/h is to leave the melter, then* $1/(1-f)$ *tons/h must enter the melter, 30 per cent of which is salty water and 70 per cent of which is salt-free ice.*
- ***Energy Balance over Melter*** *The heater provides the energy required to melt all the ice entering the melter. The melting of each pound of ice requires 144 Btu, and* $0.70/(1-f)$ *tons/h of ice enters the melter. Thus the required energy input into the heater is*

$$Q = (144\ \text{Btu/lb})(2000\ \text{lb/ton})\left(\frac{0.70}{1-f}\right)\ \text{tons/h}$$

$$= \frac{201{,}600}{1-f}\ \text{Btu/h}$$

• ***Material Balances over Washer*** *The ice passes through the washer unaltered. Hence the amount of ice that must enter the melter equals the amount of ice that must leave to enter the washer:* $0.70/(1 - f)$ *tons/h. Since the slush is 30 per cent salt solution, the amount of solution entering and leaving with the slush is* $0.30/(1 - f)$ *tons/h.*

We now set up a material balance on the salt. If s_1 *is the fraction of salt in the fresh water produced and* s_2 *the salinity of the used wash, a salt balance over the washer is of the following form:*

$$\underbrace{\frac{0.30}{1-f}0.07}_{\text{incoming slush}} + \underbrace{\frac{f}{1-f}s_1}_{\text{recycle wash}} = \underbrace{\frac{f+0.30}{1-f}s_2}_{\text{slush + used wash leaving}}$$

Now, when the slush leaving the drainer is melted, the salinity of the melt is determined by the salinity of the adhering brine as

$$s_1 = \frac{s_2[0.30/(1-f)]}{1/(1-f)} = s_2(0.30)$$

Solving these two equations for s_1*, the salinity of the fresh water formed, gives*

$$s_1 = \frac{0.021}{1 + f(7/3)}$$

Figure 6.5–3 shows the salinity of the product water and the energy consumption by the melter as a function of the fraction of fresh water recycled for washing. It is interesting to observe that this design is limited to the production of water with a salinity

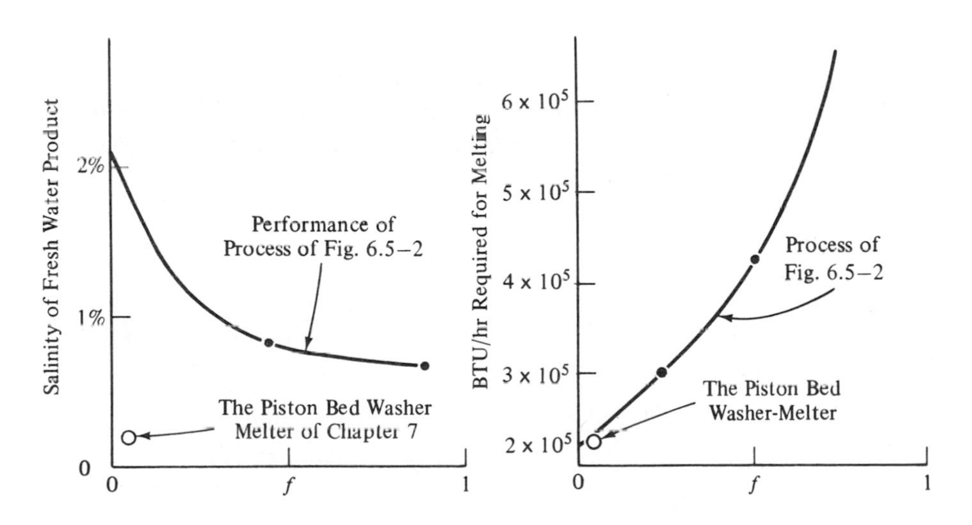

Figure 6.5–3. Predicted Performance of Process of Figure 6.5–2

greater than 0.6 per cent, and only at a great consumption of energy in the melting operation.

The piston bed washer–melter developed in Chapter 7 is a far better integration of the washing and melting tasks, leading to the almost complete washing of the ice with a 5 per cent recycle of the fresh water. From these calculations we conclude that this superior performance is not accessible to the task integration proposed in Figure 6.5–2.

6.6 HEATS OF CHEMICAL REACTION

Energy is consumed and released in great amounts by changes in chemical structure. For example, the chemical changes that occur when carbon, oxygen, and carbon dioxide undergo the reaction $C + O_2 \rightarrow CO_2$ release the heat generated during the burning of coal. We now develop skill in estimating the enthalpy changes resulting from such chemical reactions.

The *heat of reaction* ΔH_R is defined as the difference between the enthalpy of the products and the enthalpy of the reactants:

$$\Delta H_{\text{reaction}} = H_{\text{products}} - H_{\text{reactants}}$$

These data are reported at a standard temperature, usually 25°C, even though the reaction may not occur at the standard temperature.

For example, the combustion of sulfur, $S + O_2 \rightarrow SO_2$, has a heat of reaction $\Delta H_R = -70.9$ kcal/g-mole at 25°C, indicating that the SO_2 product has a lower enthalpy than the reactants S and O_2. The negative enthalpy difference indicates that heat is released by the reaction, an *exothermic reaction.*

The reaction $2\,C + H_2 \rightarrow C_2H_2$ has a heat of reaction $\Delta H_R = +54$ kcal/g-mole at 25°C, indicating that energy must be added to the reaction to raise the enthalpy of the products above those of the reactants, an *endothermic reaction.* Of course, the mere addition of energy will not ensure that the reaction occurs.

The number of chemical reactions that might occur among materials is enormously large, too large to tabulate the heat effects for all possible reactions. However, one can construct the heat of reaction for any particular reaction of interest from *heat of formation* data ΔH_f, such as that included in Table 6.5–1. The heat of formation ΔH_f is the enthalpy difference between a compound and the elements from which the compound is formed, with the data adjusted to the standard temperature.

For example, the heat of formation of methane (see Table 6.5–1) is -17.9 kcal/g-mole. This indicates that if it were possible to form methane, CH_4, from carbon, C, and hydrogen, H_2, at 25°C, the heat of formation would be

$$C + 2\,H_2 \rightarrow CH_4(g), \qquad \Delta H_f = -17.9 \text{ kcal/g-mole at 25°C}$$

As additional examples, the heat of formation of water vapor is -57.8 kcal/g-mole, and carbon dioxide gas is -94 kcal/g-mole:

$$H_2 + \tfrac{1}{2}\,O_2 \rightarrow H_2O(g), \qquad \Delta H_f = -57.8 \text{ kcal/g-mole at 25°C}$$

$$C + O_2 \rightarrow CO_2(g), \qquad \Delta H_f = -94 \text{ kcal/g-mole at 25°C}$$

Next we put these data to use. We can construct the heat of reaction for any given reaction merely by fitting together the reactions of formation and summing, algebraically, the heats of formation.

For example, the combustion of methane involves methane, oxygen, carbon dioxide, and water. What is the heat of reaction?

$$CH_4 + 2\,O_2 \rightarrow CO_2 + 2\,H_2O, \qquad \Delta H_R = ?$$

This reaction can be thought of as a composite of these individual reactions, which sum to the reaction above.

	$CH_4 \rightarrow C + 2\,H_2$	$\Delta H_R = -\Delta H_f =$	$+17.9$
	$C + O_2 \rightarrow CO_2$	$\Delta H_R = \Delta H_f =$	-94
by	$2\,H_2 + O_2 \rightarrow 2\,H_2O$	$\Delta H_R = 2\,\Delta H_f =$	-115.6
summing	$CH_4 + O_2 \rightarrow CO_2 + 2\,H_2O(g)$	$\Delta H_R =$	-191.7 kcal/g-mole

The net heat of reaction, obtained by summing the heats of reaction from the individual reactions that compose the main reaction, is $\Delta H_R = -191.7$ kcal/g-mole at 25°C.

Thus we have access to the heat effects for any possible chemical reaction by means of the heats of formation of the species that enter into the reaction. Generally, the reactions do not occur at 25°C, and temperature corrections must be made, as in Examples 6.6–1 and 6.6–2.

Example 6.6–1 Adiabatic Flame Temperature *It has been proposed to manufacture sulfur dioxide by the direct oxidation of sulfur with pure oxygen. However, cool sulfur dioxide must be recycled to the burner to lower the flame temperature to below 1000°F, so as not to damage the burner. This is shown in Figure 6.6–1. How much sulfur dioxide must be recycled?*

$$\underset{\substack{\text{gas}\\(100°\text{F})}}{O_2} + \underset{\substack{\text{liquid}\\(100°\text{F})}}{S} \rightarrow \underset{\substack{\text{gas}\\(100°\text{F})}}{SO_2}$$

$$\Delta H_R = -126{,}000 \text{ Btu/lb-mole at } 100°\text{F}$$

$$C_{p_{SO_2}} = 18 \text{ Btu/mole °F}$$

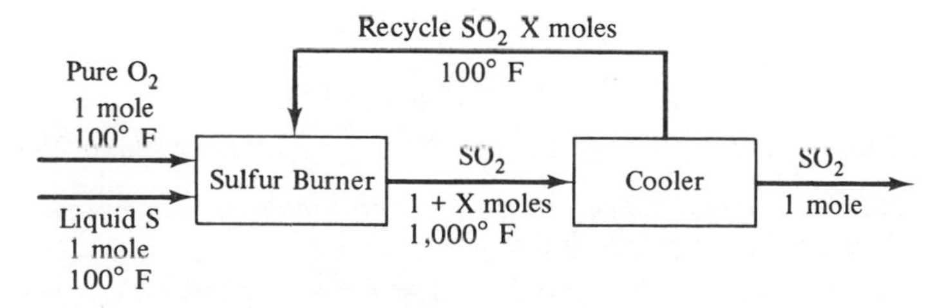

Figure 6.6–1. Use of SO_2 As Inert During Sulfur Oxidation with Pure Oxygen

An energy balance over the burner gives the following ($Q = W = 0$):

$$H_{O_2}(100°F) + H_S(100°F) + XH_{SO_2}(100°F) = (1 + X)H_{SO_2}(1000°F)$$

The definition of heat of reaction is

$$\Delta H_R = H_{SO_2}(100°F) - H_S(100°F) - H_{O_2}(100°F)$$

Thence

$$X[H_{SO_2}(1000°F) - H_{SO_2}(100°F)] = -\Delta H_R$$

$$XC_{pSO_2}(1000 - 100) = -\Delta H_R$$

$$X = -\frac{\Delta H_R}{C_{pSO_2}900} = -\frac{-126{,}000}{(18)(900)} = 7.8$$

Thus the pure oxygen should be diluted by about eight times with the SO_2 recycle to ensure a low-enough flame temperature.

Example 6.6–2 A Primitive Limestone Kiln *How costly would it be to heat limestone by the combustion of natural gas to initiate the reaction that produces quicklime and carbon dioxide?*

$$\underset{\text{(limestone)}}{CaCO_3} \xrightarrow{1000°C} \underset{\text{(quicklime)}}{CaO} + \underset{\text{(carbon dioxide)}}{CO_2}$$

Clearly, two heat effects will enter our calculation, the sensible heat to raise the temperature of the limestone to 1000°C so the equilibrium will shift toward the products, and the heat of the chemical reaction which then occurs.

Table 6.5–1 contains the necessary heat of formation data to construct the heat of reaction:

$$Ca + C + \tfrac{3}{2}O_2 \rightarrow CaCO_3, \qquad \Delta H_f = -288 \text{ kcal/g-mole at } 25°C$$

$$Ca + \tfrac{1}{2}O_2 \rightarrow CaO \qquad \Delta H_f = -152 \text{ kcal/g-mole at } 25°C$$

$$C + O_2 \rightarrow CO_2 \qquad \Delta H_f = -94 \text{ kcal/g-mole at } 25°C$$

If the first reaction is subtracted from the second two reactions, the desired reaction is constructed:

$$\begin{aligned} & -(-288) \\ & +(-152) \\ & +(\ -94) \\ CaCO_3 \rightarrow CaO + CO_2 \quad \Delta H_R = & + \quad 42 \text{ kcal/g-mole at } 25°C \end{aligned}$$

The positive heat of reaction indicates that 42 kcal/g-mole of heat must be supplied to change the limestone into slaked lime and carbon dioxide.

We now encounter a common problem. The heat of reaction data were computed at the standard conditions of 25°C. But the reaction occurs at 1000°C. These differences can be rectified using an energy balance.

Figure 6.6–2 illustrates the region over which one would make the following energy balance:

$$Q = H_{CaO}(T_2) + H_{CO_2}(T_2) - H_{CaCO_3}(T_1)$$

By definition, the heat of reaction is

$$\Delta H_R = H_{CaO}(T_1) + H_{CO_2}(T_1) - H_{CaCO_3}(T_1)$$

where $T_1 = 25°C$. Subtracting the above gives

$$Q = \Delta H_R + [H_{CaO}(T_2) - H_{CaO}(T_1)] + [H_{CO_2}(T_2) - H_{CO_2}(T_1)]$$
$$= \Delta H_R + (C_{p_{CO_2}} + C_{p_{CaO}})(T_2 - T_1)$$

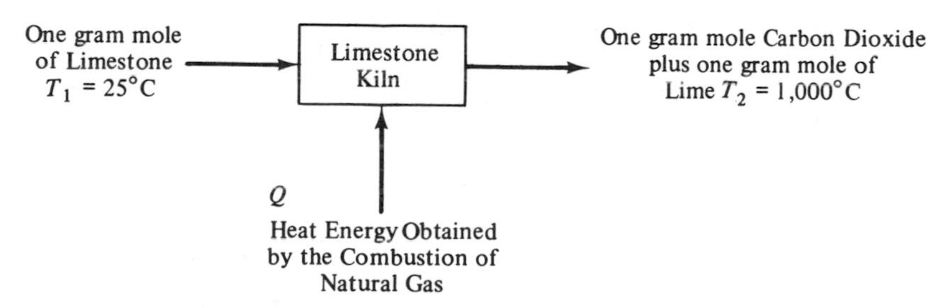

Figure 6.6–2. Energy Balance over a Primitive Limestone Kiln

Using these data,

$$\Delta H_R = 42 \text{ kcal/g-mole}$$
$$C_{p_{CO_2}} = 0.012 \text{ kcal/g-mole °C}$$
$$C_{p_{CaO}} = 0.010 \text{ kcal/g-mole °C}$$
$$T_1 - T_2 = 975°C$$

gives

$$Q = 64 \text{ kcal/g-mole}$$

To obtain the cost of this energy per ton of $CaCO_3$ processed, we assume a value of \$0.65/10^6 Btu to obtain

$$(64 \text{ kcal/g-mole})\left(\frac{1800 \text{ Btu g-mole}}{\text{lb-mole kcal}}\right)\left(\frac{\text{lb-mole}}{100 \text{ lb } CaCO_3}\right)\left(\frac{2000 \text{ lb}}{\text{ton}}\right)\left(\frac{\$0.65}{10^6 \text{ Btu}}\right)$$
$$= \$1.50/\text{ton } CaCO_3$$

In the next section we show the cost reduction that can be accomplished by proper energy management using a better limestone-kiln design. This cost easily can be cut in two.

6.7 HEAT ENERGY MANAGEMENT

In this section we show by example how task integrations are performed industrially to manage the available heat energy resources. In particular we examine the problems of supplying heat to endothermic reactions, removing energy from exothermic reactions, and handling heats of product recovery. In section 6.8 we complete the study of energy management by discussing systems for sensible heat recycle.

Heat Supply to Endothermic Reactions

The heat energy required of endothermic reactions must not only be of the right amount, it must be at the right temperature level. If we propose to melt steel, the melting being an endothermic reaction, with the energy from the combustion of coal, not only must sufficient energy be released by the combustion, but also the flame temperature must be above the melting point of steel. Such minimum acceptable temperature levels, dictated by the temperatures of melting, vaporization, chemical reaction, and so forth are common in the supplying of energy to endothermic tasks.

There are four general methods for supplying the required energy: indirect heating, electrical heating, reaction combination, and sensible heating. Each has advantages in different situations.

Indirect heating involves the transfer of heat through heat-exchange surfaces in the reactor. The external heat requirements for cracking of petroleum may be supplied by passing the petroleum in tubes back and forth through the combustion chamber of a furnace, as in Figure 6.7–1. Hot pebbles may be the source of indirect heating, as in the Wisconsin Process for nitrogen fixation, where air is converted to its oxides while being blown over the surface of very hot ceramic pebbles (see Figure 6.7–2 for a description of the Ljungstrom heater, which operates on a similar principle).

Electric heating is used to supply energy in induction furnaces that operate only on material which itself is an electrical conductor. Reactions occurring in molten salts, molten metals, beds of coke, or gas plasmas can be sustained by electrical discharge. The high cost of electrical energy makes this method of heating less attractive unless there are distinct advantages from the reduction of product contamination, process control, or other factors.

Reaction combination involves the matching of the endothermic reaction with an exothermic reaction, the heat for the one reaction being provided by the other. For example, the reaction occurring in a limestone kiln, $CaCO_3 \rightarrow CaO + CO_2$, is endothermic, and the combustion of coal, $C + O_2 \rightarrow CO_2$, is exothermic. If a mixture of coal and limestone is burned with air, the one reaction drives the other with no outside sources of heating or cooling. In a second example, the production of HCN by the reaction of NH_3 and CH_4,

$$NH_3 + CH_4 \rightarrow HCN + 3\ H_2$$

is strongly endothermic. However, if air is added to the reaction, this reaction,

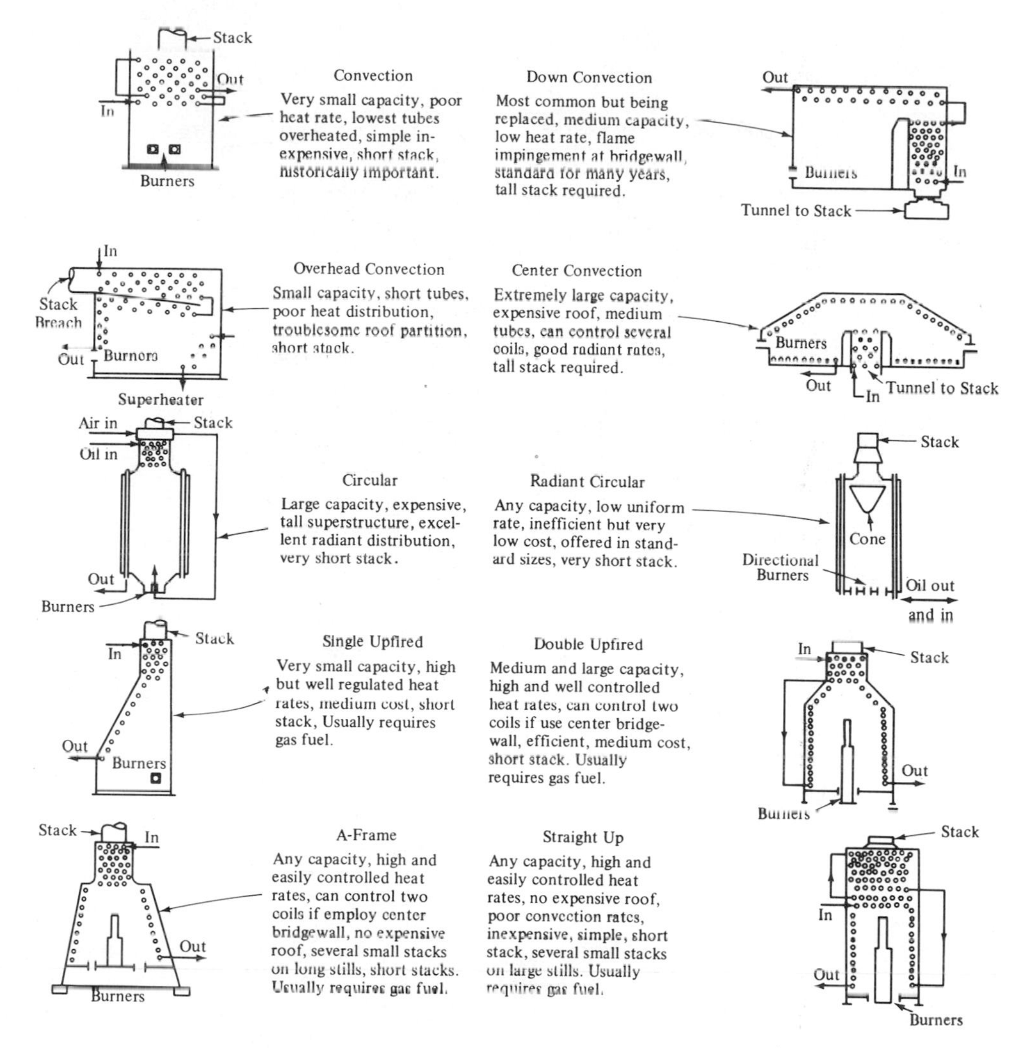

Figure 6.7–1. Petroleum Furnace Arrangements. Pipe Stills and Tubular Cracking Units
W. L. Nelson, *Petroleum Refinery Engineering*, 3rd ed., McGraw-Hill Book Company, New York, 1949, p. 526.

$3\ H_2 + 1\frac{1}{2}\ O_2 \rightarrow 3\ H_2O$, gives the net reaction, $NH_3 + CH_4 + 1\frac{1}{2}\ O_2 \rightarrow HCN + 3\ H_2O$, which requires no external heating.

In practice the mixing of endothermic and exothermic reactions is limited if the new reaction reacts with the desired products, reacts with catalysts, causes difficulties in product separation, and so forth.

Sensible heating involves the heating of the reactants well above their reaction temperature before they are mixed for reaction or before they are introduced into the

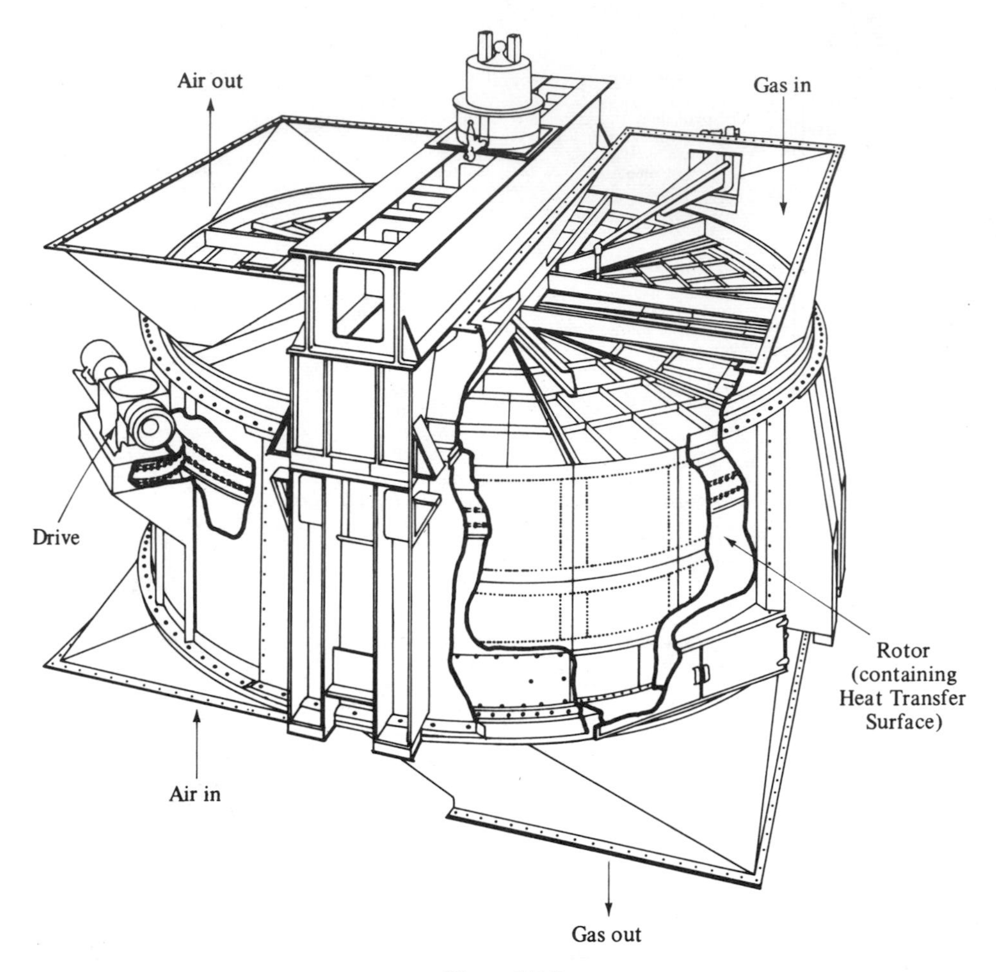

Figure 6.7–2

reactor where the catalyst is found. The heat for the reaction is then supplied by the sensible heat obtained by the cooling of the reactants.

Example 6.7–1 Reaction Combination in a Limestone Kiln *Limestone decomposes at 1000°C into lime and carbon dioxide according to the following endothermic reaction:*

$$\underset{\text{(limestone)}}{CaCO_3} \xrightarrow{1000^\circ C} \underset{\text{(lime)}}{CaO} + \underset{\text{(carbon dioxide)}}{CO_2}, \qquad \Delta H_{R_{25^\circ C}} = 42.5 \text{ kcal/g-mole}$$

It is proposed to drive the reaction by the burning of coke, an exothermic reaction:

$$\underset{\text{(coke)}}{C} + \underset{\text{(oxygen)}}{O_2} \rightarrow \underset{\text{(carbon dioxide)}}{CO_2}, \qquad \Delta H_{R_{25^\circ C}} = -94 \text{ kcal/g-mole}$$

How much coke is needed per ton of limestone?

The coke-to-limestone ratio depends on the energy management that can be designed into the kiln. If it were possible to design a kiln in which the carbon dioxide and the lime leave at room temperature (25°C), the energy requirements per ton of limestone would be

$$M(H_{CO_2} + H_{CaO} - H_{CaCO_3})_{25^\circ C} = M\,\Delta H_{R_{25^\circ C}}$$

where

$$M = (1 \text{ ton})\left(2000 \frac{\text{lb}}{\text{ton}}\right)\left(\frac{\text{lb-moles}}{100 \text{ lb}}\right) = 20 \text{ lb-moles } CaCO_3$$

and

$$\Delta H_R = \left(42.5 \frac{\text{kcal}}{\text{g-mole}}\right)\left(1800 \frac{\text{Btu/g-mole}}{\text{kcal lb-mole}}\right) = 76{,}500 \text{ Btu/lb-mole}$$

The heat energy requirements are

$$(20 \text{ lb-moles/ton})\left(\frac{76{,}500 \text{ Btu}}{\text{lb-mole}}\right) = 1{,}530{,}000 \text{ Btu/ton } CaCO_3$$

if the reaction products leave at ambient temperature.

However, if the solids leave at 1000°C instead of room temperature, the 20 lb-moles of CaO are heated from 25°C to 1000°C. The sensible heat energy of the CaO is

$$MC_p\,\Delta T = \left(\frac{20 \text{ lb-moles}}{\text{ton}}\right)\left(\frac{8 \text{ Btu}}{\text{lb-mole °F}}\right)(975°\text{C})\left(\frac{1.8°\text{F}}{°\text{C}}\right)$$
$$= 280{,}000 \text{ Btu/ton } CaCO_3$$

for a total energy requirement of 1,530,000 + 280,000 = 1,810,000 Btu/ton $CaCO_3$.

Should both the lime and the carbon dioxide leave at 1000°F, the additional energy consumed in the heating of the 20 lb-moles of CO_2 *is*

$$MC_p\,\Delta T = (20\text{lb-moles})\left(\frac{12 \text{ Btu}}{\text{lb-mole °F}}\right)(975°\text{C})\left(\frac{1.8°\text{F}}{°\text{C}}\right)$$
$$= 420{,}000 \text{ Btu/ton } CaCO_3$$

for a total of 1,810,000 + 420,000 = 2,230,000 Btu/ton $CaCO_3$.

The energy for these endothermic operations is generated by the combustion of coke. Each ton of coke corresponds to

$$(2000 \text{ lb/ton})\left(\frac{\text{lb-moles}}{12 \text{ lb}}\right) = 166 \text{ lb-moles/ton coke}$$

Thus, for each ton of coke burned 166 lb-moles of coke are consumed along with 166 lb-moles of O_2, and generated are 166 lb-moles of CO_2. Since air is about 20 per cent oxygen and 80 per cent nitrogen, 166(80/20) = 664 lb-moles of nitrogen are brought in with the oxygen.

If the kiln is designed so that the effluent gases leave at room temperature, the heat generated by the combustion is

$$\begin{aligned} -M_C \, \Delta H_R &= -\left(166 \, \frac{\text{lb-moles}}{\text{ton coke}}\right)\left(-94 \, \frac{\text{kcal}}{\text{g-mole}}\right)\left(1800 \, \frac{\text{Btu g-mole}}{\text{lb-mole kcal}}\right) \\ &= 28{,}000{,}000 \text{ Btu/ton coke} \end{aligned}$$

However, if the kiln is designed so the effluent gases leave at 1000°C, a significant part of this energy is tied up in the sensible heat of the 166 moles of CO_2 and 664 lb-moles of N_2. This reduces the total amount of heat released by the combustion by

$$\begin{aligned} & M_{\text{CO}_3} C_{p\text{CO}_2}(1000 - 25°\text{C}) + M_{\text{N}_2} C_{p\text{N}_2}(1000 - 25°\text{C}) \\ &= \left(166 \, \frac{\text{lb-moles}}{\text{ton coke}}\right)\left(12 \, \frac{\text{Btu}}{\text{lb-mole°F}}\right)(975°\text{C})\left(\frac{1.8°\text{F}}{°\text{C}}\right) \\ &\quad + \left(664 \, \frac{\text{lb-moles}}{\text{ton coke}}\right)\left(8 \, \frac{\text{Btu}}{\text{lb-mole°F}}\right)(975°\text{C})\left(\frac{1.8°\text{F}}{°\text{C}}\right) \\ &= 12{,}850{,}000 \text{ Btu/ton-coke} \end{aligned}$$

This leaves a net available energy of

$$28{,}000{,}000 - 12{,}850{,}000 = 15{,}150{,}000 \text{ Btu/ton-coke}$$

Now by assuming that the energy released by the coke burning either leaves with the kiln products as sensible heat or as the result of the endothermic reaction, we can estimate the coke requirements. For example, if all the materials leave the kiln at 1000°C, the heat energy required by the endothermic reaction has been computed to be 2,230,000 Btu/ton $CaCO_3$, and the heat released by the coke burning as 15,150,000 Btu/ton-coke. The ratio of these two is the required ratio of limestone to coke:

$$\frac{\text{tons limestone}}{\text{ton coke}} = \left(\frac{15{,}150{,}000 \text{ Btu/ton coke}}{2{,}230{,}000 \text{ Btu/ton CaCO}_3}\right) = 6.7$$

Table 6.7–1 summarizes the limestone-to-coke ratio for all possible ways of running the kiln in which the limestone and coke are charged together, and in which the effluents

TABLE 6.7–1 Limestone-to-Coke Ratio for Limiting Kiln Designs

		Gaseous Effluent	
		Cold (25°C)	Hot (1000°C)
Solids Effluent	Cold (25°C)	18	7.8
	Hot (1000°C)	15	6.7

either leave at room temperature or at the maximum reaction temperature. Notice how important it is to cool the effluent gases; over twice as much coke is required in the cases when the gases leave hot.

Figure 6.7–3 shows a commercial limestone kiln in which the coke is burned separately and the hot combustion gases are fed to the calcination reactor. The ash from coke burning will not contaminate the lime in this design. In other designs the limestone and coke are blended as a solids feed, and air is blown through the solids. In such commercial operations less than 10 tons of limestone are calcined per ton of coke. This is on the lower limit of our calculations, the higher coke consumption being caused by heat losses from the kiln to the surroundings and the excess air used to ensure complete combustion of the coke.

Heat Removal from Exothermic Reactions

The heat released by exothermic reactions may be used to drive other endothermic reactions or separation processes, or may leave the processes as sensible heat in cooling water or hot stack gases. Three general flow sheets to be discussed for supporting exothermic reactions involve adiabatic reactors followed by product coolers or else internally cooled reactors.

Adiabatic reactors involve the absorption of the exotherm by the sensible heat of the reactants, products, and diluents present in the reactor. The product stream reaches the *adiabatic reaction temperature.* Should the adiabatic reaction temperature be excessively high, internal reactor cooling must be used, rather than the external product cooling envisioned. Example 6.6–1 illustrates the kind of calculation used to determine the operation of adiabatic reactions.

Internal cooling of reactors is accomplished by transferring heat through heat-exchange surfaces built directly in the reactor, by cycling part of the reaction mass through an external cooler, or by allowing part of the reaction mass to boil and condensing the vapors formed. In all these cases the heat of reaction generally ends up in water circulating through coolers and condensers. Figure 6.7–4 shows some arrangements for pulling the heat out of the interior of reactors.

Cooling by reaction combination is the inverse of the heating-by-reaction combination discussed earlier. When a bed of coal is burned in an updraft of air, the temperatures reached may cause the ash to melt, fouling the furnace. However, by

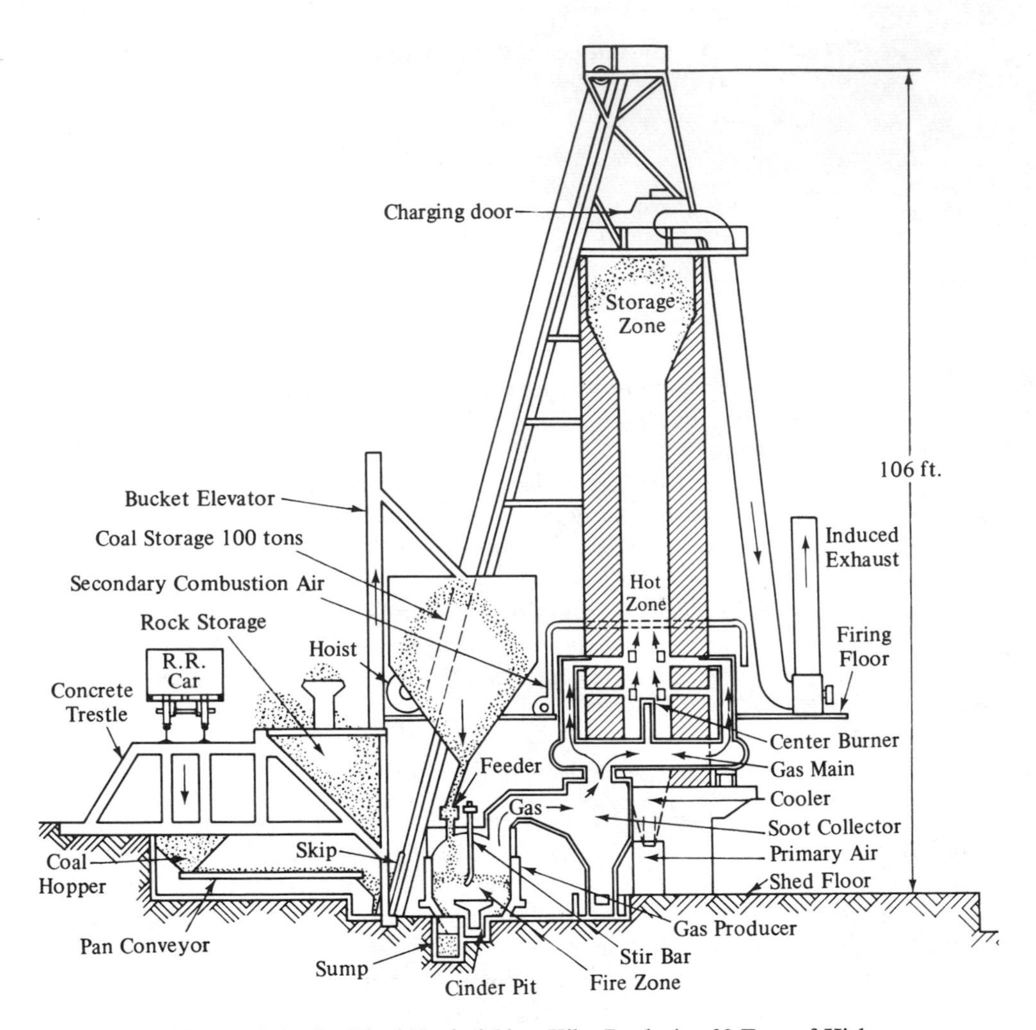

Figure 6.7–3. Gas-Fired Vertical Lime Kiln, Producing 80 Tons of High Calcium Lime per Kiln per Day
Victor J. Azbe, St. Louis, Missouri—patent granted.

injecting water into the burner, the following endothermic reaction lowers the temperature:

$$C + H_2O \rightarrow CO + H_2$$

Then, the CO and H_2 are burned above the bed of coal by the injection of additional air. In this way excessively hot spots are smoothed out.

Heats of Product Recovery

In the recovery of the products of chemical reaction considerable amounts of energy transfer occur. Usually steam is the source of the heat energy, which, after its use in separation, is rejected in the process cooling water.

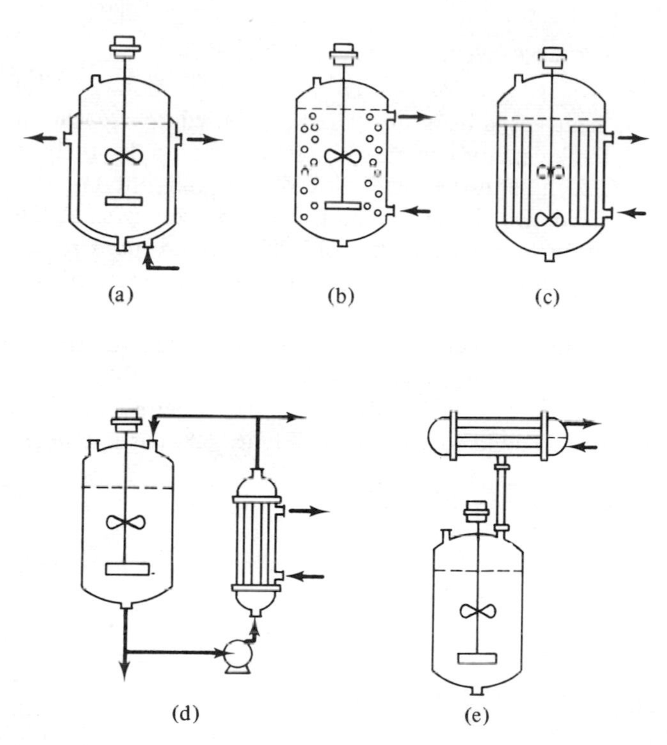

Figure 6.7–4. Heat Transfer to Stirred Tank Reactors
(*a*) Jacket; (*b*) internal coils; (*c*) internal tubes; (*d*) external heat exchanger; (*e*) external reflux condenser.

In *simple condensation* of the product that appears in a vapor phase in the reactor off gases, no steam is required, only cooling water to condense the simple product. For example, zinc metal (mp 419°C, bp 907°C) is condensed out of a retort gas effluent (50 per cent zinc and 50 per cent CO) by passing the vapor through a water-cooled condenser. The condenser may generate steam, which can be used elsewhere in the process. Clearly, simple condensation can only be accomplished when boiling points of the species are widely different.

Example 6.7–2 Steam Generation by Zinc Condensation *How valuable might be the steam generated by the condenser on a zinc retort that recovers 1 ton of zinc per day?*

The heat of vaporization of zinc (see Table 6.5–1) is 27.43 kcal/g-mole. This amounts to the removal from the zinc vapor of the following amount of energy:

$$\left(\frac{\text{1 ton zinc}}{\text{day}}\right)\left(\frac{\text{27.4 kcal}}{\text{g-mole}}\right)\left(\frac{\text{1800 Btu g-mole}}{\text{lb-mole kcal}}\right)\left(\frac{\text{2000 lb}}{\text{ton}}\right)\left(\frac{\text{lb-mole}}{\text{65 lb}}\right) = 1.55 \times 10^6 \text{ Btu/day}$$

At $0.30/1 million Btu, the value of the steam is less than $0.50/ton of zinc condensed. Rather than bother with the external sale of steam, usually the heat energy generated by zinc condensation leaves as sensible heat of cooling water.

6.8 HEAT-EXCHANGE NETWORKS

Networks of heat exchangers are commonly used to recycle energy within a process, avoiding the escape of energy with effluent materials. If the process runs at high temperatures, such as in the distillation of seawater, the hot effluents are used to heat the colder feed. On the other hand, if the process runs at low temperature, such as in desalination by freezing, the cold effluents are used to cool the warmer feed. The sequencing of heat-exchange operations is an interesting aspect of task integration.

In a heat exchanger, the device most commonly used for thermal energy task combination, two fluids pass on opposite sides of a conducting surface. As a consequence of the second law of thermodynamics, heat energy transfers through this surface from the warmer fluid to the cooler. Figure 6.8–1 is a detailed drawing of a

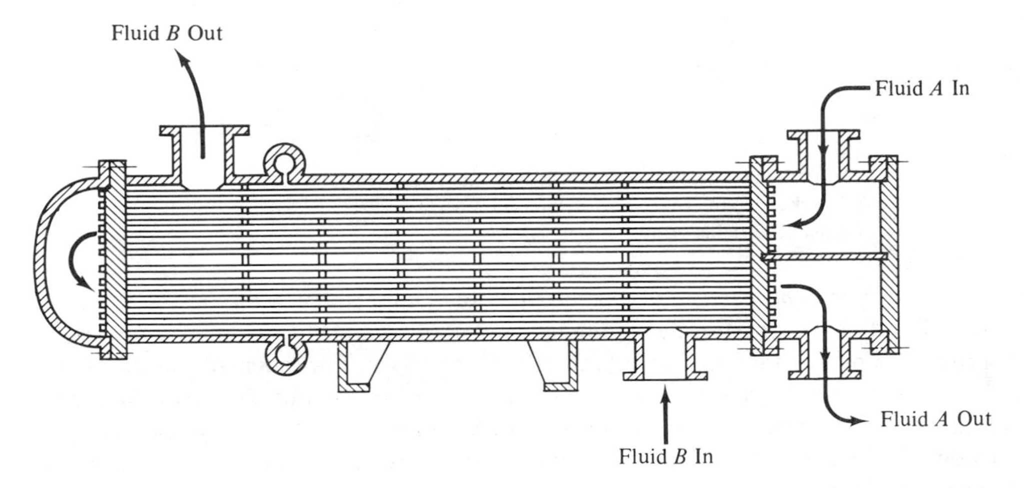

Figure 6.8–1. Shell-and-Tube Heat Exchanger

double-pass shell-and-tube heat exchanger. One fluid passes down and back through the tubes, and the other fluid passes on the outside of the tubes. The amount of heat exchanged depends on the flow rates, temperature differences, and thermal properties of the fluids, as well as the design of the heat exchangers, in particular the heat-exchange surface area.

For our purposes we picture a heat exchanger in a simpler form, as in Figure 6.8–2, ignoring most of the details that contribute to the advanced study of heat transfer. In cocurrent operation the hot and cold streams pass through the exchanger in the same direction, and in countercurrent operation the streams flow in opposite

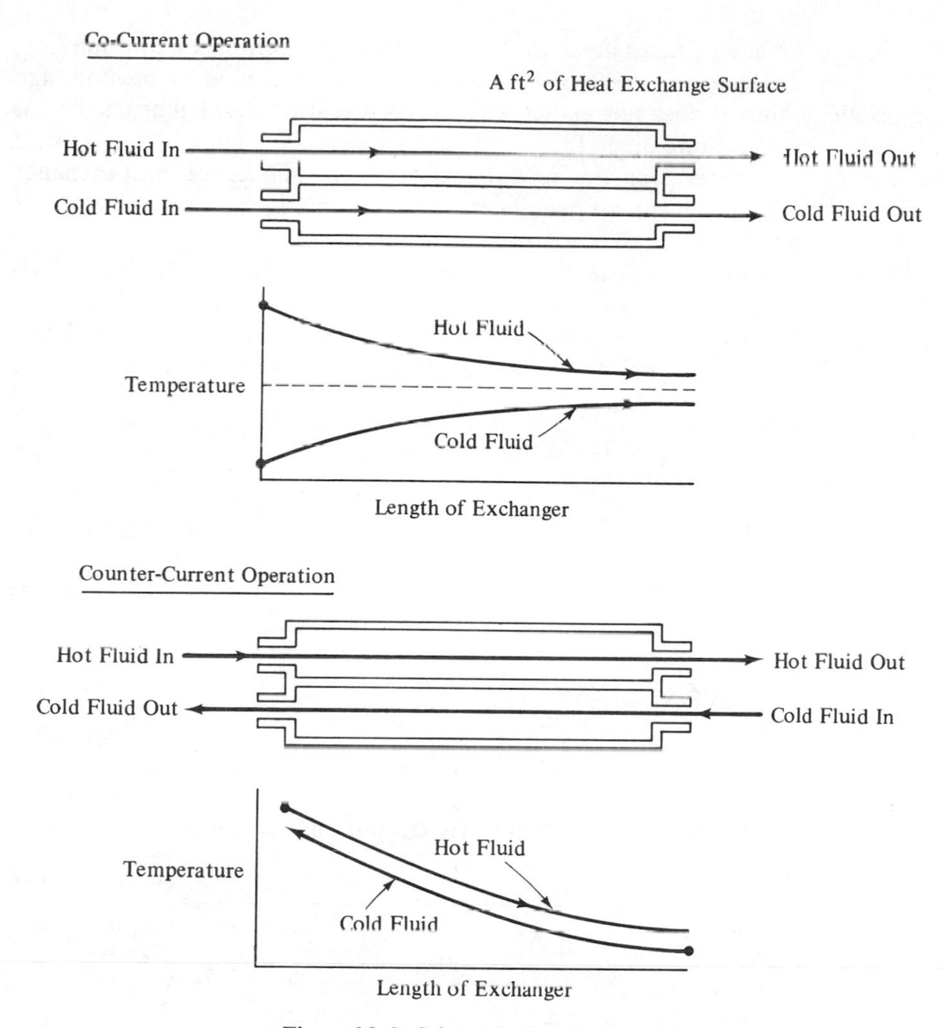

Figure 6.8–2. Schematic Heat Exchanger

directions. The direction of flow has a significant effect on the performance of the exchanger.

Notice in the cocurrent exchange how the temperatures of the fluids approach each other as the hot fluid transfers heat to the cold fluid. Regardless of the size of the heat exchanger, the hot and cold fluids are driven toward intermediate temperature. On the other hand, observe the temperature profile in countercurrent operation. The cold fluid temperature is driven toward the highest hot fluid temperature and the hot fluid temperature is driven toward the coldest cold fluid temperature. This observation leads to a heuristic:

> **Individual heat exchangers are most effective when internal flow of hot and cold fluids is countercurrent.**

If one of the energy tasks involves isothermal heat production or consumption, such as the phase-changing tasks of boiling, condensing, freezing, or melting, and some reactions, then it does not matter whether cocurrent or countercurrent flow is used. Also, considerations other than heat transfer may lead the engineer toward cocurrent flow; however, for the predominately large number of heat-exchange situations, the countercurrent heuristic holds. We presume in the remainder of this chapter that countercurrent internal flow is used.

The total rate at which heat is transferred between the two fluids, Q Btu/h, is proportional to the heat exchange area, A ft^2, and the average temperature difference between the fluids, $(\Delta T)_{av}$°F. The proportionality constant is called the overall heat transfer coefficient, U Btu/h-°F-ft^2, the range of values of which are given in Table 6.8–1:

$$Q = UA(\Delta T)_{av} \qquad \text{Btu/h}$$

where

$$(\Delta T)_{av} = \frac{(T_{hot}^{in} - T_{cold}^{out}) + (T_{hot}^{out} - T_{cold}^{in})}{2}$$

This arithmetic average ΔT is sufficiently accurate for our purposes, although the log-mean average, $(\Delta T)_{LM}$, is better when precision is required.

$$(\Delta T)_{LM} = \frac{(T_{hot}^{in} - T_{cold}^{out}) - (T_{hot}^{out} - T_{cold}^{in})}{\text{natural logarithm}\ \dfrac{(T_{hot}^{in} - T_{cold}^{out})}{(T_{hot}^{out} - T_{cold}^{in})}}$$

Table 6.8–1 Typical Values for Overall Heat Transfer Coefficient

SITUATION	U(BTU/H-°F-FT^2)
Gas–gas exchange	2–20
Liquid–liquid exchange	10–1000
Boiling liquids or condensing vapors	200–20,000

The heat transferred from the hot fluid is picked up by the cold fluid, as the following equation indicates:

$$Q = M_{hot}C_{p_{hot}}(T_{hot}^{in} - T_{hot}^{out}) = M_{cold}C_{p_{cold}}(T_{cold}^{out} - T_{cold}^{in})$$

where, of course,

$$T_{hot}^{in} > T_{cold}^{in}$$

$$T_{hot}^{in} > T_{hot}^{out} > T_{cold}^{in}$$

$$T_{hot}^{in} > T_{cold}^{out} > T_{cold}^{in}$$

Looking at these equations, we observe several things:

1. The amount of heat transferred Q increases as the area of the heat-transfer surface A increases.
2. The maximum amount of heat transferable is the minimum of

$$M_{hot}C_{p_{hot}}(T^{in}_{hot} - T^{out}_{hot})$$

and

$$M_{cold}C_{p_{cold}}(T^{out}_{cold} - T^{in}_{cold})$$

3. The maximum Q is reached when $T^{in}_{hot} = T^{out}_{cold}$ or $T^{out}_{hot} = T^{in}_{cold}$, at infinite heat-transfer area.

We conclude then that any increase in heat transfer must be paid for by an increase in heat-exchanger size and that the upper limit on energy transfer is determined by the lesser sensible energy change desired of the two streams.

Example 6.8–1 Cooling Hot Oil *A hot oil stream is to be cooled from 200 to 100°F using cooling water available at 60°F. The data on the streams are as follows:*

	HOT OIL	COOLING WATER
Flow rate: M lb/h	10,000	6000
Heat capacity: C_p Btu/lb °F	0.8	1.0
Input temperature °F	200	100
Output temperature °F	60	?

The overall heat transfer coefficient for the heat exchanger is $U = 100$ Btu/h °F ft^2. How much heat can be exchanged, what is the outlet temperature of the cooling water, and how large a heat exchanger is required?

The rate at which heat is to be removed from the oil is

$$\begin{aligned} Q &= M_{oil}C_{p_{oil}}(T^{in}_{oil} - T^{out}_{oil}) \\ &= (10{,}000 \text{ lb/h})(0.8 \text{ Btu/lb °F})(200 - 100\text{°F}) \\ &= 0.8 \times 10^6 \text{ Btu/h} \end{aligned}$$

That energy must be picked up by the cooling water:

$$\begin{aligned} Q &= M_{water}C_{p_{water}}(T^{out}_{water} - T^{in}_{water}) \\ 0.8 \times 10^6 &= (6000 \text{ lb/hr})(1.0 \text{ Btu/lb °F})(T^{out}_{water} - 60) \end{aligned}$$

Solving gives

$$T^{out}_{water} = 193\text{°F}$$

The average temperature difference between the fluids during heat exchange is

$$(\Delta T)_{av} = \frac{(T_{oil}^{in} - T_{water}^{out}) + (T_{oil}^{out} - T_{water}^{in})}{2}$$

$$= (200 - 193) + (100 - 60)$$

$$= 23.5°F$$

The surface area of the heat exchanger is

$$A = \frac{Q}{U(\Delta T)_{av}}$$

$$= \frac{0.8 \times 10^6 (Btu/h)}{100(Btu/h\text{-}°F)(23.5°F)}$$

$$= 334 \text{ ft}^2$$

We see in this example that the cooling-water temperature must approach to within 7°F of the hot-oil temperature of 200°F to accomplish the heat exchange required. In practice one might use more cooling water to increase the $(\Delta T)_{av}$ and reduce the required heat-exchange area. Usually, a minimum-approach temperature of 20°F results from an economic balance between heat-exchanger cost and cooling-water cost.

It is convenient to picture the heat exchange, as in Figure 6.8–3, for the problem of cooling stream *A* by heating streams *B* and *C* (Table 6.8–2). Plotted vertically is the stream temperature and horizontally the product of the flow rate and the heat capacity. The area enclosed is the rate at which heat energy must be exchanged with the streams to accomplish the desired temperature change, the heat duty. The horizontal scale is a floating scale, and the positioning of the streams on that scale is arbitrary. We place the streams to be heated and the streams to be cooled on separate graphs.

The first observation made is that the total area of the blocks for the streams to be heated must equal the total area of the blocks for the streams to be cooled, if the

TABLE 6.8–2 Stream Heat-Exchange Problem

STREAM	INPUT TEMP (°F)	OUTPUT TEMP (°F)	MC_p (BTU/H-°F)
A	300	100	10,000
B	100	200	5,000
C	50	100	5,000

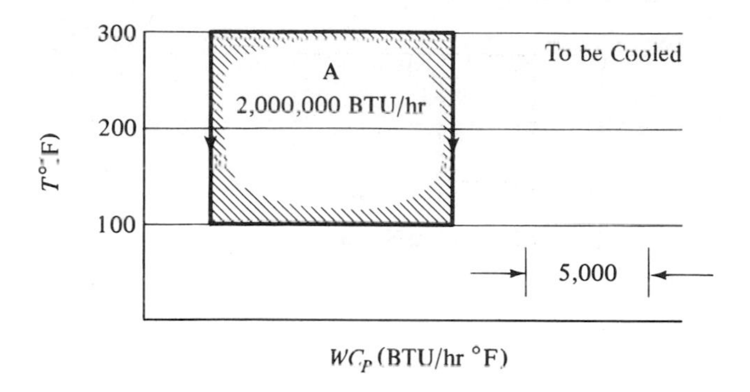

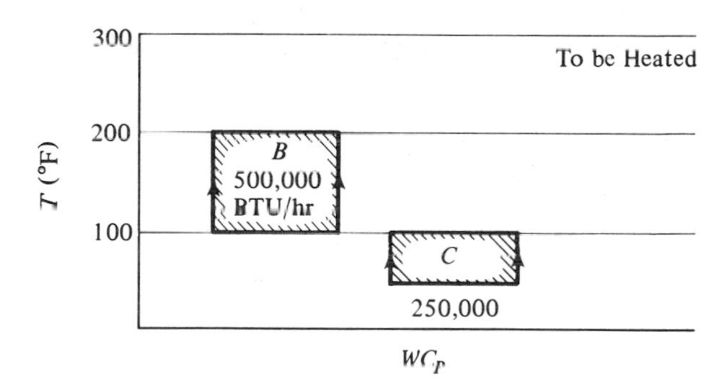

Figure 6.8–3. Problem of Table 6.8–2

heating and cooling are to be accomplished by heat exchange alone. In Figure 6.8–3 the heating duty is 750,000 Btu/h, whereas the cooling duty is 2,000,000 Btu/h. Auxiliary cooling must be used, amounting to 1,250,000 Btu/h.

A number of options are available for matching streams A, B, and C in heat exchange. One possibility is to divide stream A into two streams, the one part containing the 750,000 Btu/h to be exchanged with streams B and C, and the other part containing the 1,250,000 Btu/h to be removed by auxiliary cooling. This plan is shown in Figure 6.8–4. The 500,000 Btu/h in stream B is exchanged with the top part of stream A', reducing its temperature from 300 to 167°F. The remaining energy in stream A' is exchanged with stream C.

A second option is to switch the exchange with B and C to obtain the heat-exchange network shown in Figure 6.8–5. The temperature differences at which the last exchanger must operate are impossible in this case. The inlet temperature for B is 100°F and the outlet temperature for A is 100°F, a condition that can only be reached at infinite area. In practice one would lower the infinite area of the last exchanger and complete the cooling of stream A by auxiliary cooling. Or better yet, go back to the network in Figure 6.8–4.

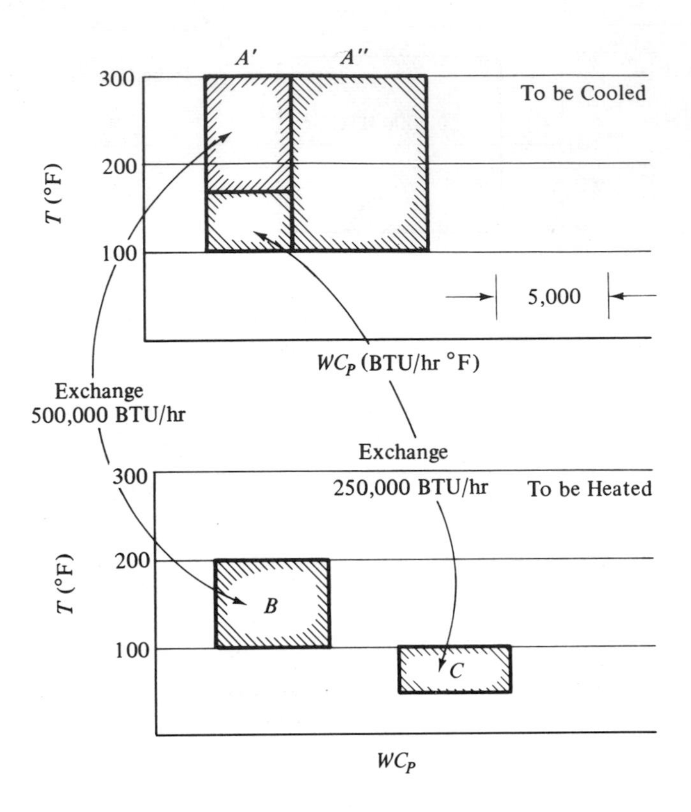

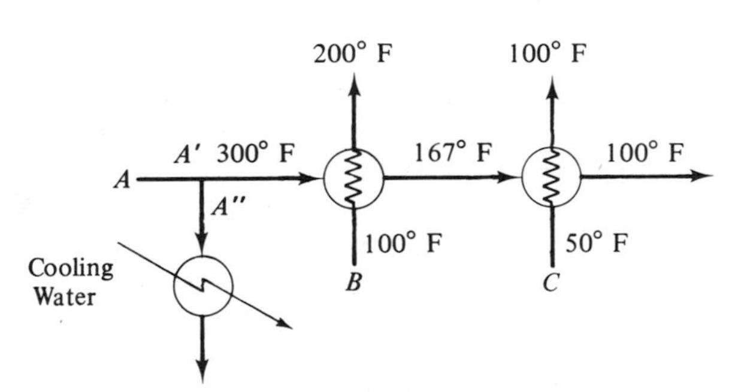

Figure 6.8–4. Splitting and Cooling of Stream *A*

A third option for handling this problem would be to leave stream *A* intact, and exchange energy with stream *B*, stream *C*, and cooling water, in that order. This is shown in Figure 6.8–6.

How many other networks can you visualize? The number of alternative designs becomes overwhelming in most problems involving more than just two streams, and the engineer requires some sort of orderly approach to heat-exchange network synthesis.

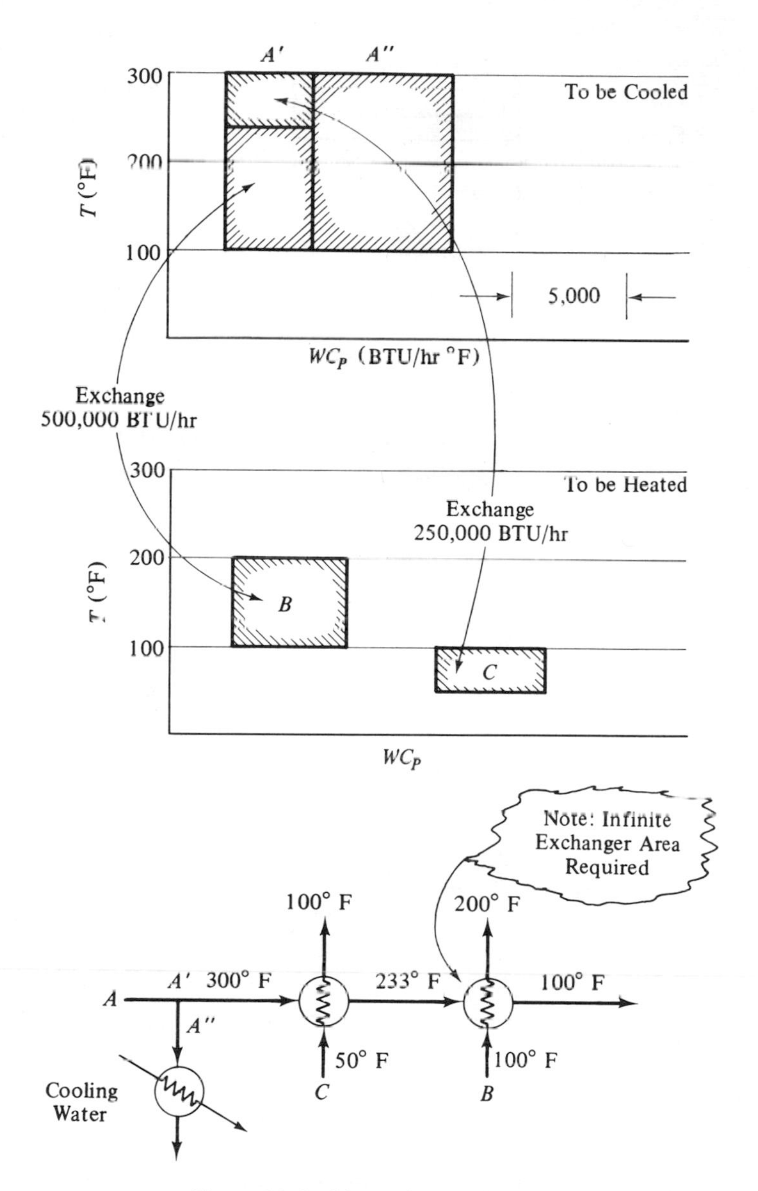

Figure 6.8–5. Alternative Configuration

Heuristics for Network Synthesis

The synthesis of heat-exchanger networks economically to recycle energy within a process has been the subject of a number of research investigations. The aim is to find the configuration of heat exchangers, such as those illustrated in Figure 6.8–7, that accomplishes the required heat exchange with the lowest total cost, including the

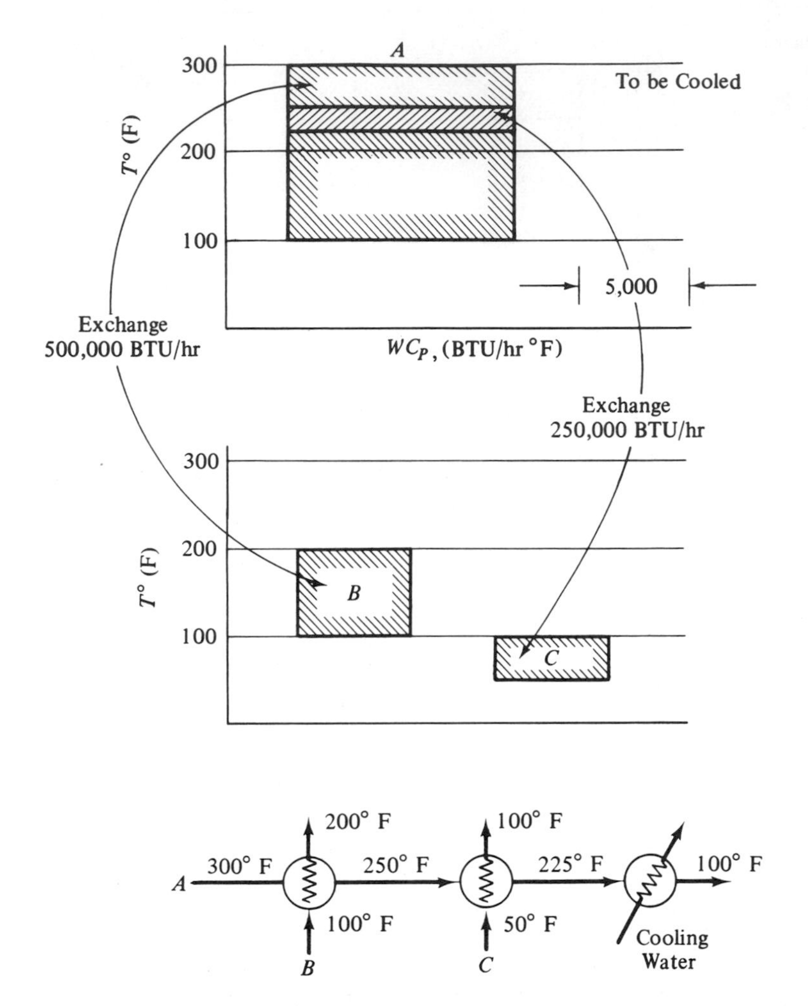

Figure 6.8–6. Third Alternative Heat-Exchange Network

investment in the heat exchangers, the auxiliary coolers and heaters, and the purchase of steam and cooling water.

Of the many heuristics developed for the synthesis of heat-exchange networks, we shall consider three. The first is the establishment of a minimum exchanger. It has been pointed out earlier that the area, and hence the cost of an exchanger, is inversely proportional to the temperature differences. If the temperatures of the two streams are getting close together, a point is reached where it is more economical to perform the remainder of the energy tasks with other integrations or external utilities, rather than increase the size of the exchanger precipitously. In many cases, the economic trade-off point occurs at a minimum temperature difference of 15 to 20°F. Therefore, a useful heuristic is:

Figure 6.8–7. Network of Heat Exchangers Integrating Energy Usage in a Petroleum Refinery. (Courtesy of Western Supply Co.)

Do not specify heat exchange between two streams such that the temperature difference at either end is below the minimum-approach temperature.

Figure 5.6–1 implied that the cost of processing increases with excursions from ambient temperature. This is also true for the operating cost of heating and cooling utilities. For utilities these costs usually occur in discrete steps. Steam may only be available for heating at several temperatures. Similarly, cooling water, brine, glycol, propane, or other refrigerants will be available only at characteristic temperatures. Therefore, propose exchangers that will allow auxiliary heating to be done at the *lowest* possible temperature and auxiliary cooling at the *highest* possible temperature, so that auxiliary heating and cooling are done as close to ambient as possible. This is especially important when an alternative heat-exchange integration would require an auxiliary utility from a less expensive source. This leads to two more useful heuristics:

Consistent with the minimum-approach temperature, propose heat exchange between the *hottest* stream to be cooled with the *warmest* stream to be heated.

Alternatively:

> **Consistent with the minimum-approach temperature, propose heat exchange between the *coldest* stream to be heated with the *coldest* stream to be cooled.**

Example 6.8–2 Four-Stream Heat-Exchange Network *In a proposed process there are four thermal energy tasks whose existing and required temperatures, flow rates, and heat capacities, assumed constant over the temperature range, are given below. Synthesize a heat integration network.*

STREAM	T_{in} (°F)	T_{out} (°F)	M (LB/H)	C_p (BTU/LB °F)
A	140	320	20,600	0.7
B	320	200	27,800	0.6
C	240	500	23,200	0.5
D	480	280	25,000	0.8

Steam utility available at 250, 350, and 550°F.
Cooling water available at 80°F.
Brine available at 30°F.

The energy content–temperature diagram for this problem with the flow heat capacity for each stream is given in Figure 6.8–8. Applying the temperature-difference heuristic, we shall set the minimum-approach temperature at 20°F.

Looking at the hottest stream to be cooled, stream D, we see that its existing temperature is 480°F. Applying the minimum-approach temperature, the hottest any cold stream may be heated by integration is 460°F; since stream C is required to be heated to 500°F, we have no choice but to use auxiliary 550°F steam, as shown in Figure 6.8–9.

Now we propose a match between the hottest stream to be cooled, stream D, and the warmest stream to be heated, the remainder of stream C. The heat available in stream D is

$$(480 - 280°\text{F})(20{,}000\text{Btu/h °F}) = 4.00 \times 10^6 \text{ Btu/h}$$

The heat required by the remainder of stream C is

$$(460 - 240°\text{F})(11{,}600\text{Btu/h °F}) = 2.56 \times 10^6 \text{ Btu/h}$$

Thus all the remainder of stream C can be matched against stream D. We must use the hottest portion of stream D, because the hottest part of stream C is at the minimum-approach temperature difference from the hottest part of stream D. After the match, the remainder of stream D will have the temperature

$$480 - \frac{2.56 \times 10^6 \text{ Btu/h}}{20{,}000 \text{ Btu/hr °F}} = 362°\text{F}$$

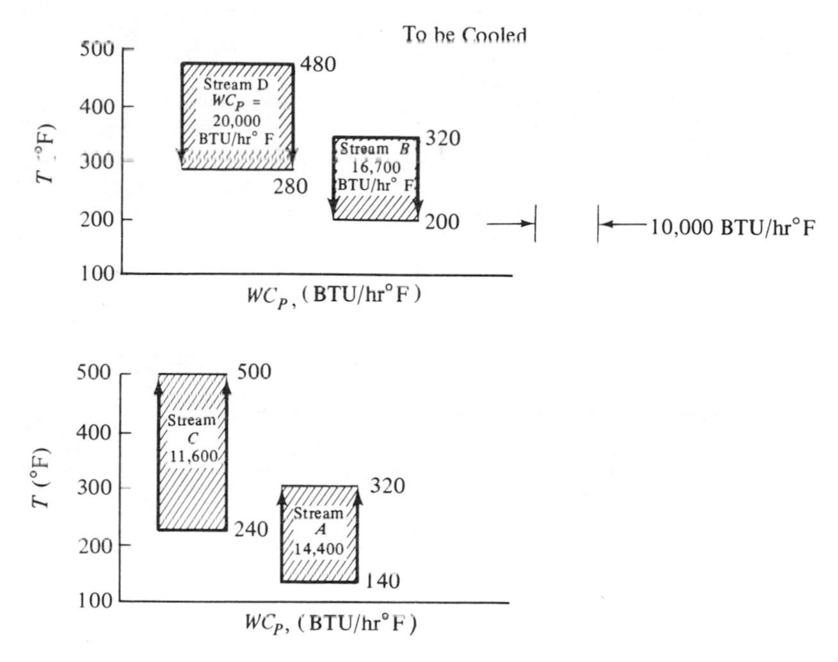

Figure 6.8–8. Energy Content–Temperature Diagram

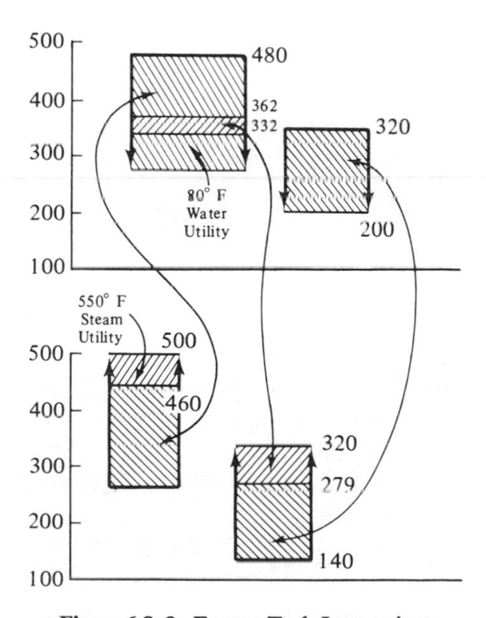

Figure 6.8–9. Energy Task Integrations

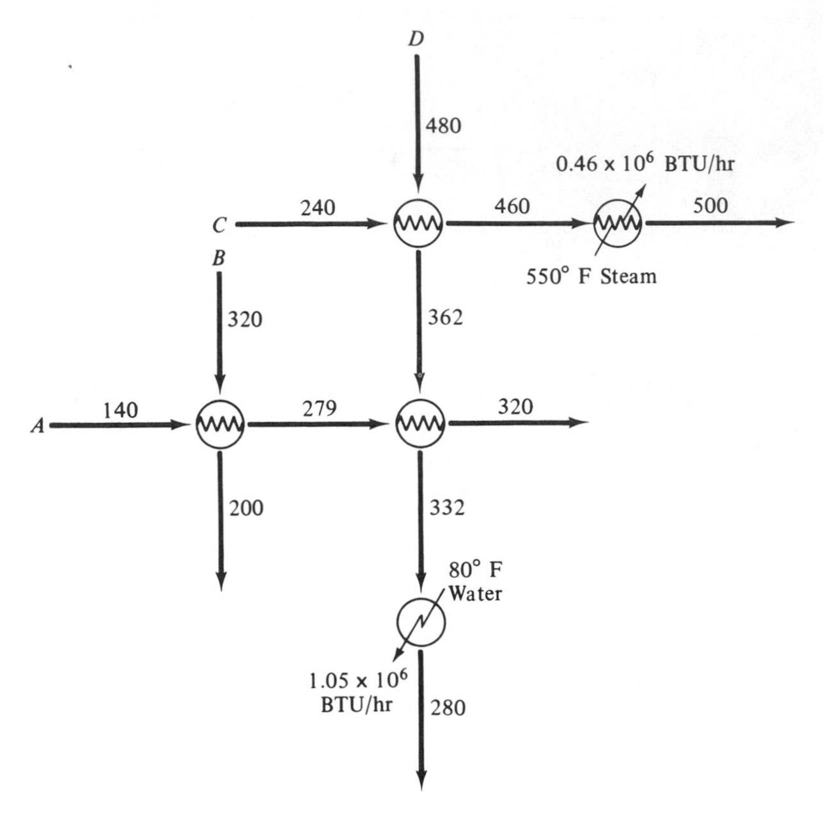

Figure 6.8–10. Four-Stream Heat-Exchange Network

We now might propose a second match between the coldest stream to be heated, stream A, with the coolest stream to be cooled, stream B. The heat required by stream A is

$$(320 - 140°\text{F})(14{,}400\text{Btu/h °F}) = 2.59 \times 10^6 \text{ Btu/h}$$

The heat available from stream B is

$$(320 - 200°\text{F})(16{,}700\text{Btu/h °F}) = 2.00 \times 10^6 \text{ Btu/h}$$

All of stream B can be used to heat stream A. The match cannot be made with the hottest part of stream A, because the minimum-approach temperature constraint would be violated. Therefore, the match must be made with the coolest portion of stream A, as shown in Figure 6.8–9. The new stream A temperature will be

$$140 + \frac{2.00 \times 10^6}{14{,}400} = 279°\text{F}$$

The remainder of stream D and the remainder of stream A may now be matched.

The heat required by stream A is

$$(320 - 279)14{,}400 = 0.59 \times 10^6 \text{ Btu/h}$$

The heat available from stream D is

$$(362 - 280)20{,}000 = 1.64 \times 10^6 \text{ Btu/h}$$

All of stream A is matched to stream D. Again, to avoid the minimum-approach temperature, the hottest portion of stream D is used. Its temperature is now

$$362 - \frac{0.59 \times 10^6}{20{,}000} = 332°\text{F}$$

The remainder of stream D is cooled with utility water, as in Figure 6.8–9. The complete heat-exchange network is shown in Figure 6.8–10. If the heuristics had been applied in different order at the second match, and the warmest part of stream D had been matched against the warmest remaining stream to be heated, stream A, then a slightly different network, but with exactly the same minimal utility requirement, would have been generated. The details of this alternative network, Figure 6.8–11, are left as an exercise.

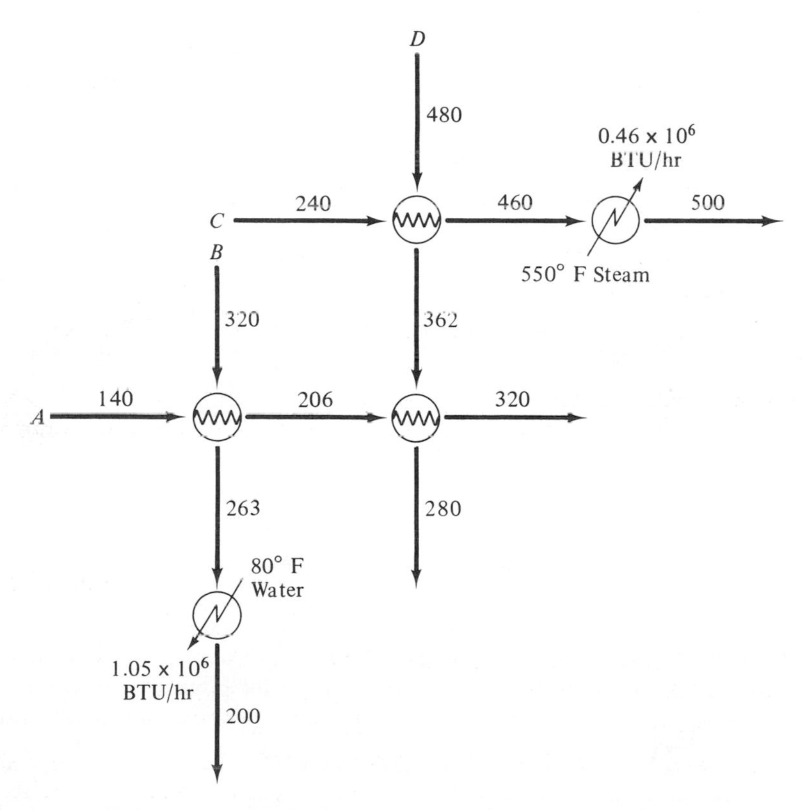

Figure 6.8–11. Alternative Four-Stream Heat-Exchange Network

6.9 CONCLUSIONS

The amount of raw material required in processing is an immutable function of the reaction path. Little can be done to economize there, unless a reaction sequence is found that gives access to cheaper raw materials. The research chemist works on this aspect of process discovery.

However, quite a bit can be done to change the energy consumption in processing. This is done by the clever reuse of energy, by having the heating and cooling of materials do many tasks before the energy is rejected to the surroundings.

In this chapter we learned to estimate the heat requirements for processing involving sensible heat effects, changes of phase, and changes in chemical composition. This is the basic material required to discover the broad features of energy management.

In the next chapter we consider a process in which energy management is the dominant feature, the recovery of fresh water from the sea. The raw material is free of charge, and the cost of the fresh water is determined solely by the engineer's skill in energy management.

REFERENCES

Useful texts on energy management include

D. M. HIMMELBLAU, *Basic Principles and Calculations in Chemical Engineering*, Prentice-Hall, Inc., Englewood Cliffs, N.J., 1967.

J. C. WHITWELL and R. K. TONER, *Material and Energy Balances*, McGraw-Hill Book Co., New York, N.Y. 1973.

H. P. MEISSNER, *Processes and Systems in Industrial Chemistry*, Prentice-Hall, Inc., Englewood Cliffs, N.J., 1971.

For more on heat exchanger network synthesis see

K. F. LEE, A. H. MASSO, and D. F. RUDD, *Industrial and Engineering Chemistry Fundamentals*, **9**, 48, 1970.

PROBLEMS

1. **ICE MELTING BY STEAM CONDENSATION.** In Chapter 7 we develop a process for obtaining fresh water from the sea by freezing. In the process ice is melted by condensing steam. How many pounds of ice can be melted per pound of steam condensed?

2. **THERMAL POLLUTION OF THE ENVIRONMENT.** The release of large amounts of hot water into a lake or river, say, from a power plant, can cause severe ecological damage. For this reason upper limits are set on the temperature of effluent streams, and

these are maintained either by the precooling of the effluent water or by the dilution of the hot effluent with cooler river water.

If 10 million gallons/day of 160°F water must be released from a power plant, how much 60°F river water must be brought into the power plant for dilution purposes? The water leaving the power plant must not exceed 80°F.

3. **COOLING DISTILLATE WITH PROPANE.** A stream of 35°API distillate is to be cooled from 200 to 100°F, using a liquid propane stream that is to be heated from 0 to 100°F. How many pounds per hour of 35°API distillate can be cooled per pound of propane heated?

4. **FREEZING FISH.** Fish fillets are frozen by a spray of liquid nitrogen as they pass continuously on a conveyer belt. If the process is to freeze 1000 lb/hr of fish, how much liquid nitrogen is required?

   The fish are to leave at −30°F and used nitrogen vapor at −10°F. In lieu of data on the thermal properties of fish, assume the fish to be 50 per cent water and 50 per cent solids. The solids are assumed to have the thermal properties of cellulose. This problem illustrates how an order-of-magnitude solution can be obtained on relatively little data. In practice, on the order of 100 to 200 gal/h of nitrogen is required.

5. **THERMITE WELDING.** The highly exothermic reaction between powdered aluminum and powdered iron oxide generates molten iron and aluminum oxide:

$$2\,Al + Fe_2O_3 \rightarrow Al_2O_3 + 2\,Fe$$

   This reaction is used to weld steel rails and other large pieces of metal. If the initial powder mixture is at room temperature, what is the maximum temperature of the reaction products? Is this sufficient to generate molten iron?

6. **GAS DESULFURIZATION TO REDUCE POLLUTION.** Pure sulfur can be formed from hydrogen sulfide present in natural gas by a controlled partial oxidation reaction:

$$3\,H_2S(g) + \tfrac{3}{2}\,O_2(g) \rightarrow 3\,S(s) + 3\,H_2O(g)$$

   Such a sweetening process is useful to reduce the air pollution caused by the formation of SO_2 when the natural gas is burned for heating purposes. Calculate the heat that must either be added or removed from such a process in Btu per pound of sulfur recovered.

7. **SULFUR SOLIDIFICATION.** Liquid sulfur is cooled and solidified on a continuously moving stainless-steel belt. The cooling is accomplished by a spray of cool water (65°F) that boils and forms steam, which is vented off the process. How much cooling water is required for a process that solidifies 10 tons of sulfur per day? What is the cost of cooling water in dollars per ton of sulfur?

8. **CARBON MONOXIDE AS A KILN FUEL.** Rather than use coal to fire the limestone kiln of Example 6.7–1, a waste carbon monoxide is to be used. For this purpose, how many pounds of carbon monoxide is equivalent to 1 pound of coal?

9. **ENERGY STORAGE BY CHECKER BRICK REGENERATORS.** The preheating of combustion air in open-hearth furnaces, ingot-soaking pits, glass-melting tanks, by-product coke ovens, heat-treating furnaces, and the like is universally carried out in checker brick regenerators made of fireclay, chrome, or silicon bricks. The flue gases

from the process are vented through a cooled regenerator, thereby heating the brickwork and storing the heat energy. The fresh air to be heated passes through other hot regenerators. By proper cycling of the heating and cooling cycles, the heat energy in the vent gases is transferred to the incoming combustion air. How many dollars worth of heat energy is stored in a regenerator when 100 tons of brickwork are heated from 1000 to 2000°F?

10. **SPRAY-DRIED COFFEE.** Estimate the cost of heat energy to spray dry the coffee extract in the process of Figure 3.3–4. Express your answer in dollars per pound of soluble coffee powder. The major energy consumption is from water vaporization.

11. **COST OF SUGAR EVAPORATION.** In Problem 4 of Chapter 3, a sugar solution is concentrated in an evaporator. Estimate the total cost of heating and cooling per ton of concentrated sugar solution formed. Assume that the energy involved in vaporization and condensation dominates.

12. **BENZOIC ACID BY EVAPORATION.** Benzoic acid in a dilute water solution is to be recovered in solid form. One plan is to heat the water solution to evaporate the water, and the second plan is to extract the benzoic acid with benzene and then evaporate the benzene. Determine the heat addition rate to the evaporators for each of these processing plans if 1000 gal/h of solution containing 0.05 mole of benzoic acid per gallon is to be processed. What is the savings in heating cost in dollars per hour? How much benzoic acid is lost? Benzene?

Plan 1 Direct Evaporation

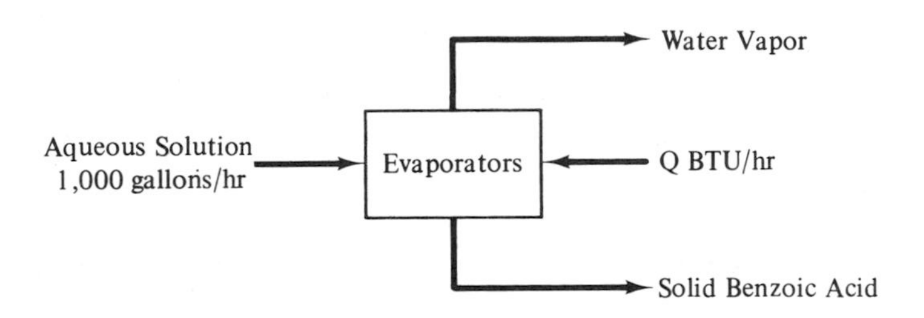

Plan 2 Extraction–Evaporation

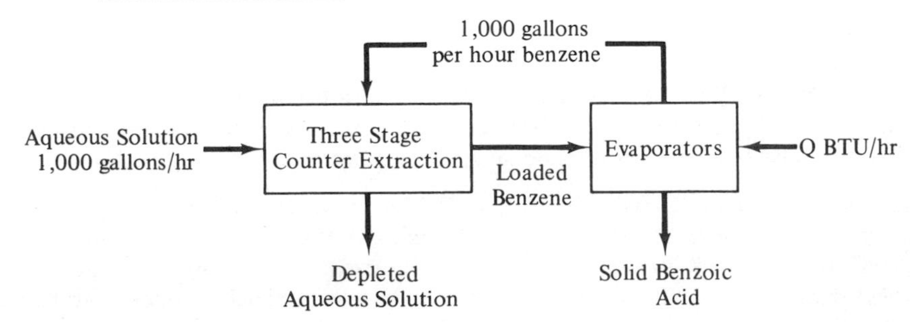

Hint. The three-stage countercurrent extraction problem has been solved as Example 4.4–2. Assume that the heat addition rate equals the energy required to vaporize the water and the benzene. Solubility of benzene in water is found in library.

13. **HEATING VALUE OF COAL.** What is the cost of energy produced by the combustion of $10/ton coal? Assume the coal to be essentially carbon. The combustion products leave the furnace at 600°F, and 20 per cent more air is to be used than required by the stoichiometry of the reaction. How does the cost compare to that of Table 6.3–1?

14. **OXYGEN–ACETYLENE TORCH.** What is the maximum temperature that can be obtained by a torch fueled with acetylene and pure oxygen? What effect on the flame temperature would the use of air instead of oxygen have?

15. **ACETONE REACTOR PREHEATING.** Acetone is to be manufactured from isopropanol by the following reaction:

$$\underset{\text{(isopropanol)}}{C_3H_7OH} \xrightarrow{\text{catalyst}} \underset{\text{(acetone)}}{C_3H_6O} + \underset{\text{(hydrogen)}}{H_2}$$

The isopropanol is to be preheated to 600°F and injected into a catalyst-filled reactor, and in the reactor the reaction goes to completion. It is proposed to use the reactor effluents as a source of energy to preheat the cold isopropanol feed. Does this make any sense? What energy management procedures do you suggest?

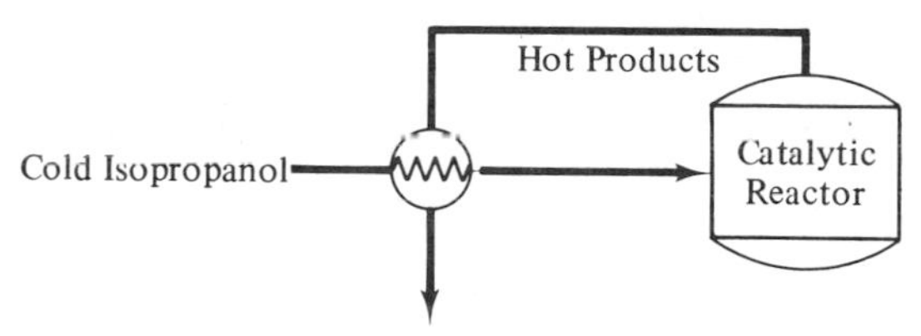

16. **ECONOMIZERS FOR FREEZING DESALINATION PROCESS.** A process is visualized to recover fresh water from the sea by the partial freezing of seawater. The ice crystals formed are salt free, and when washed free of adhering brine can be melted to form the fresh water. For each ton of seawater brought into the process $\frac{1}{2}$ ton of a brine concentrate leaves and $\frac{1}{2}$ ton of fresh water is formed.

    If the brine leaves at 28°F, the fresh water leaves at 32°F, and the seawater enters at 60°F, how much of a reduction in product cost could be accomplished if there were some way of recovering all the refrigerative capacity of the effluent streams to precool the incoming seawater? Express your answer in dollars per thousand gallons of fresh water.

    Synthesize the network of heat exchangers that best accomplishes the desired energy exchanges. In this situation a minimum-approach temperature of as little as 2°F is often used, since the energy economy is a very critical factor. Synthesize the network for a 10°F minimum-approach temperature. What change in water cost occurs?

17. **HEAT-EXCHANGE NETWORK SYNTHESIS.** A wide variety of problems in heat-exchange network synthesis can be formed by removing one or more of the streams

from the example problems used in the text. Form and solve as many problems as you wish from the data given in Example 6.8–2.

18. **CRYOGENIC SHREDDING OF JUNK CARS.** The following item was reported in *Chemical Engineering*, Feb. 22, p. 65, 1971.

> Cryogenics makes it easier to recover steel from junked cars, according to Liege, Belgium, raw-materials trader George & Co. The firm has developed a procedure whereby cars (minus their tires) are conventionally compressed into cube-shaped bundles, which are then chilled to about −190 F. to −310 F. by liquid-nitrogen-induced cold. At these temperatures, the ferrous metals present turn very brittle but the nonferrous components do not. When the bundles next go to a shredder, the ferrous components therefore chip into coin-sized pieces that can be easily separated from the rest of the junk. These chips are suitable feed for electric furnaces. And, says George & Co., much less residual copper accompanies the chips than is the case with steel from conventional junked-car reclaiming. About 4 gal. of liquid nitrogen is needed per 100 lb. of scrap.
>
> The Belgian company and Klockner & Co., Duisburg, Germany, have invested more than $600,000 in developing a prototype plant. Meanwhile, a full-scale plant to process 60,000 tons/yr. is to be built in Liege this year, and a similar one in Germany.

What is your estimate of liquid nitrogen consumption? How does it compare to 4 gal/100 lb?

7

Fresh Water by Freezing

In the winter of the year 1729, I exposed beer, wine, vinegar, and brine in large open vessels, to the frost; which, thus, congeal'd nearly all the water of these liquors into a soft, spongy kind of ice, and united the strong spirits of the fermented liquors; so that by piercing the ice, they might be pour'd off and separated from the water, which diluted them before freezing; and the more intense the frost is, the more perfect this separation: whence we see that cold unfits water to dissolve alcohol, and the salt of vinegar: and it is probable, that the utmost possible cold in nature would deprive water of all its dissolving power.

Hermann Boerhaave

7.1 INTRODUCTION

The scope of each of the preceding chapters was limited in turn to a specific aspect of process discovery. Such limitations are necessary and desirable to present a compact package of knowledge that can be digested without too much difficulty. In practice, no field is as clean and orderly as textbooks tend to describe it to be. This is certainly true of the field of process discovery.

For one thing, few, if any, real problems are equally balanced among the several topics presented in the earlier chapters. Some problems are dominated by the chemistry, others by task sequencing, still others by a critical separation, and many by the integration of energy and material use. Furthermore, an experimental program usually parallels process discovery, for rarely is sufficient information to be found in the literature on phenomena to be exploited in a new area of processing. Finally, the early discovery stages of engineering are not sufficiently detailed to arrive at just one process, and several alternatives survive to undergo the detailed engineering studies necessary to determine economic and engineering feasibility.

These things are now brought into focus by tracing the development of a process to obtain fresh water from brackish water. We risk the appearance of contrived spontaneity by condensing the years of work of a number of engineers into a scenario of process discovery. However, it is important for the reader to come close to the act of discovery, even if the discovery is only one simulated by the authors.

This chapter begins with the observation that the *ice crystals formed in brackish water are salt free*, and ends with a process now commercially available for the production of fresh water. This engineering problem is dominated by task integration problems.

In Chapter 8 we approach a discovery problem dominated by chemistry, species allocation, and task sequencing problems. The problem is that of producing the chemical chlorodecane for detergent manufacture.

7.2 Background on Desalination

In this section the current status of desalination is reviewed to set the stage for the engineering of a freezing process.

In many areas of the world population growth strains the available supplies of water for drinking, agriculture, and industry. Enormous amounts of water are to be

TABLE 7.2–1 Major Constituents of Seawater in Parts per Million

Chloride, Cl	18,980
Sodium, Na^+	10,561
Sulfate, SO_4^{2-}	2,649
Magnesium, Mg^{2+}	1,272
Calcium, Ca^{2+}	400
Potassium, K^+	380
Bicarbonate, HCO_3^-	142
Bromide, Br^-	65
Other salts	34

Salinity of Other Water Sources

Source	Salinity (ppm)
West Texas (well)	1000–6700
Utah (river)	1000–4000
Great Salt Lake	270,000
Caspian Sea	14,000
Lake Erie	180
Seawater	35,000

found in the sea and in other brackish water sources. However, these supplies are contaminated by salts and minerals, and are useless unless an economic means of purification can be discovered.

Economic is the critical modifier of the phrase "means of purification." Municipal water in the United States costs considerably less than $0.50/1000 gallons, which is $0.12/ton or only $0.00006/lb. Even fresh water barged into the Virgin Islands costs only $1.20/ton. This then is no ordinary processing problem: the processing costs are measured in thousandths of a cent per pound of product.

Furthermore, the processing capacity must be great. The water usage in the average American household is on the order of 100 gal/person/day. Based on this average, a community of 100,000 would require *10 million* gallons of water per day. Of course, in areas of limited water supply the average household use would be less.

Finally, the raw material for these processes is highly contaminated. The U.S. Public Health Service recommends that the salinity of drinking water be less than 500 ppm (parts per million by weight). Seawater is 3.5 per cent salt or 35,000 ppm. Table 7.2–1 shows the major salts that contaminate seawater, and the salinity of several other brackish water sources.

The state of the art in desalination is most advanced for multistage flash distillation and electrodialysis. Worldwide in 1968 there were over 700 desalination plants

TABLE 7.2–2 Methods of Desalination

I Distillation processes—when salt water is boiled the salt remains behind
  a. Long-tube vertical multieffect distillation
  b. Multistage flash distillation
  c. Multieffect, multistage distillation
  d. Vapor compression distillation

II Membrane processes—water passes through a membrane and salt does not
  a. Electrodialysis
  b. Transport depletion electrodialysis
  c. Reverse osmosis

III Humidification processes—vapor space above warm seawater contains water but not salt
  a. Solar humidification
  b. Diffusion humidification

IV Freezing processes—salt does not participate in ice crystal formation
  a. Direct freezing
  b. Secondary refrigerant freezing

V Chemical processes—water and salt have different chemical properties
  a. Ion-exchange method
  b. Hydrate formation

in operation, and 323 of these were in the United States. The largest single plant in the United States is the 2.6 million gallon/day plant supplying Key West, Florida. A 7.5 million gallon/day plant was recently completed at Rosarito, Mexico. The cost for fresh water from the sea using the technology available in 1952 was $4/1000 gal; by 1970 this had been reduced to $0.70/1000 gal; and in the near future costs are expected of $0.40/1000 gal for very large plants. Some of the existing processes are mentioned in Table 7.2–2.

Rather than examine these different processes in detail, we focus on the discovery of one of these, that involving desalination by freezing. The mental activities involved are the object of this chapter, and the process itself is merely a by-product of the study presented. We chose the freezing process primarily because of the simplicity of the freezing phenomenon.

7.3 FACTS ON FREEZING

Both the water vapor formed upon boiling seawater and the individual ice crystals formed by freezing are salt free. However, the heat of vaporization of water is about 1000 Btu/lb and the heat of crystallization is only 144 Btu/lb. Seven times less heat must be transferred to form salt-free ice crystals than to form salt-free water vapor. This is one of the factors that attracts attention to the freezing process. Does this great difference in energy requirements have engineering significance, or is the currently well-advanced distillation technology also the inherently better technology?

Unfortunately, the freezing of salt water generates a slurry of ice crystals contaminated with brine, which does not separate naturally. The brine adhering to the surface of well-drained ice amounts to from 20 to 40 per cent of the weight of the actual ice.

These then are the central facts on freezing upon which we begin the discovery of a desalination process:

1. Individual ice crystals formed in brackish water are salt free.
2. Brine adheres to the ice crystals in an amount of from 20 to 40 per cent of the ice weight.
3. The heat of crystallization of ice is 144 Btu/lb.

7.4 FREEZING, WASHING, AND MELTING

The central task sequence upon which to base a freezing process is obvious. The brackish water is to be cooled to its freezing point, partially frozen, the ice freed from the brine, and the ice melted to form fresh water. All this must be done at a target cost of about $0.50/1000 gallons of water formed.

How might the freezing be done? Normally, refrigeration is accomplished by vaporizing a fluid that has the desired boiling point. The heat is removed as the heat of vaporization of the refrigerant. Butane boils at 31°F, so we might begin by thinking in terms of a butane refrigerant. Seawater begins to freeze at about 28°F, so a slight reduction in operating pressure will be required to cause the butane to boil at the required lower temperature.

Figure 7.4–1 suggests itself as a feasible sequence of operations. How costly might this be?

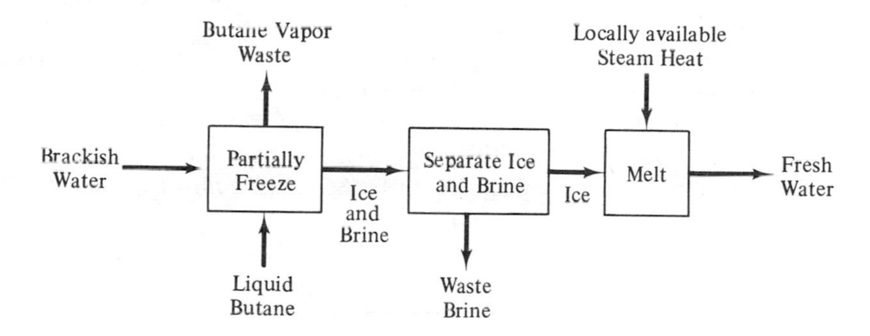

Cost of Butane Alone is Five Hundred Times the Target of $0.50 per Thousand Gallons. Steam to Melt the Ice is also Much too Costly.

Figure 7.4–1. A First Attempt at Process Discovery Is Wildly Expensive.

Butane costs about $0.03/lb and has a heat of vaporization of 165 Btu/lb compared to the heat of crystallization of water of 144 Btu/lb. The cost of the butane necessary to power this freezing plan can be estimated quite easily.

$$\left(\frac{1000 \text{ gal}}{\text{of water}}\right)\left(\frac{8 \text{ lb}}{\text{gal}}\right)\left(\frac{144 \text{ Btu}}{\text{lb ice}}\right)\left(\frac{\text{lb butane}}{165 \text{ Btu}}\right)\left(\frac{\$0.03}{\text{lb butane}}\right) = \$230$$

Just to freeze the ice we are running into costs of $230/1000 gal. Some major modifications are in order.

How costly is it to use steam to melt the ice? Low-pressure steam costs about $0.30/1 million Btu. On this basis the steam cost to melt the ice is

$$(1000 \text{ gal})\left(\frac{8 \text{ lb}}{\text{gal}}\right)\left(\frac{144 \text{ Btu}}{\text{lb water}}\right)\left(\frac{\$3 \times 10^{-7}}{\text{Btu}}\right) = \$0.35$$

The idea of using steam to melt the ice is just not economically feasible if we must purchase the steam, for the steam cost is too large a fraction of the target water cost of $0.50/1000 gal.

7.5 TASK INTEGRATION ATTEMPTED

Our attention now is directed to the problem of task integration. We must drastically reduce the butane and steam costs, or else abandon this concept.

Butane is easily liquified. We might recycle the vented butane. Liquifaction of gases is accomplished by compressing the gas and removing the heat of vaporization by some kind of heat exchange, usually with cooling water. Furthermore, we might reduce the cost of the steam needed to melt the ice by burning a little bit of the butane to provide the energy. Figure 7.5–1 is the result of this first attempt at task integration.

Let us test the economics of this butane recovery scheme. To compress the butane to the point where it will liquify upon cooling and expanding requires about 10 Btu/lb of butane compressed. With energy costs on the order of \$0.50/1 million Btu, the compression costs are calculated to be \$5 $\times$ 10^{-6}/lb of butane liquified.

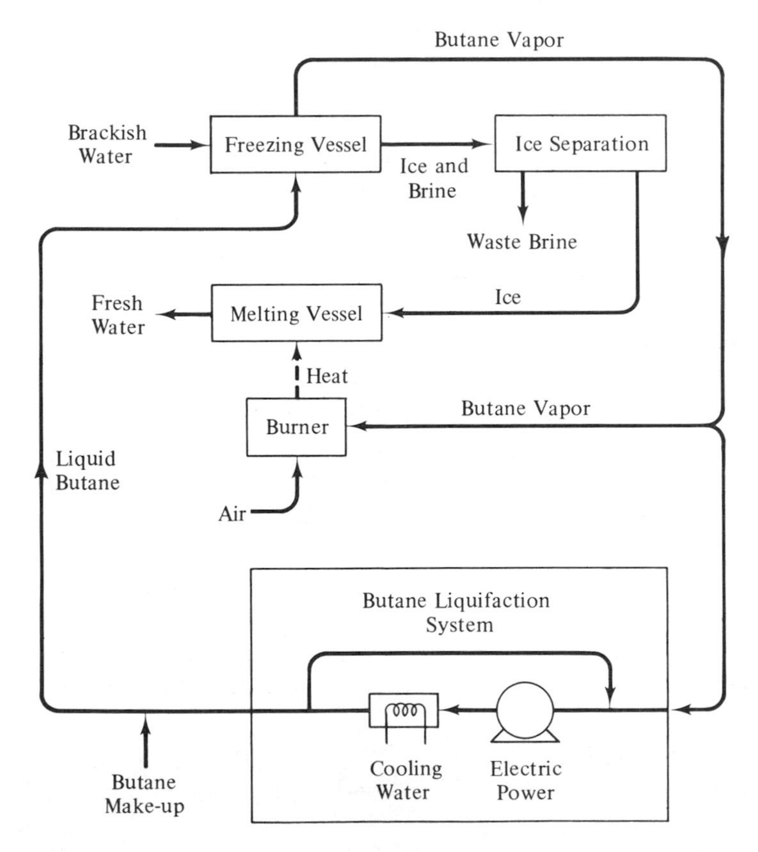

Figure 7.5–1. A First Attempt at Task Integration
Butane recovery cost is not unreasonable, but ice melting plan is too expensive.

Converting this cost to dollars per thousand gallons of fresh water formed gives

$$\left(\frac{\$5 \times 10^{-6}}{\text{lb butane compressed}}\right)\left(\frac{144 \text{ lb butane}}{165 \text{ lb ice}}\right)\left(\frac{8000 \text{ lb}}{1000 \text{ gal}}\right) = \$0.04/1000 \text{ gal}$$

This rough calculation does not screen out the idea of recycling the butane. We shall keep this in the plan for a while at least. Of course, the cost of the cooling water needed during liquifaction has not been included.

What about the use of butane to provide the energy to melt the ice? The heating value of butane is 20,000 Btu/lb burned. The cost of burning sufficient butane is easily computed:

$$\left(\begin{matrix}1000 \text{ gal}\\ \text{of water}\end{matrix}\right)\left(\frac{8 \text{ lb}}{\text{gal}}\right)\left(\frac{144 \text{ Btu}}{\text{lb melted}}\right)\left(\frac{1 \text{ lb butane}}{20{,}000 \text{ Btu}}\right)\left(\frac{\$0.03}{\text{lb butane}}\right) = \$1.70$$

The idea of melting the ice with a butane-powered burner is out of the question.

In summary, in Figure 7.5–1 the idea of recycling the butane for the freezing of the ice is economically viable so far, but the melting phase is far too costly. Additional task integration is needed, or maybe the plan is inherently too costly.

7.6 TASK INTEGRATION IMPROVED

Every Btu pulled out of the water to form ice must be placed back into the ice to form water. This is a one-to-one exchange that invites task integration. The energy removed from the water to form ice is picked up by the butane and is contained in the butane vapor leaving the freezer. To change the butane vapor back into butane liquid, we add energy by the compression and remove energy by the cooling of the compressed vapors. A total energy balance indicates that more than enough energy must be removed from the butane during liquefaction than is needed to melt the ice. The melting process itself can be integrated into the butane liquifaction cycle.

Furthermore, the liquid butane and water are not miscible, so the compressed butane might just as well be flashed directly into the melting tank and the liquid decanted. Figure 7.6–1 is the much simpler and probably much less costly plan resulting from these two task integrations.

However, something critical has been omitted. In computing the butane costs we only considered the heat of crystallization of the ice. What about the cost of cooling the incoming brackish water to its freezing point? Might this cost make the process unattractive?

Suppose that the brackish water is to be concentrated from a 3.5 per cent salt solution, say, to a brine of 7 per cent concentration. If S is the amount of brackish water brought into the process, F the amount of fresh water formed, and B the amount of brine formed, material balances give the following:

$$S = B + F \qquad \text{total material balance}$$

$$0.035S = 0.07B \qquad \text{salt balance}$$

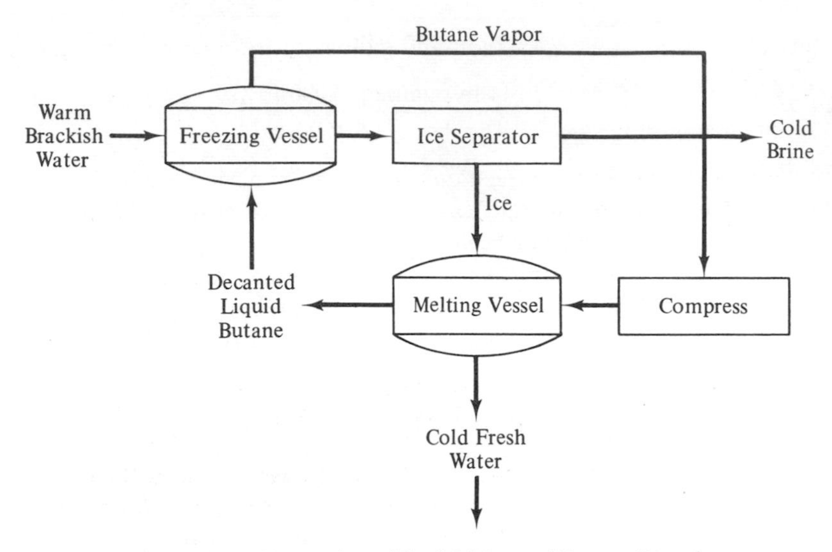

Figure 7.6–1. Integration of Ice Melting and Butane Recycle

Solving gives

$$\frac{S}{F} = 2$$

In this situation twice as much brackish water must be cooled to the freezing point as fresh water formed.

If the brackish water is available at 60°F and must be cooled to 30°F prior to freezing, the energy requirements for precooling amount to

$$\left(\frac{2 \text{ lb brackish water}}{\text{lb fresh water}}\right)\left(\frac{1 \text{ Btu}}{\text{lb °F}}\right)(60 - 30)\text{°F} = \left(\frac{60 \text{ Btu}}{\text{lb of fresh water}}\right)$$

Now since the energy used in freezing is 144 Btu/lb, the precooling load will increase the power costs during refrigeration by $\frac{60}{144}$ or 40 per cent. We conclude that the precooling of the brackish water is an important problem with which we must be concerned. If it were only 4 per cent, our time might be better spent elsewhere.

The fresh water leaves cold, as does the used brine in the process outlined in Figure 7.6–1. This is unnecessary, for these cold streams could be used to precool the incoming brackish water. Certainly the energy these cold streams can pick up is on the same order of magnitude as the energy needed to precool the brackish water. The combined flow rates of the fresh water and the brine must equal the flow rate of the incoming brackish water. The brackish water is to go from ambient temperatures to 30°F, and the fresh water and the brine are both available at 30°F and should leave

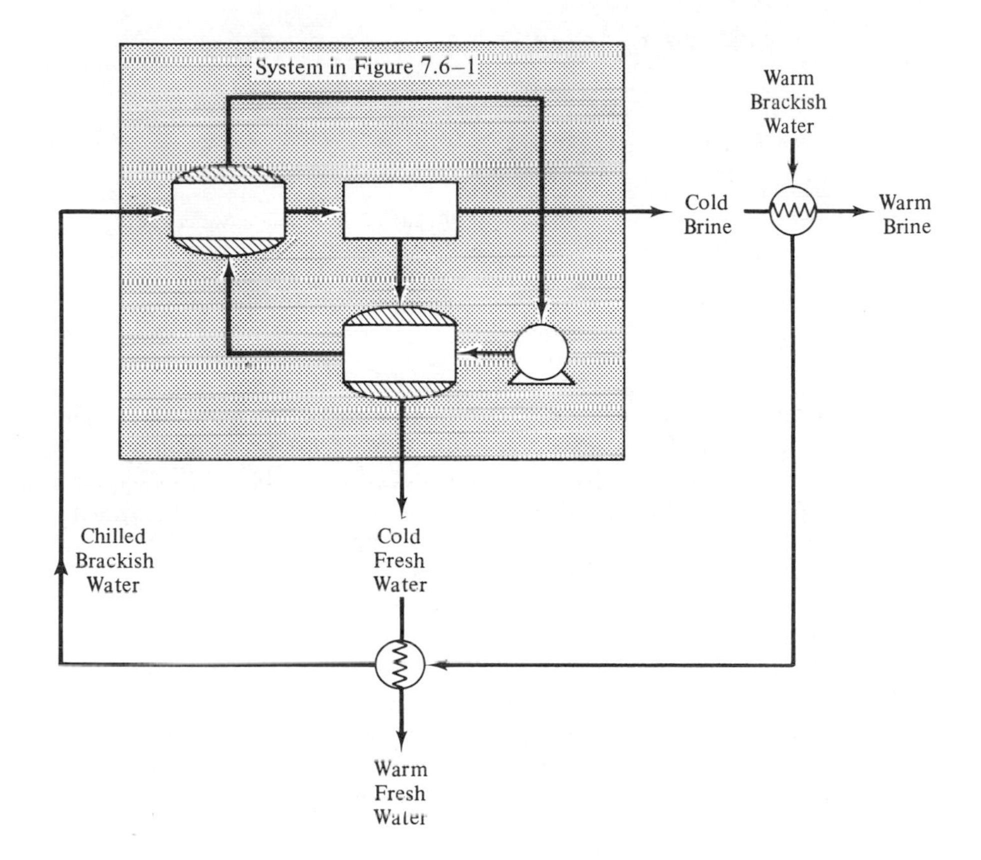

Figure 7.6–2. Task Integration to Chill Incoming Brackish Water Using Heat-Exchanger Network

the process at near ambient temperatures. The two heat exchangers that appear in Figure 7.6–2 should be useful in reducing costs.

7.7 Choice of Refrigerant Examined

Figure 7.6–2 is a roughed-out processing plan based on the use of butane as the refrigerant. Doubtless a different processing scheme would have been obtained if the refrigerant choice had been a material with a cost and properties considerably different than butane. Butane was selected initially because it boils at about the freezing point of water and is immiscible with water. Do these criteria necessarily lead to the discovery of an economic process?

Table 7.7–1 lists several alternative refrigerants, each with advantages and disadvantages. We seek a refrigerant that does not interact with water, is relatively

TABLE 7.7–1 Properties of Refrigerant Candidates

Compound	$\Delta H_{vaporization}$ (Btu/lb)	Vapor Pressure at 30°F (psia)	Comments
Butane	164	14	Relatively insoluble in water
Isobutane	156	25	Forms a hydrate with water
Ammonia	172	60	Very soluble in water
Water	1000	0.07	Only compound not foreign to processing

cheap or easily recoverable, is easily bolied and reliquified at temperatures near the freezing point of water, and has a relatively high heat of vaporization so that the amount of refrigerant to be recycled is low. In Table 7.7–1 water is unique. Water is the only substance that is not foreign to our processing problem. In Chapter 5 the heuristic was presented: *All other things being equal, shy away from separations that require the use of a species not normally present in the processing.* With water, all other things are not equal, since the pressure must be reduced greatly to cause water to vaporize at the freezing point of water. However, since it is not foreign, let us examine the choice of water as the refrigerant.

Figure 7.7–1 is a phase diagram for pure water and for seawater. At high pressures and low temperatures the water is present as a solid, at low pressures and high temperatures as a vapor, and at moderate temperatures and pressures as a liquid. The line that forms the interface between the vapor and liquid region sweeps out the boiling point of water as a function of pressure. Along this line the liquid

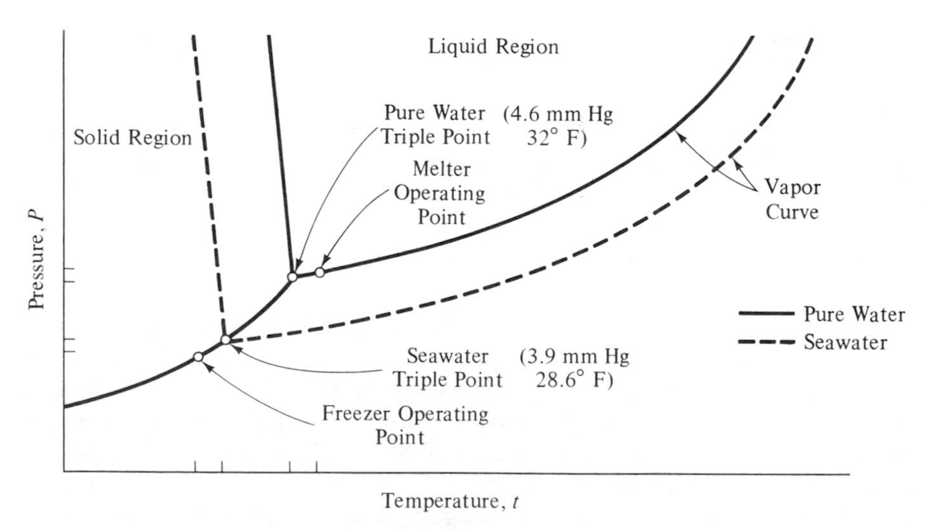

Figure 7.7–1. Phase Diagram for Pure Water and Seawater

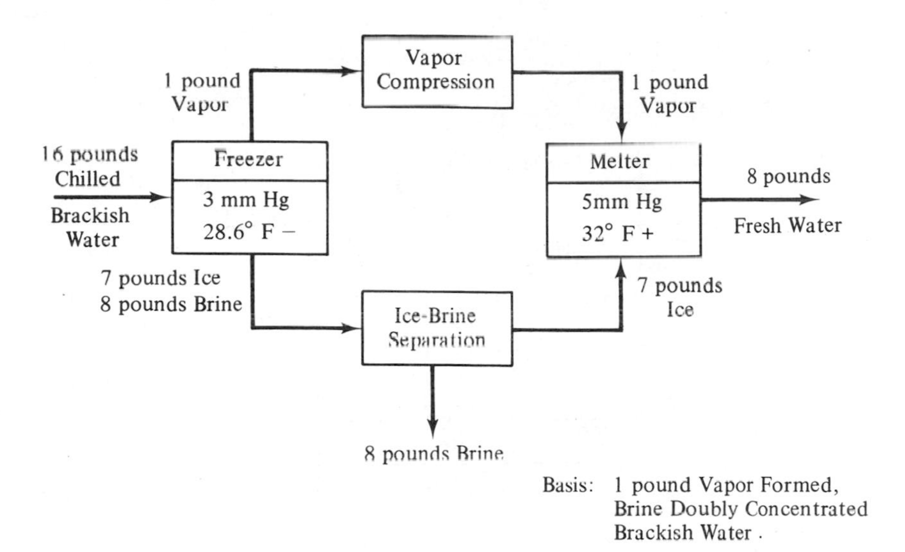

Figure 7.7–2. Task Sequence for Vacuum Freezing–Vapor Compression Desalination

and vapor can exist simultaneously. The *triple point* is defined by the unique temperature and pressure under which the vapor, liquid, and solid can exist simultaneously. For seawater the triple point occurs at 3.9 mmHg and 28.6°F, and for pure water 4.6 mmHg and 32°F.

By reducing the pressure above seawater at its triple point, the water vaporizes, and the withdrawal of the heat of vaporization causes ice to form. For each pound of water vaporized about 7 pounds of ice forms; this is the ratio of the heat of vaporization to the heat of crystallization. After the ice is freed of the brine and brought into contact with the pure vapor, melting occurs if the pressure is above the triple point for the pure water. The operating points of the freezer and melter shown in Figure 7.7–1 are each on the proper side of the respective triple points.

Figure 7.7–2 outlines the task sequence based on this vacuum freezing–vapor compression concept. The basis taken in this figure is the vaporization of 1 lb of water, and the concentration of seawater from 3.5 per cent salt to a brine of 7.0 per cent salt. The flow rates shown are consistent with ideal operation under these conditions.

The design by Colt Industries, Inc., of the equipment necessary to implement the freezing–melting portion of this task sequence is shown in Figure 7.7–3. The freezer–melter consists of a single vessel with a central cylinder above which the main compressor is mounted. The vapor is drawn from the seawater in the lower part of the vessel and compressed into the upper part of the vessel into which the clean ice is placed. There the melting takes place and the fresh water is formed.

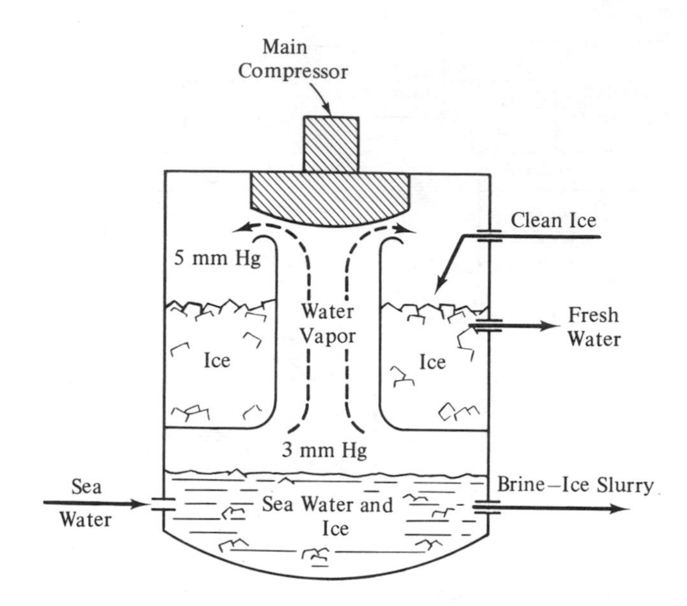

Figure 7.7–3. Schematic Design of an Early Freezer–Melter by Colt Industries, Inc.

7.8 INTEGRATION OF A PISTON BED WASHER

Our attention so far has been centered on the freezing and melting operations. Now the question of ice–brine separation is considered. First, how important a problem is the separation? A material balance gives considerable insight into the problem.

Brine amounting to 20 to 40 per cent of the weight of the ice adheres to the surface of the ice crystals. To estimate a lower bound on the importance of ice washing, suppose that the adherence only amounts to 20 per cent. And suppose that the brine is a result of concentrating seawater to a 7.0 per cent salt brine. If 1 lb of wet ice is melted, 0.20 lb of the water will come from the adhering brine, and this will contain $(0.20)(0.07) = 0.014$ lb of salt. Thus the unwashed ice when melted produces a water with 1.4 per cent salt (14,000 ppm), which is well above the salinity limits of 500 ppm set on potable water. We conclude that the brine must be removed from the ice before melting.

The development of an effective ice-washing procedure is an interesting story in itself; one of the first large-scale freezing processes put on stream did not produce

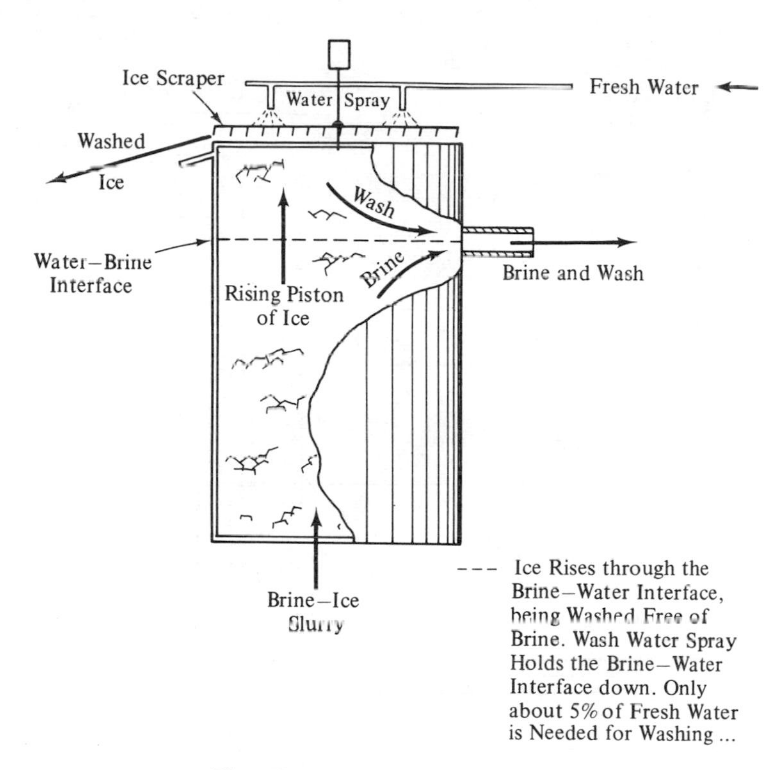

Figure 7.8–1. Piston Bed Ice Washer

fresh water at near the expected rate, since the washing of the ice consumed too much fresh water. However, piston bed washing methods are well developed now, as shown in Figure 7.8–1. The piston bed consists of a tall cylinder into the bottom of which the ice–brine slurry is pumped. The ice rises as a piston through the cylinder, and on the top of the piston of ice fresh, cold water is sprayed. The brine is removed through ports partway down the cylinder, along with the fresh water that has washed down through the ice. The washed ice is scraped off the top of the cylinder by rotating knives. Such a device when properly designed has been found to use only about 5 per cent of the fresh water in the washing operation.

Figure 7.8–2 is a schematic diagram of the desalter Model 100A1 of Colt Industries, Inc., in which the freezer–melter unit and the ice-washing unit are placed side by side. Additional equipment is needed to deaerate the incoming brackish water, and to provide the additional refrigeration to balance any heat leaks into the system and the heat generated by the compressor and pumps. Figure 7.8–3 is a photograph of this 100,000 gallon/day unit mounted on a railroad car ready to be shipped to the site of operation. A number of these vacuum freezing–vapor compression units are in operation around the world (see Figure 7.8–5).

A newer design shown in Figure 7.8–4 combines the freezing–washing–melting operations into one vessel, resulting in greater economy. In the next section we show

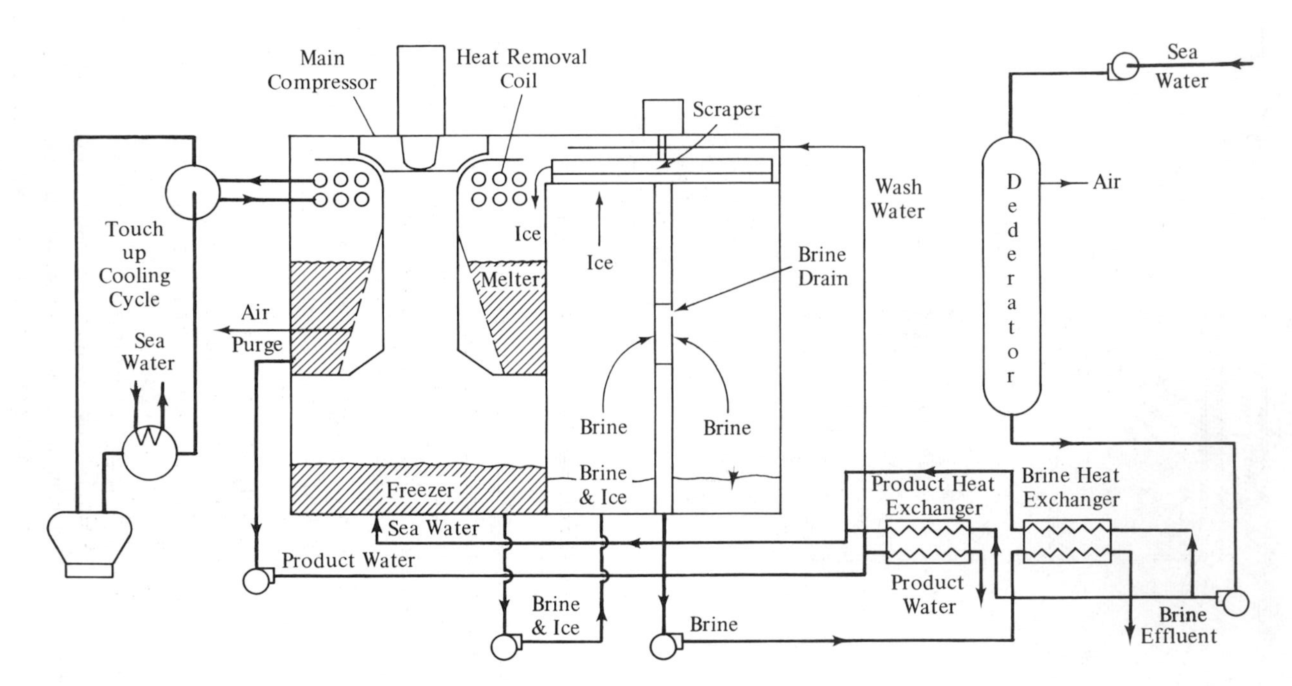

Figure 7.8–2. Schematic for the Complete Vacuum Freezing–Vapor Compression Unit Pictured in Figures 7.8–3 and 7.8–4

Figure 7.8–3. Portable 100,000 Gal/Day VF–VC Desalting Process, Model 100A1, Colt Industries, Inc., Beloit, Wisconsin

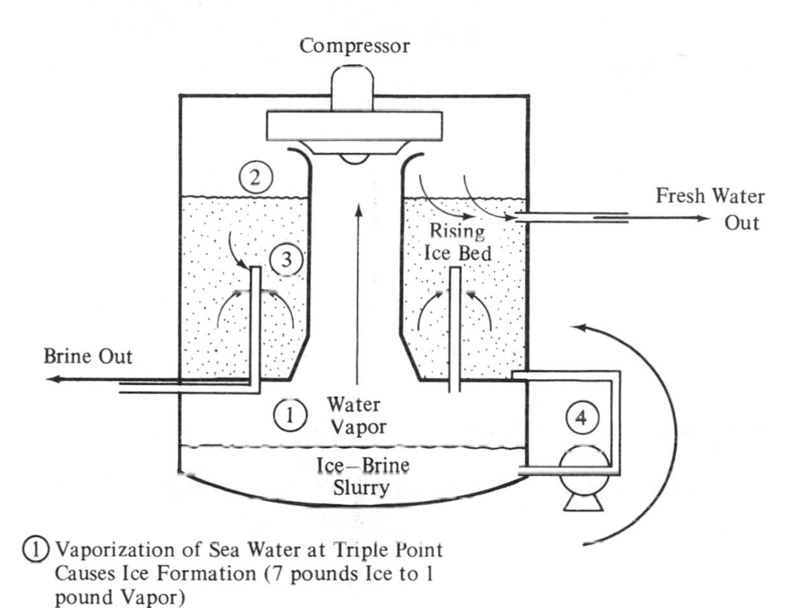

① Vaporization of Sea Water at Triple Point Causes Ice Formation (7 pounds Ice to 1 pound Vapor)

② Vapor Condenses on Clean Ice Surface to Form Fresh Water (7 pounds Ice Melted to 1 pound Vapor Condensed)

③ 5% of Fresh Water Formed is Used to Wash Rising Piston of Ice. Brine and Wash Water Leave Process

④ Ice–Brine Slurry Pumped to Bottom of Ice Wash Column.

Figure 7.8–4. Freezing–Washing–Melting Operations Consolidated into One Piece of Equipment, a New Design

Figure 7.8–5. Vacuum Freezing–Vapor Compression Test Unit in Place at Elat, Israel

how this task-integrated process fits into the design of a 500,000 gal/day desalting module, and how these modules are stacked to form a 5 million gal/day desalting plant.

7.9 DESIGN AND ECONOMICS OF A COMMERCIAL MODULE

In the previous sections we have seen how the concept of VF–VC desalting processes might come about. In this section we present the engineering detail and economics of a 500,000 gal/day module developed by Colt Industries, Inc., of Beloit, Wisconsin. The module is small enough so that it can be factory assembled and shipped to the desalting site, and desalting capacity greater than 500,000 gal/day can be achieved by connecting a number of the modules in parallel. An economic study is presented of a plant using 10 of these modules to achieve a 5 million gal/day capacity. We have witnessed the beginning stages of this process, and now we look at the final specifications of the process. An enormous amount of engineering effort goes into the transformation of our rough concept into an operating system. The transition from the material in the previous sections to the material in this section is great and cannot be included here, but it is desirable to finish this case study of process discovery with a completed design.

The desalting module shown in Figure 7.9–1 is 65 ft long, 10 ft wide, and 33 ft high. It consists of two factory-assembled skids joined together at the site. The lower

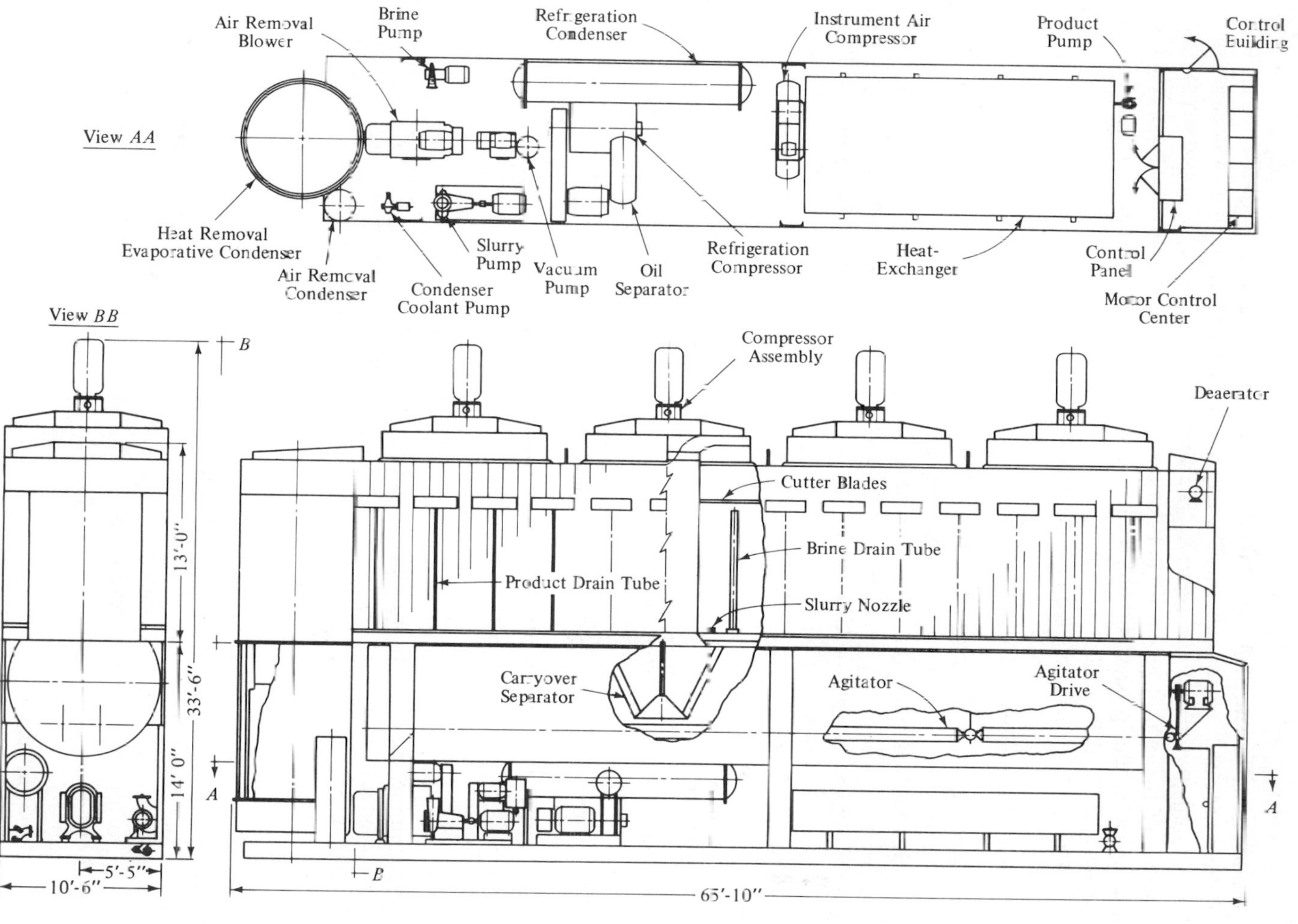

Figure 7.9–1. 500,000 Gal/Day VF–VC Desalter Module of Colt Industries, Inc.

skid contains the main subframe, the lower portion of the main pressure vessel, and all the other auxiliary equipment. The upper skid contains the melter–condenser and ice-washing portions of the process. The four compressors are shipped separately because of the overall height limitations during shipping. When these packages are assembled at the processing site, Colt VC–VF Module 500A1 is formed with a capacity of 500,000 gal/day of fresh water.

The modular elements are now described.

Heat Exchanger

The heat exchanger for precooling the feed brackish water consists of thin aluminum sheets approximately 29 in. wide and 238 in. long, which contain stamped-in rows of hemispherical dimples and shaped ports at each end. These are stacked so that the dimples space the sheets a uniform distance. By varying the gasket arrangements, the three fluids, warm brackish water, cold brine, and cold fresh water, can be forced to exchange heat so that the incoming brackish water is cooled to less than 32°F. The heat-exchange section is 30 in. high, 20 ft long, and 10 ft wide, and is mounted on the lower skid.

Deaerator

Air must be removed from the incoming seawater so that the proper vacuum can be maintained in the freezing unit. After the brackish water has been precooled, it passes to a 7-ft-high vessel attached to the main frame, filled with porous packing and maintained at a vacuum suitable to cause the dissolved air to leave the water. The deaerated water flows by gravity to the freezer.

Freezer

The lower section of the main vessel is the freezing chamber. It is divided into four stages connected in series, and the ice-slurry effluent from one stage passes on to the next. Agitators stir the slurry to increase the surface area and prevent the formation of ice on the vessel walls. The freezer is a 10-ft-diameter semicylindrical vessel 50 ft long.

Vapor Compressor

The most important single item in the process is the vapor compressor. Four identical compressors are mounted above the melter. These are a specially designed radial-bladed centrifugal unit powered by a 100-hp 3600-rpm motor. The flow rate through a single unit is approximately 400,000 ft^3/min of water vapor at a compression ratio of 1.5.

Counterwasher–Melter

The ice is washed and melted in the region surrounding the four compressor inlet tubes. The ice slurry is pumped there from the fourth freezing stage, and the brine is removed through porous tubes located in the center of the ice pack. Melting occurs at the top of the pack along paths formed in the ice by rotating knife blades. The fresh water is removed through screened openings located along the sides of the vessel.

Carryover Separator

This prevents brine drops from being carried over with the vapor from the freezing section to the melting section.

Heat Removal System

A small amount of refrigeration is needed to remove the excess water vapor formed by heat leaks into the system and by the heat generated by compression and pumping operations. This is accomplished by a self-enclosed ammonia refrigeration cycle.

Air-Removal System

The air not removed by the feed-water deaerator is released during the freezing operation, and this is removed by a special vacuum pumping operation.

Material and Energy Balance

The complete material and energy balance for a single module is shown in Figure 7.9–2. There one finds the specifications of all the process flows, including both material and energy flow.

Economics

Table 7.9–1 shows the cost breakdown for a water-processing plant consisting of 10 of these modules and having a total capacity of 5 million gal/day. The investment

TABLE 7.9–1 Economics of a 5 Million Gal/Day Plant of 10 Modules

I	Capital costs	$5,500,000
II	Operating cost	$410,000/year
	Equipment amortization	$435,000/year
III	Water cost ($/1000 gal)	
	Capital cost	0.26
	Electric power	0.19
	Labor	0.04
	Other	0.02
	Total	$0.53/1000 gal

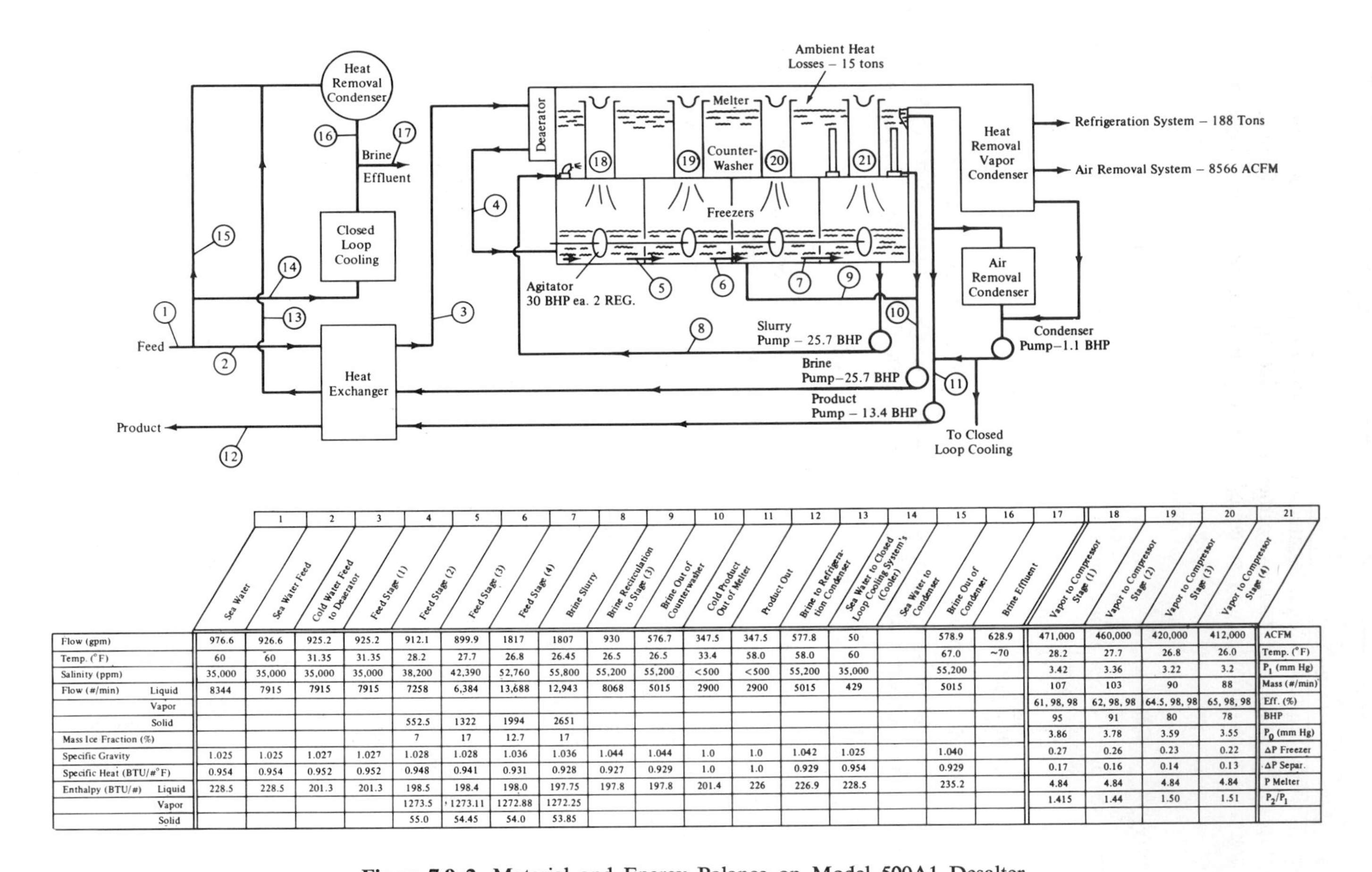

	1 Sea Water	2 Sea Water Feed	3 Cold Water Feed to Deaerator	4 Feed Stage (1)	5 Feed Stage (2)	6 Feed Stage (3)	7 Feed Stage (4)	8 Brine Slurry	9 Brine Recirculation to Stage (3)	10 Brine Out of Counterwasher	11 Cold Product Out of Melter	12 Product Out	13 Brine to Refrigeration Condenser	14 Sea Water to Closed Loop Cooling System's (Cooler)	15 Sea Water to Condenser	16 Brine Out of Condenser	17 Brine Effluent
Flow (gpm)	976.6	926.6	925.2	925.2	912.1	899.9	1817	1807	930	576.7	347.5	347.5	577.8	50		578.9	628.9
Temp. (°F)	60	60	31.35	31.35	28.2	27.7	26.8	26.45	26.5	26.5	33.4	58.0	58.0	60		67.0	~70
Salinity (ppm)	35,000	35,000	35,000	35,000	38,200	42,390	52,760	55,800	55,200	55,200	<500	<500	55,200	35,000		55,200	
Flow (#/min) Liquid	8344	7915	7915	7915	7258	6,384	13,688	12,943	8068	5015	2900	2900	5015	429		5015	
Vapor																	
Solid					552.5	1322	1994	2651									
Mass Ice Fraction (%)					7	17	12.7	17									
Specific Gravity	1.025	1.025	1.027	1.027	1.028	1.028	1.036	1.036	1.044	1.044	1.0	1.0	1.042	1.025		1.040	
Specific Heat (BTU/#°F)	0.954	0.954	0.952	0.952	0.948	0.941	0.931	0.928	0.927	0.929	1.0	1.0	0.929	0.954		0.929	
Enthalpy (BTU/#) Liquid	228.5	228.5	201.3	201.3	198.5	198.4	198.0	197.75	197.8	197.8	201.4	226	226.9	228.5		235.2	
Vapor					1273.5	1273.11	1272.88	1272.25									
Solid					55.0	54.45	54.0	53.85									

	18 Vapor to Compressor Stage (1)	19 Vapor to Compressor Stage (2)	20 Vapor to Compressor Stage (3)	21 Vapor to Compressor Stage (4)
ACFM	471,000	460,000	420,000	412,000
Temp. (°F)	28.2	27.7	26.8	26.0
P_1 (mm Hg)	3.42	3.36	3.22	3.2
Mass (#/min)	107	103	90	88
Eff. (%)	61, 98, 98	62, 98, 98	64.5, 98, 98	65, 98, 98
BHP	95	91	80	78
P_0 (mm Hg)	3.86	3.78	3.59	3.55
ΔP Freezer	0.27	0.26	0.23	0.22
ΔP Separ.	0.17	0.16	0.14	0.13
P Melter	4.84	4.84	4.84	4.84
P_2/P_1	1.415	1.44	1.50	1.51

Figure 7.9–2. Material and Energy Balance on Model 500A1 Desalter (Seawater Feed)

in equipment is about $5,500,000 and the fresh water costs $0.53/1000 gal. It is interesting that of this water cost $0.26 is attributed to the prorated cost of the equipment, $0.19 to the electricity to run the process, and $0.04 to the operating labor.

7.10 Conclusion

The case study examined here illustrated the application of task integration and equipment selection, and showed how, in a particular problem, one of the several areas of process discovery may dominate. In any situation where large amounts of material must be processed at low cost, the source of the raw material and the efficiency of processing may be critical. Here the raw material is available at no cost, so the efficiency of processing is of critical importance.

To see the end result of these deliberations, it was necessary to jump over much of the more specialized engineering to show the constructed vacuum freezing–vapor compression process module and to show the results of an economic study. Solely on the basis of the material in this text we cannot distinguish between the butane process and the VC–VF process; both are viable alternatives that would survive to the more advanced stages of development.

In situations where processing efficiency is not so important, process discovery may take on a completely different appearance. The development of a process to oxygenate blood during open-heart surgery might focus on process reliability and simplicity. The development of a process for the production of chlorodecane, in the next chapter, focuses on reaction-path synthesis, species allocation, and task selection. Part of the skill of the engineer is in sensing the central problem upon which to pay attention.

REFERENCES

An interesting early survey of water manufacture is

E. R. Gilliland, "Fresh Water for the Future," *Industrial and Engineering Chemistry*, **47**, 2410, 1955.

For more detail see

K. S. Spiegler, *Salt Water Purification*, John Wiley & Sons, Inc., New York, 1962.

J. McDermott, *Desalination by Freeze Concentration*, Noyes Data Corp., New York, 1971.

H. F. Wiegandt, "Saline Water Conversion By Freezing," *Advances in Chemistry*, **27**, 82, 1960.

R. Popkin, *Desalination*, Praeger Publishers, Inc., New York, 1968.

The enormous research and development effort sponsored by the *Office of Saline Water* is reported and summarized in the annual document to the President:

Saline Water Conversion Reports, Government Printing Office, Washington, D.C., 1952–present.

See also

Proceedings of the First International Symposium on Water Desalination, Vols. I–III, U.S. Department of the Interior, Office of Saline Water, Government Printing Office, Washington, D.C., 1965.

PROBLEMS

Here we provide some fragmentary information from which desalination processes might be discovered. This information is to be used as a basis of class discussion to see if, in the period of three or four class meetings, some interesting processes evolve. We find it useful for the students to take over the class period as an informal workshop. Some standard source of physical property data is useful.

1. **SOLVENT EXTRACTION.** Triethylamine dissolves about 30 per cent by weight of water at 20°C and only 2.5 per cent at 50°C.

2. **HYDRATE PROCESS.** Propane is capable of forming a solid hydrate, with 17 molecules of water, although liquid propane dissolves only very small amounts of water. At about 5 atm the hydrate precipitates out at about 2°C in seawater, and about 20 per cent by weight of brine adheres to the solid hydrate.

8

Detergents from Petroleum

What you have been obliged to discover by yourself leaves a pathway in your mind which you can use again when the need arises.

Lichtenberg,
Aphorismen

8.1 Introduction

In this chapter we apply the principles of process discovery to the development of part of the technology to convert crude oil into detergents. No new ideas are presented; we apply what we already know. Our attention is focused on the problem of converting a kerosene fraction of the crude oil into an intermediate material, a chlorinated hydrocarbon, which then fits into the larger campaign of detergent production. In this section we become acquainted with the processing framework in which we are to innovate.

The main active ingredient in detergents are specially designed molecules that alter the properties of oil–water interfaces. These surface-active molecules, or *surfactants*, are constructed to have part of the molecule oil soluble and part water soluble. The oil-soluble part of a surfactant dissolves in oils and greases, and the water-soluble part dissolves in water, thereby solubilizing oils and greases. Sodium dodecylbenzenesulfonate is one of the most common surfactants used in detergent mixtures.

$$\underbrace{C_{12}H_{25}\text{-}C_6H_4\text{-}}_{\text{hydrocarbon part tends to be oil soluble}}\ \underbrace{SO_3^-Na^+}_{\text{salt part tends to be water soluble}}$$ (sodium dodecylbenzenesulfonate)

These surfactants are not found in nature and must be constructed from readily available material, such as crude oil.

Figure 8.1–1 shows the acceptance of detergent blends as cleaning agents, and Table 8.1–1 presents the composition of several typical blends.* In 1970 over 5 billion lb of detergents were used in the United States, and the production capacity of a typical detergent synthesis process was 100 million lb/year. A target cost of some $0.20/lb of detergent blend gives some idea of the production costs toward which we must aim.

The chemistry that gives access to suitably inexpensive raw materials is quite straightforward. One reaction path begins with the chlorination of kerosene to produce a chlorinated hydrocarbon. The chlorinated hydrocarbon is used to attach the kerosene molecule to benzene. This product is then sulfonated to attach the water-soluble ion, and the product is neutralized to form a surfactant powder.

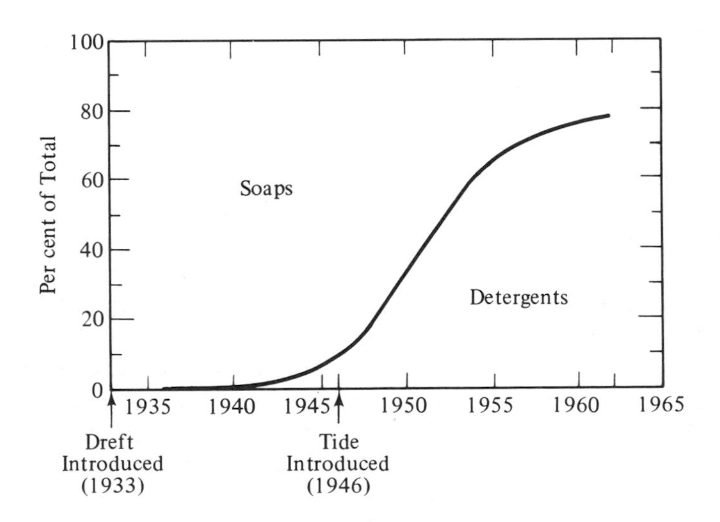

Figure 8.1–1. Relative Production of Detergents and Soap

* The amount of phosphates in detergent blends will be drastically reduced in the near future. It is now realized that phosphates contribute to the early death of lakes by overfertilization.

TABLE 8.1–1 Basic Composition of Three Types of Dry Detergents (Granules)

Ingredient	Function	Ingredients on Dry-Solids Basis (wt %)		
		Light-Duty High Sudsers	Heavy-Duty Controlled Sudsers	Heavy-Duty High Sudsers
Surfactants				
Organic active, with suds regulators	Removal of oily soil, cleaning	25–40	8–20	20–35
Builders				
Sodium tripolyphosphate and/or tetrasodium pyrophosphate	Removal of inorganic soil, detergent building	2–30	30–50	30–50
Sodium sulfate	Filler with building action in soft water	30–70	0–30	10–20
Soda ash	Filler with some building action	0	0–20	0–5
Additives				
Sodium silicate having 2.0 $\leq SiO_2/Na_2O \leq$ 3.2	Corrosion inhibitor with slight building action	0–4	6–9	4–8
Carboxymethyl cellulose	Antiredeposition of soil	0–0.5	0.5–1.3	0.5–1.3
Fluorescent dye	Optical brightening	0–0.05	0.05–0.1	*ca.* 0.1
Tarnish inhibitors	Prevention of silverware tarnish	0	0–0.02	0–0.02
Perfume and sometimes dye or pigment	Aesthetic, improved product characteristics	0.1[a]	0.1[a]	0.1[a]
Water	Filler and binder	1–5	2–10	3–10

Source: J. R. Van Wazer, *Phosphorus and Its Compounds*, Vol. 2, John Wiley & Sons, Inc. (Interscience Division), New York, 1961, p. 1760.

[a] Approximate.

Chlorination of kerosene:

$$\underset{\text{(kerosene)}}{C_{12}H_{26}} + \underset{\text{(chlorine)}}{Cl_2} \xrightarrow{\text{light}} \underset{\text{(chlorododecane)}}{C_{12}H_{25}Cl} + \underset{\text{(hydrogen chloride)}}{HCl}$$

reaction with benzene:

$$C_{12}H_{25}Cl + \underset{\text{(benzene)}}{C_6H_6} \xrightarrow{\text{catalyst}} \underset{\text{(alkylated benzene)}}{C_{12}H_{25}C_6H_5} + HCl$$

sulfonation:

$$C_{12}H_{25}C_6H_5 + \underset{\text{(sulfuric acid)}}{H_2SO_4} \rightarrow \underset{\text{(a surfactant ion)}}{C_{12}H_{25}C_6H_4SO_3^-} + H_2O$$

The extensive use of detergents can cause serious ecological problems, since eventually the 5 billion lb used annually will work its way into the nation's water resources. It is important for this reason that the surfactants be very pure and active, so that little need be used for cleaning, and that the molecules quickly degrade biologically into ecologically innocuous molecules. To prohibit the use of detergents would have a serious effect on public health. To alter the molecules so they behave properly is the reasonable course of action.

In this chapter we examine the development of technology to convert decane, a lower molecular hydrocarbon, into chlorodecane, which then can be used as a building block during the construction of larger molecules, such as the surfactants discussed in this section. Reaction-path synthesis, species allocation, and task selection dominate this problem. We show how *two* or *three* promising processes are isolated from nearly *one thousand* processing alternatives.

8.2 REACTION-PATH SCREENING

Monochlorodecane is not available to any appreciable extent in natural products. Therefore, if it is to be used in the construction of a surfactant or other special molecule, it must be synthesized from readily available material. Since the sales price of detergents is to be measured in cents per pound, the chlorodecane production costs must also be measured in cents per pound. We must find the chemistry that gives access to such inexpensive raw materials.

Monochlorodecane consists of a long hydrocarbon chain with a chlorine atom attached on the end:

$$\underset{\text{(monochlorodecane)}}{C_{10}H_{21}Cl}$$

Common sources of the chlorine molecule are molecular chlorine (Cl_2) and hydrogen chloride (HCl). Sources of the hydrocarbon are normal decane ($C_{10}H_{22}$), decene ($C_{10}H_{20}$), and decanol ($C_{10}H_{22}OH$). Reaction paths that give access to these materials include the following:

Path 1:

$$\underset{\text{(decane)}}{C_{10}H_{22}} + Cl_2 \xrightarrow{\text{light}} \underset{\text{(monochlorodecane)}}{C_{10}H_{21}Cl} + HCl$$

with the side reaction

$$C_{10}H_{21}Cl + Cl_2 \rightarrow \underset{\text{(dichlorodecane)}}{C_{10}H_{20}Cl_2} + HCl$$

Path 2:

$$\underset{\text{(decene)}}{C_{10}H_{20}} + HCl \xrightarrow{\text{catalyst}} \underset{\text{(monochlorodecane)}}{C_{10}H_{21}Cl}$$

Path 3:

$$\underset{\text{(decanol)}}{C_{10}H_{21}OH} + HCl \rightarrow \underset{\text{(monochlorodecane)}}{C_{10}H_{21}Cl} + H_2O$$

All these reactions produce the desired molecule, monochlorodecane, or MCD as we shall call it. We must examine the economics to see if there are essential differences. Table 8.2–1 gives typical costs of the materials.

TABLE 8.2–1 Costs of Raw Materials

	\$/LB-MOLE	\$/LB
Decane	4.80	0.034
Decene	12.00	0.086
Decanol	14.00	0.089
Chlorine	1.77	0.025
Hydrogen chloride	1.00	0.027

We notice that decane is by far the cheapest source of the hydrocarbon part of the molecule. The difference is nearly a factor of 3, dominating the relatively small difference in the sources of the chlorine molecule.

Table 8.2–2 summarizes the total reactant costs for the several reaction paths. Of course, this early economic screening is based on the assumption that the side reaction can be suppressed in reaction path 1, thereby preventing the unwanted conversion of the monochlorodecane into the dichlorodecane.

Based on this preliminary economic screening, the direct chlorination of decane stands out from the other reaction paths. The raw materials are over twice as expensive for the other routes. Let us see what can be done to engineer a process to support the direct chlorination chemistry.

TABLE 8.2–2 Total Reactant Costs Used for Early Economic Screening

REACTION	\$/LB-MOLE MCD	\$/LB MCD
Path 1	6.57	0.037
Path 2	13.00	0.073
Path 3	15.00	0.084

8.3 BROAD OUTLINE OF A PROCESS

The preliminary reaction-path screening isolated this chemistry as a probable basis of a commercial process:

$$\underset{\text{(decane)}}{\text{DEC}} + \underset{\text{(chlorine)}}{Cl_2} \xrightarrow{\text{light}} \underset{\text{(monochlorodecene)}}{\text{MCD}} + \underset{\text{(hydrogen chloride)}}{\text{HCl}}$$

$$\underset{\text{(monochlorodecane)}}{\text{MCD}} + \underset{\text{(chlorine)}}{Cl_2} \xrightarrow{\text{light}} \underset{\text{(dichlorodecane)}}{\text{DCD}} + \underset{\text{(hydrogen chloride)}}{\text{HCl}}$$

These reactions occur when chlorine is dissolved in decane and exposed to light. We must know more about the reaction, especially the extent to which the second reaction destroys the monochlorodecane.

Let us anticipate what might happen commercially. To prevent the production of the DCD, the reactor must not contain large amounts of MCD or Cl_2. This can be accomplished by feeding only small amounts of Cl_2 into the reactor and quickly converting it into MCD. The unwanted side reaction can be starved by swamping the reaction with DEC, thereby removing the Cl_2 needed to convert the MCD into DCD. This qualitative reasoning is confirmed by the pilot-plant studies below.

Figure 8.3–1 shows a pilot-plant reactor used to gather the data needed to confirm our qualitative reasoning on reactor performance. In this small process no attempt is made to achieve economic operation, for the only product is data on the reaction. Reaction data for two feed ratios are shown; they confirm our initial impression of the effects on product distribution. In the excess decane case a large amount of DEC and some Cl_2 remain in the output of the reactor. Can this waste be disposed of economically, say, by using it as a fuel? What are the raw material costs?

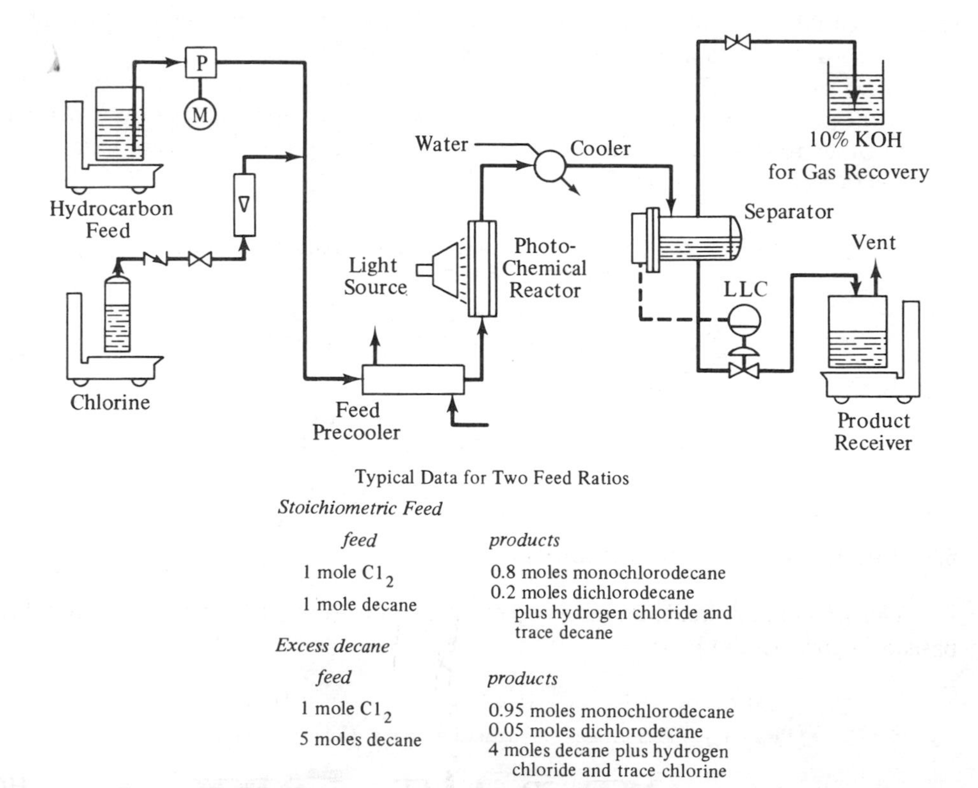

Figure 8.3–1. Pilot Plant from Which Reaction Data Can Be Obtained

$$\left(\frac{1}{0.95\text{ mole MCD}}\right)\left[(5\text{ lb-moles DEC})\left(\frac{\$4.80}{\text{lb-mole}}\right) + (\text{lb-mole Cl}_2)\left(\frac{\$1.77}{\text{lb-mole}}\right)\right]$$

$$= \frac{\$27.00}{\text{lb-mole MCD}} = \$0.15/\text{lb MCD}$$

Hence, just to carry out the reaction, the costs are almost three times as high as for the screening calculation of Table 8.2–2. How can this additional cost be avoided?

Examine the reactor operation at the stoichiometric feed ratio. Note the amount of dichlorodecane produced. The raw material cost for this situation is

$$\left(\frac{1}{0.8\text{ mole MCD}}\right)\left[\left(\frac{1\text{ lb}}{\text{mole }n\text{-decane}}\right)\left(\frac{\$4.80}{\text{lb-mole}}\right) + (\text{lb-mole Cl}_2)\left(\frac{\$1.77}{\text{lb-mole}}\right)\right]$$

$$= \frac{\$8.23}{\text{lb-mole MCD}} = \$0.0465/\text{lb MCD}$$

The use of stoichiometric feed ratios to the reactor reduces the raw-material cost but produces a large amount of overchlorinated material. How can we take advantage of the low amount of by-product produced with excess DEC without incurring the large raw-material cost?

The effluent from the reactor in the excess decane case is a mixture of HCl, Cl_2, DEC, MCD, and DCD. If the species required in the input to the reactor (Cl_2 and DEC) could be separated from this mixture, they could be recycled. This would recover 4 moles of DEC, reducing the feed requirements of 5 moles to only 1 mole. If this separation could be accomplished, the feed costs in the excess-decane case could be brought down to those of the stoichiometric case, and still low DCD production would be achieved.

Figure 8.3–2 outlines the material flow that enables us to run the reaction at excess DEC and still not waste the unreacted DEC.

To test the economics of this recycle idea we must determine how the separation of the mixture is to be performed. If the separation costs are high compared to the raw material savings, other alternatives will have to be developed.

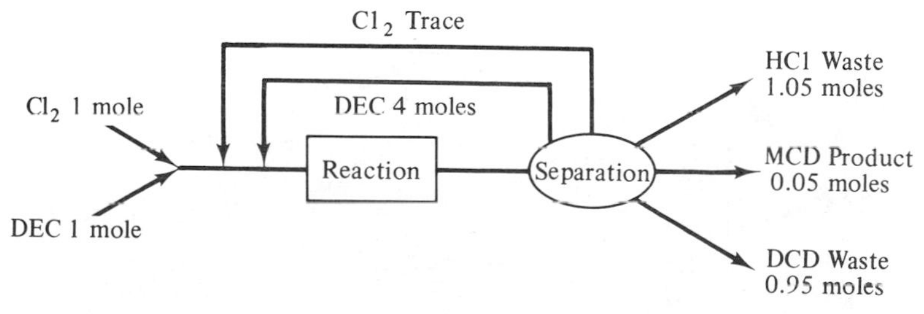

Figure 8.3–2. First Attempt at Species Allocation
Reaction operates at excess DEC to limit DCD production. Recycle mitigates waste of unreacted DEC.

8.4 BASES FOR SEPARATION

To operate the reactor at an excess DEC, and also to recycle the unreacted DEC, we are forced into the problem of separating from large amounts of DEC smaller amounts of HCl, MCD, and DCD. To separate mixtures, we look for differences in physical and chemical properties among the species to be separated. Table 8.4–1 contains the basic physical property data. For this mixture it is clear that differences in boiling points will be useful.

TABLE 8.4–1 Physical Properties and Relative Molar Amounts

	MOL WT	MP (°C)	BP (°C)	SOLUBILITY IN WATER (G/LITER)	RELATIVE AMOUNT
Monochlorodecane, MCD	177	−50	215	—	0.95
Dichlorodecane, DCD	211	−40	241	—	0.05
Decane, DEC	142	−30	174	—	4.00
Chlorine, Cl_2	71	−101	−34	25	Trace
Hydrogen chloride, HCl	36.5	−111	−85	380	1.05

The boiling-point order is shown in Figure 8.4–1. Notice that there is a wide range of boiling points and that all the boiling-point differences are relatively large. The low-boiling mixtures may cause problems, as they will be costly to condense. However, the large differences in boiling points are attractive, and separation by distillation will be considered as the initial move in the development of the separation scheme to support the reaction path.

It must be realized that by limiting attention only to those properties in Table 8.4–1 our options are severely limited. Perhaps there is some other property difference waiting to be seen, waiting for its chance to revolutionize an industry.

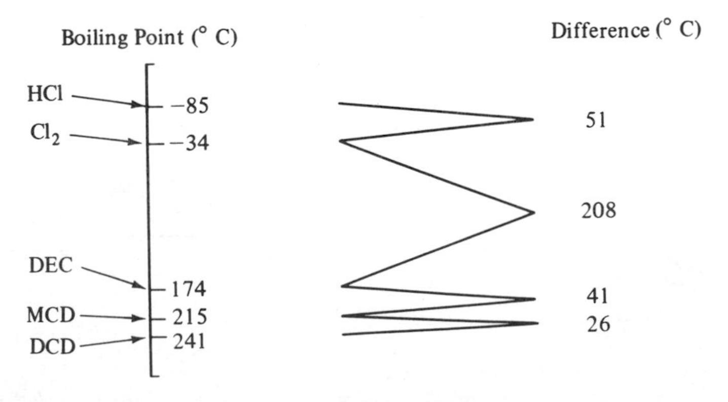

Figure 8.4–1. Boiling-Point Differences Are Attractive

8.5 INITIAL TASK SEQUENCE

If differences in boiling points are to be used as a basis for separation, the question arises, in what order should the separating be made? For this separation problem there are five possible separation sequences (Figure 8.5–1) if the chlorine and decane

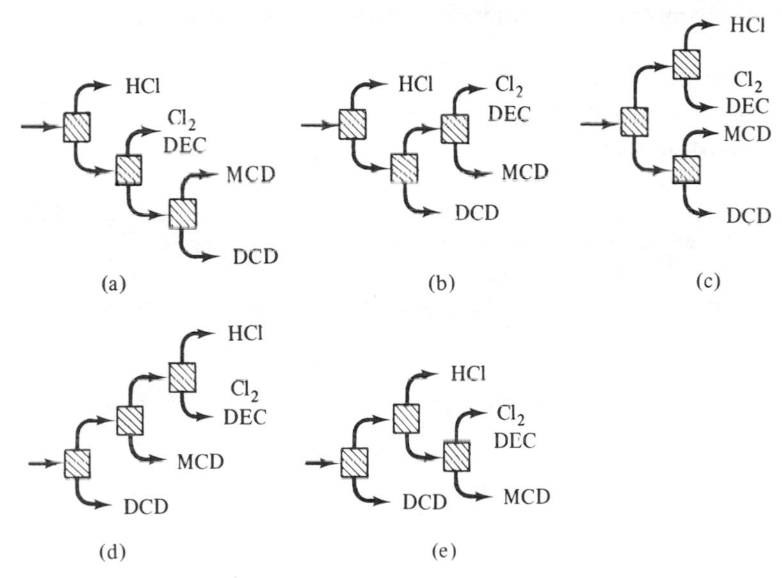

Figure 8.5–1. All Possible Separation Sequences Based on Boiling-Point Differences If Cl_2 and DEC Are Kept Together

are kept together during the separations. How can we decide which of these separation sequences is best?

If we seek to minimize the total amount of material processed, this heuristic is useful: *Remove the most plentiful species early*. The concentrations of the species in the reactor effluent are shown in Table 8.4–1. It is obvious from the magnitude of the Cl_2–DEC fraction that it should be removed early, but it is in the midst of the property ordering shown in Figure 8.4–1. Is it better to remove the HCl first or the MCD – DCD to get at the Cl_2–DEC fraction? The amount of material processed will not differ much in either case. Hence, the best structures, with respect to the amount processed, are (a) and (c) in Figure 8.5–1.

Structures (a) and (c) lead to low amounts of material processed.

One heuristic often used in distillation system design is *remove low-boiling species first*. The basis for this rule is in the increased costs associated with condensing low-boiling components. Since these species are difficult to condense, it is often best to remove them early in a separation to avoid repeated processing. For our separation problem structures, (a) and (b) satisfy this rule.

Structures (a) and (b) lead to low cooling costs.

Another rule developed in Chapter 5 states that *the most difficult separations go last*. Of the separations involved here the smallest boiling-point difference is for the MCD–DCD separation. If the boiling-point difference is used as the criterion for difficulty of separation, this separation should come last. This is true for structures (a) and (c).

Structures (a) and (c) reduce processing costs by minimizing the amount of material processed during the most costly separation.

Also we aim to remove *the corrosive material* early. Hydrogen chloride is the most corrosive and structures (a) and (b) remove it first.

Structures (a) and (b) reduce costs by removing corrosive material early.

The monochlorodecane ought to be an overhead distillation product to insure its purity.

Structures (a) and (c) have the MCD as distillates; this insures product purity.

Hence, by applying these simple rules it appears that structure (a) is favored for a number of reasons.

We have in this analysis, however, made the assumption that the chlorine and decane remained together. This seems like a reasonable assumption at first because they are both reactor feeds, and to separate them would require one more distillation (for a total of four instead of three). However, the large property difference between chlorine and *n*-decane attracts our attention. Boiling-point differences of this magnitude suggest that a simple phase separation may be all that is required to obtain a sufficiently clean split between Cl_2 and DEC. This might considerably simplify the removal of HCl from the mixture.

In structure (a) a distillation was required to remove the HCl from the Cl_2 + DEC fraction. This involves condensing HCl, which, because of its low boiling point, required high operating pressures and low-temperature coolants. If the Cl_2 is allowed to be removed with the HCl, a simple phase separator may be substituted for the distillation. A possible structure for this sequence is illustrated in Figure 8.5–2.

However, this may only postpone the separation problem. The chlorine still may have to be removed from the HCl. This observation may or may not be accurate depending on how the reactor is operated. If large excesses of DEC are utilized and long reaction times maintained, the amount of chlorine in the reactor effluent may be small enough to be discarded with the HCl. Should this not occur, the problem of removing chlorine from HCl must be solved.

In summary, sequence (a) in Figure 8.5–1 and, for low chlorine concentrations, Figure 8.5–2 look promising.

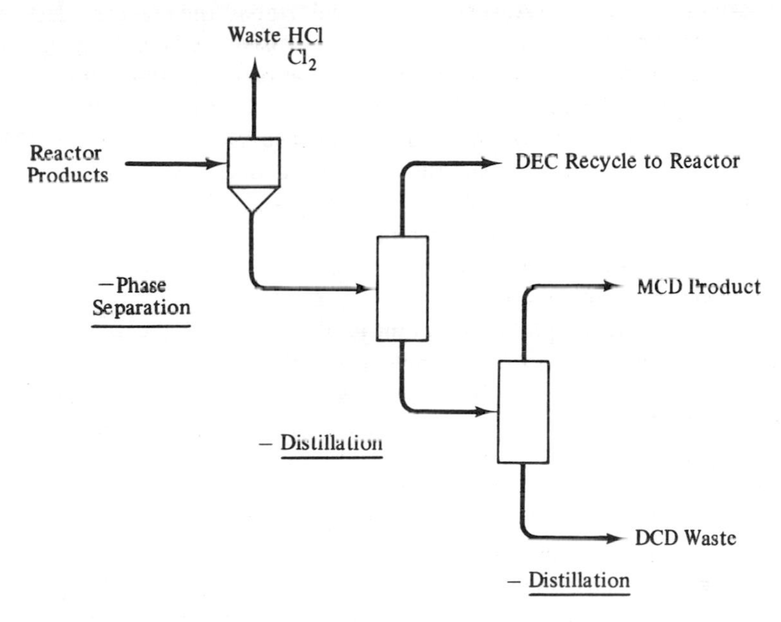

Figure 8.5–2. Low Boiling Points of Cl_2 and HCl Suggest a Simple Phase Separation
The number of distillations is reduced from three to two. However, this does not meet our species allocation, since now Cl_2 must be separated from HCl. This opens new separation possibilities not included in Figure 8.5–1.

8.6 IF CHLORINE CONCENTRATION IS HIGHER

In cases of low conversion there will be some amount of Cl_2 in HCl; hence expensive chlorine will be wasted (see Figure 8.5–2). The boiling-point difference between these two low-boiling compounds is large enough to consider distillation. However, the low boiling points make distillation difficult, since it would then be necessary to use low-temperature refrigerants for condensing and high pressures to increase the condensation temperature.

Absorption of the chlorine in a suitable solvent might be a convenient means for removing it from the HCl. This separation method has the advantage that the HCl does not have to be condensed. Even better than a solvent would be a reagent that would preferentially react with the Cl_2 in the HCl–Cl_2 mixture to form an easily separable product. Can you think of any reactions that do this?

Assume temporarily that absorption is to be used to remove the Cl_2. What solvent should we use? Water? At high water temperatures HCl is quite soluble in H_2O, whereas Cl_2 is not (*2.5* g of Cl_2/liter of water at 100°C and 760 mmHg of Cl_2 partial pressure, and *380* g of HCl/liter of water at 100°C and 760 mmHg of HCl partial pressure). Hence it may be feasible to operate an absorption unit as shown in Figure 8.6–1. Notice, however, that the chlorine leaving this unit is saturated with water. Should this water be returned with the chlorine to the reactor inlet? No. Any water in the reactor would form an extremely corrosive mixture with the HCl. It is probably best to keep any water out of most of the process equipment.

The introduction of water would cause another separation problem. As suggested in Chapter 5, indirect separation methods often cause many additional problems. To remove water from the chlorine stream, a standard drying system utilizing concentrated sulfuric acid could be used. Water is strongly absorbed by concentrated sulfuric acid, whereas chlorine is not. A possible processing system is shown in Figure 8.6–2. Introducing water into the system causes several auxiliary problems; perhaps these can be avoided with another solvent.

A hydrocarbon with high boiling point might be useful. The low vapor pressure of the solvent would reduce losses of solvent into the HCl vapors. The selectivity or ability of solvent to remove chlorine would depend on the relative vapor pressures of Cl_2 and HCl. The chlorine could be recovered from the solvent by distillation. What sources of hydrocarbon are available? It is certainly feasible to buy a special hydrocarbon on the open market. What about solvents in the process itself?

Could DEC, MCD, or DCD serve as solvents? If all that is required is a high-boiling liquid, they could serve well. Which one should we use? Since the chlorine is destined for the reactor inlet, it might be convenient to use DEC (which is also allocated to the reactor inlet) as the solvent. This combined role for DEC of reactant and solvent eliminates a separation column. Figure 8.6–3 illustrates a process utilizing the concept. Notice how the recycle DEC is combined with feed *n*-decane to supply the solvent for the HCl–Cl_2 separation.

If MCD was selected as the solvent, the flow sheet shown in Figure 8.6–4 would result. Note that the effluent from the absorber containing chlorine and MCD is sent to the *n*-decane–MCD separation unit. The chlorine comes overhead in this unit with

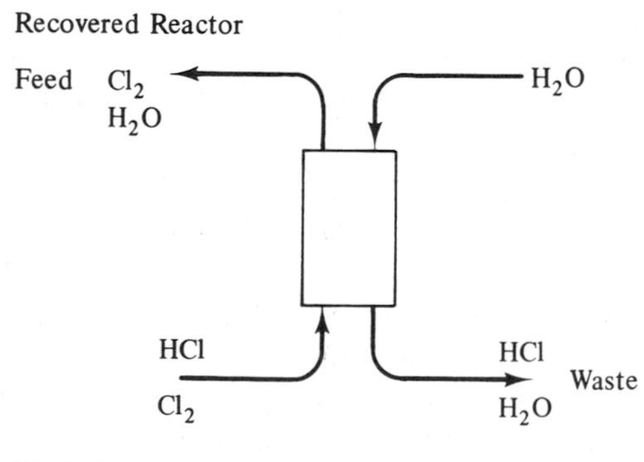

Figure 8.6–1. Exploiting Solubility Differences in Water to Recover Cl_2 from Cl_2–HCl Waste Stream in Figure 8.5–2

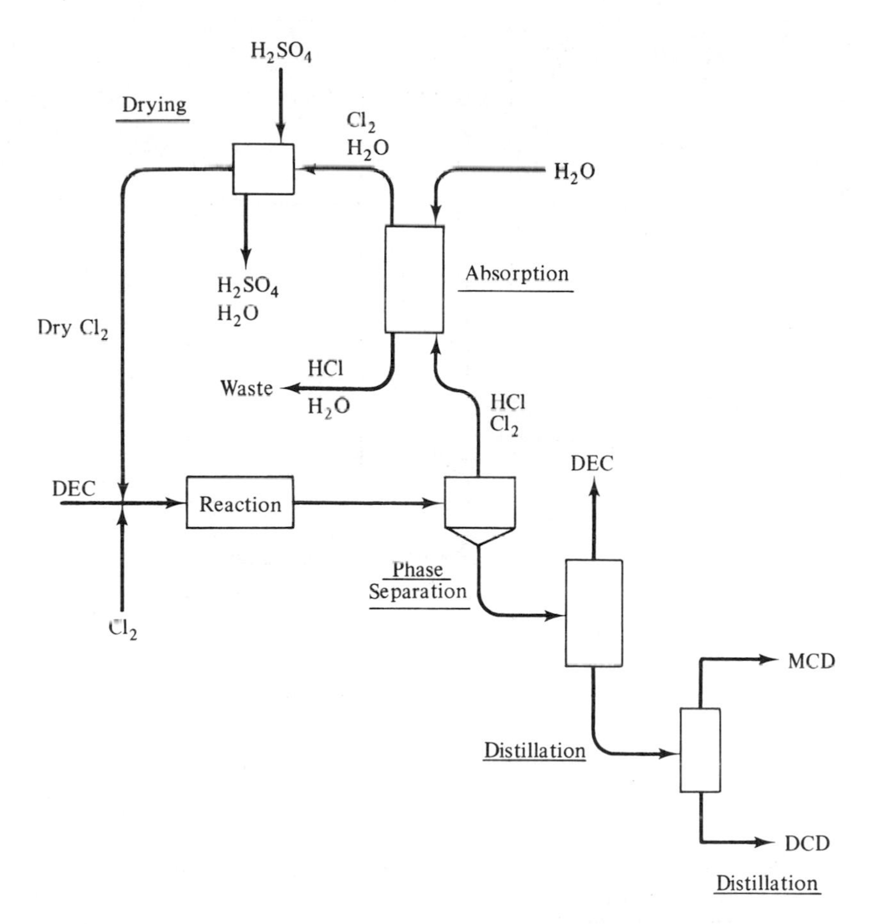

Figure 8.6–2. Separation Sequence Using Water Absorption of HCl and Using H_2SO_4 for Drying Cl_2
The use of water as a solvent causes problems typical of any indirect separation. Water is foreign to the processing.

the DEC. This flow sheet suffers from the large amounts of MCD that recycle through the absorber, DEC separator, and the MCD separator.

8.7 FOCUSING ON PLAUSIBLE PROCESSING PLANS

Let us retrace our steps during this synthesis to see how the number of alternative flow sheets has been reduced by using simple analysis techniques and heuristics. The problem as initially stated was—produce MCD. Three reactions and perhaps several combinations of these were discovered to utilize readily available raw materials. One

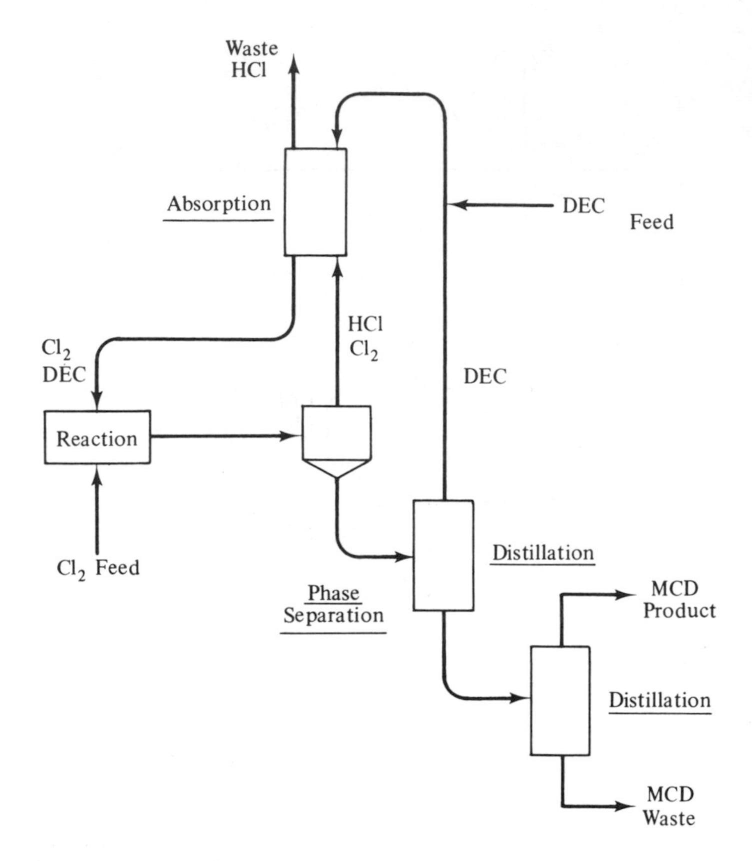

Figure 8.6–3. Recycle *n*-Decane and Feed *n*-Decane Are Used to Drive the HCl–Cl_2 Separation
An indirect separation has been made direct.

reaction was selected for further evaluation by performing a material balance on the reactions and analyzing the economies of each scheme.

Several species allocations were then generated and evaluated for this reaction path. The allocation selected gave rise to a separation problem, which for distillation alone had five alternative sequences. For two bases of separation, say, distillation and extraction, there are 30 alternatives. When one of the separations required a solvent, at least another five alternatives for each sequence were included.

Using these estimates of the number of alternatives at each level, an estimate of the total number of structures can be made: (3 reactions)(2 species allocations)(30 separation sequences) × (5 solvent choices) = *900* alternative structures. Hence, even before confronting the problems of task integration or equipment selection, there are 900 possible flow sheets we might consider. However, using evaluations and heuristics at each level of synthesis, we have reduced the number that we consider

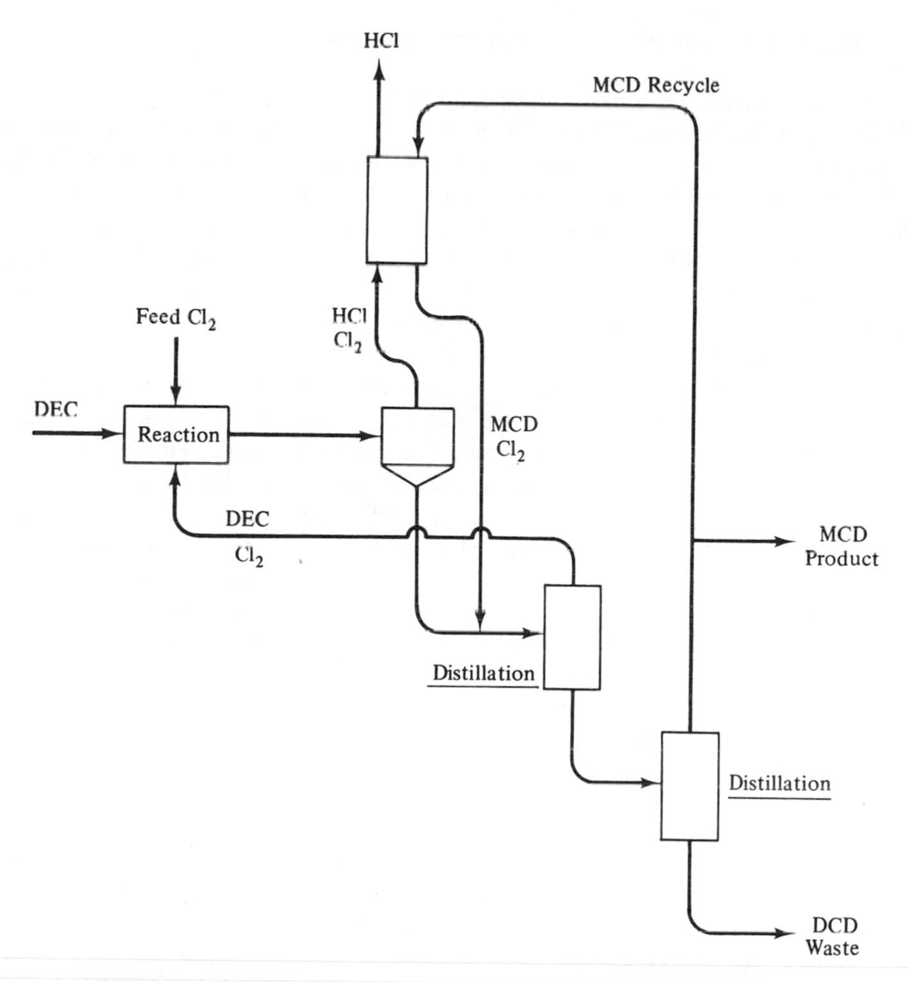

Figure 8.6–4. Using MCD As Solvent for HCl–Cl_2 Separation Places a High Load on the Two Distillations

likely to be best to five or six. This tremendous reduction in the number of alternatives that must be analyzed in detail is the great advantage of the heuristics. This advantage is not without drawbacks. When the decision was made to not investigate the decene or decanol reactions, two-thirds of the alternatives were eliminated. If the processing costs to operate the decane reaction are high, the raw-material advantage it enjoyed may not make it better than, say, the decene reaction. Hence when making evaluations of alternatives, if two are not significantly different, they should both be carried along. These competing process concepts can only be distinguished accurately by the detailed cost estimates outlined in the next section.

8.8 DETAILED ANALYSIS OF COMPLETE DESIGNS

In the previous sections we have followed the probable development of several MCD processing concepts. In this section we present the detailed flow sheets and evaluations of one of these concepts. First, we examine in detail the process shown in Figure 8.8–1, which is based upon the processing observation that the amount of chlorine in the HCl stream is small and can be discarded. This corresponds to the processing concept of Figure 8.5–2, after the task integrations resulting in the addition of pumps, coolers, heaters, integrated heat-exchange network, and distillation equipment. An enormous amount of effort goes into transforming our rough concepts into detailed specifications.

At the level of design shown in Figure 8.8–1 we have specified, in detail, the kind of equipment and its sequence in the process. It remains to determine how this equipment is constructed and operated. For example, there remains freedom to select the length of the reactor, the diameter and height of the distillation columns, the pressure and temperature for the reactor, and the feed ratio of decane to chlorine. The selection of these variables is not easy, because the variables interact with each other. For example, setting the reactor length partially determines the composition of the reactor effluent, which influences the distillation column design.

This difficult optimization problem is normally solved by developing detailed mathematical models describing the performance of the process equipment. These models are then solved for many levels of the design variables to determine which combination of reactor length, column pressure, and reactor feed ratio is best. The organized manipulation of these variables to achieve the best design is called *optimization.*

We now describe the processing details shown in Figure 8.8–1. Decane flows from a storage tank through pump P-1 to a mixing tee, where it mixes with recycled decane. The decane is mixed with chlorine, which has been stored as a liquid in storage tank T-1. The mixture of chlorine and *n*-decane is heated to 200°F in a steam jacketed heat exchanger prior to entering the reactor. The reaction mixture flows through glass tubes, which are irradiated with ultraviolet light. Cooling water is circulated around the tubes to remove the heat of reaction.

As the mixture reacts, HCl gas is evolved and a two-phase mixture containing HCl, Cl_2, DEC, MCD, and DCD flows out of the reactor. The gas–liquid mixture is separated in a simple phase separator. The gas phase containing HCl with a small amount of chlorine exits from the top of the separator and is sent to waste recovery. Liquid containing DEC, MCD, and DCD flows from the bottom of the separator. This liquid is heated by heat exchange with the overhead streams from columns C-1 and C-2. In column C-1, which is 80 ft high, 7 ft in diameter, and contains 39 trays, the DEC is separated from the MCD and DCD. Following condensation in heat exchangers E-2 and E-3, the DEC stream is split in two. Eighty per cent of the liquid is returned to the column as reflux; the remainder is recycled through pump P-2 and cooler E-1 to the reactor inlet. MCD and DCD liquid flow from the bottom of column C-1, where the stream is split into two streams; one, 80 per cent of the original

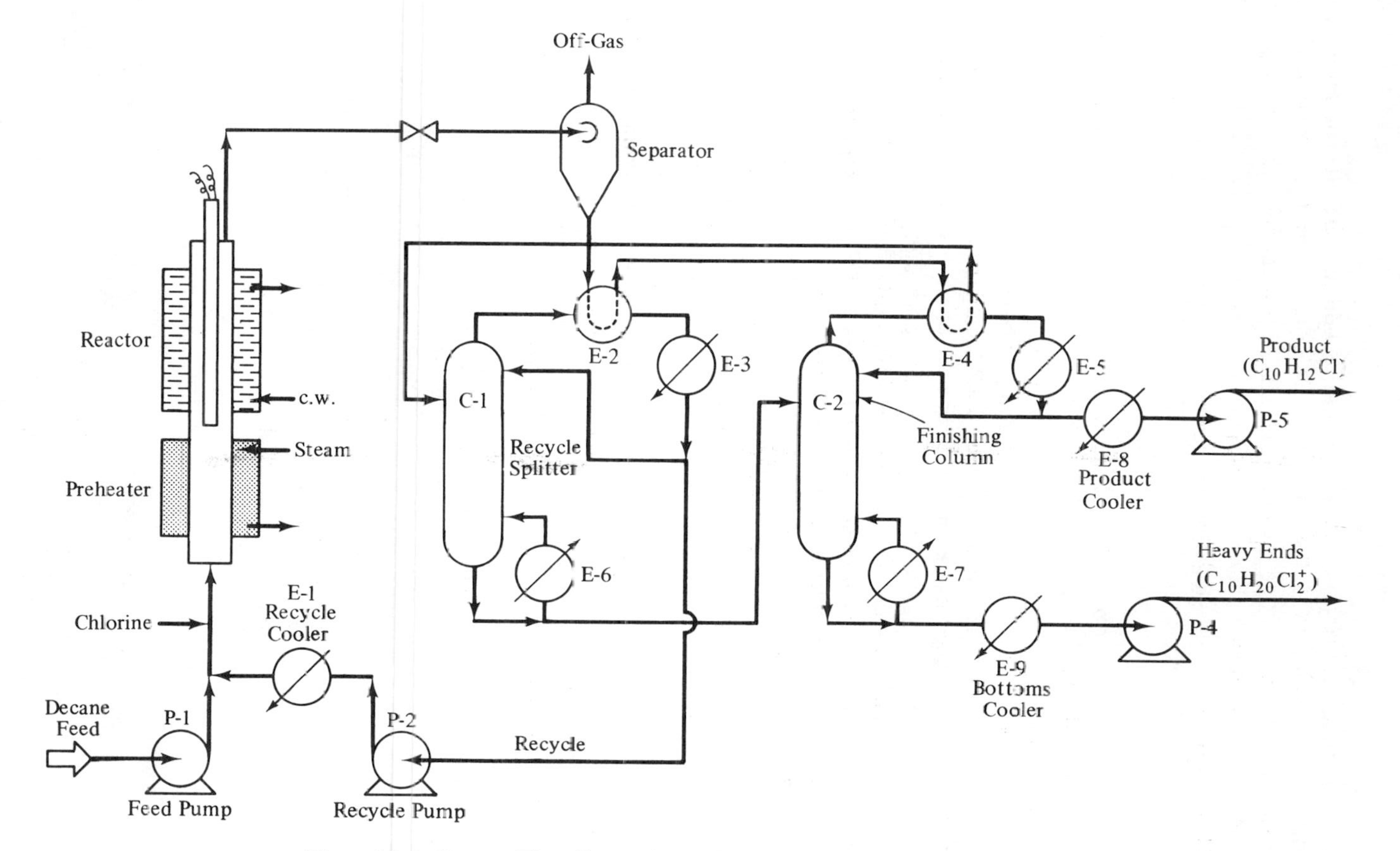

Figure 8.8–1. Process Flow Sheet Direct Chlorination of *n*-Decane (R. R. Hughes)

stream, is vaporized and returned to the column as boil-up; the other is forwarded to column C-2. In column C-2 the MCD and DCD are separated. Column C-2 is 100 ft high, 5 ft in diameter, and contains 50 trays. It is made of carbon steel. DCD flows from the bottom of column C-2 and is pumped to storage. MCD is taken overhead, cooled, and also pumped to storage.

Economics

Table 8.8–1 shows the cost breakdown for a monochlorodecane plant having a total capacity of 100 million lb/year. The investment in process capital is about $4,000,000, and the product monochlorodecane costs $0.05/lb. The raw-material costs contribute $0.035/lb, or 71 per cent of the cost of the product. Note that Hughes used slightly lower raw-material costs than we (see Table 8.2–2). Similar detailed engineering and economic studies have been made on a wide variety of processes, and the results of these studies are interesting, since they give us an opportunity to see which of the several processing plans we develop prove to be competitive.

Figure 8.8–2 shows eight flow sheets that Powers and Rudd examined in detail to compute the profit as a function of reactor conversion and price of chlorine. We recognize flow sheet 1 as that of Figures 8.8–1 and 8.5–2. Flow sheet 3 is that of Figure 8.6–4, and flow sheet 6 was developed in Section 8.6. The economic studies summarized in Figure 8.8–3 indicate that of the eight flow sheets, these three dominate under certain conditions.

Let us examine the reasons for the dominance of these processes. Flow sheet 1 has a lower profit at low conversions of the limiting reagent chlorine due to the loss of chlorine from the system. As higher conversions of chlorine are achieved, the losses of chlorine decrease and the profit increases. Also, as the value of the wasted chlorine increases, flow sheet 1 becomes less attractive.

TABLE 8.8–1 Estimated Economics of Producing 100 Million Pounds Per Year of Monochlorodecane by Process of Figure 8.8–1 (R. R. Hughes)

I. Capital costs	$4,000,000
II. Annual operating costs	$4,400,000/yr
Amortization of capital	$600,000/yr
III. Product cost	$0.05/lb of MCD

	$/LB	% OF TOTAL
Capital costs	0.006	12
Raw materials	0.035	71
Utility	0.003	5
Labor	0.005	10
Other	0.001	2
Total	0.050	

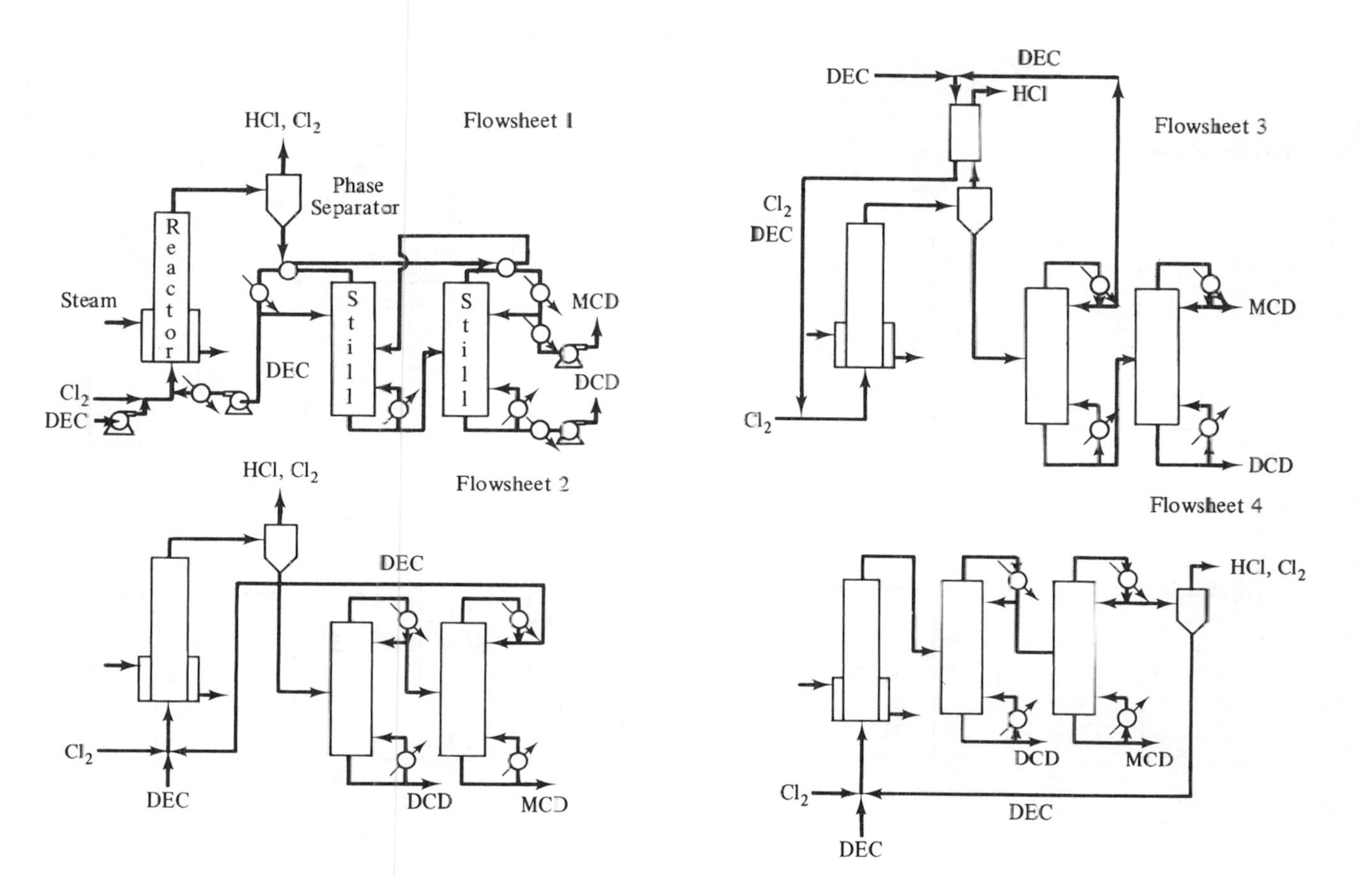

Figure 8.8–2. Eight Process Flow Sheets, the Economics of Which Are Summarized in Figure 8.8–3

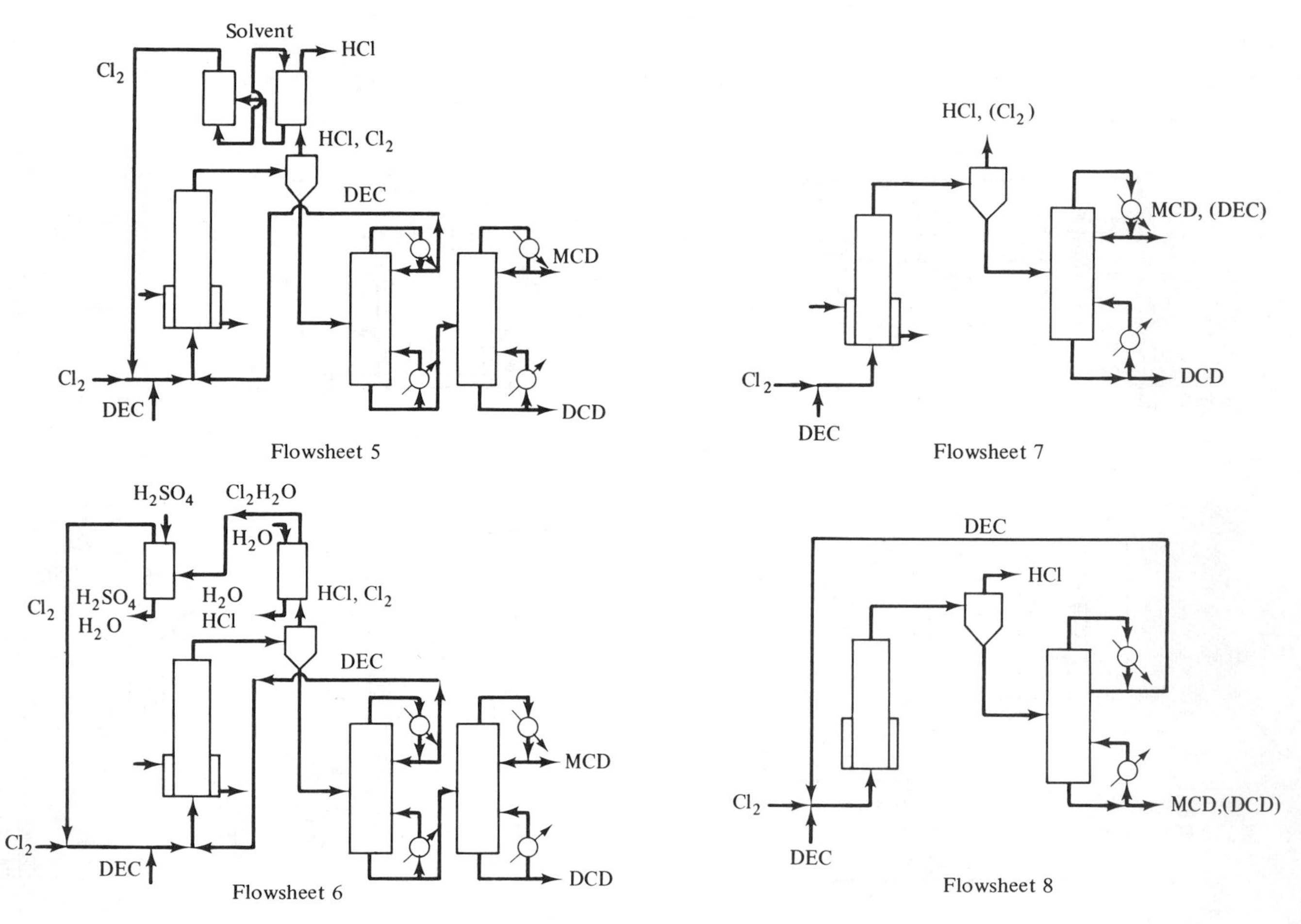

Figure 8.8–2 (*continued*)

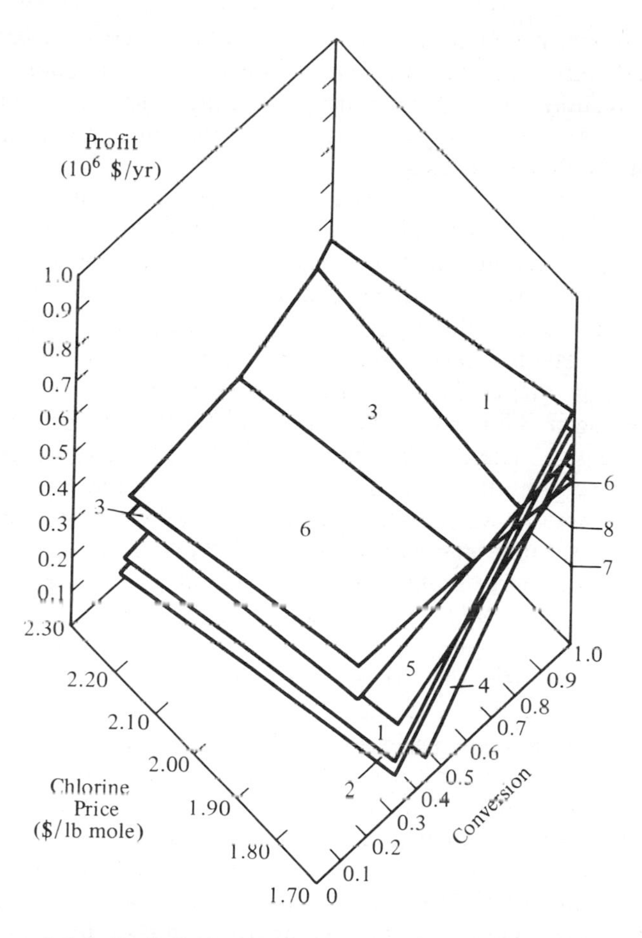

Figure 8.8 3. Profit Surfaces for Eight Monochlorodecane Processes

Flow sheet 2, which has the separation sequence for the last two columns in flow sheet 1 reversed, is less attractive than flow sheet 1 for all values of conversion. The decane is the largest component in the reactor effluent, and is also the lowest-boiling component leaving the phase separator. The failure to remove decane early causes the increased cost in this case.

Flow sheet 3 involves the use of decane as a solvent to remove chlorine from the vapors leaving the phase separator. Since the chlorine not converted in the reactor is recycled, this flow sheet is not nearly as sensitive to conversion or chlorine cost as are flow sheets 1 and 2. The maximum profit for flow sheet 3 is not as high as the optimum for flow sheet 1, but it is much less sensitive to the level of conversion. If uncertainty in the reactor design is large, the selection of flow sheet 3 may be wise. The addition of a separator to recover unconverted chlorine adds to the equipment and operating costs, but gives much greater flexibility in operation.

Flow sheet 4, which utilizes the reverse-separation sequence of flow sheet 1, is decidedly less attractive. The repeated processing of the low-boiling components, particularly HCl, causes very high equipment and utility costs. The loss of chlorine, as in flow sheets 1 and 2, makes the venture profit for this process dependent on reactor conversion and chlorine cost.

Flow sheet 5 shows the same independence of conversion as flow sheet 3. The profit is less due to solvent costs and the additional separator.

A profit nearly as high as that for flow sheet 3 is obtained for flow sheet 6. Flow sheet 6 utilizes water to remove hydrogen chloride from the vapors leaving the phase separator. Wet chlorine returning to the reactor inlet is dried by contacting concentrated sulfuric acid. The low cost of the solvent, water, and the fact that the water need not be regenerated make this alternative very attractive. However, the corrosive nature of the hydrogen chloride–water mixture and the sulfuric acid–water mixture requires more expensive materials of construction. The danger of water entering the reactor must also be considered. If the reactor and separators are constructed of carbon steel, any traces of water in the presence of the hydrogen chloride reaction product would lead to rapid failure of these units due to corrosion.

Flow sheet 6 has a higher profit than flow sheet 3 at low conversion. This is due to the large amounts of decane solvent required in flow sheet 3 for low chlorine conversions.

Flow sheets 7 and 8 are the special cases when conversion is complete. Flow sheet 7 has the advantage of only requiring two separators and no recycle. The high conversions using stoichiometric feed ratios lead, however, to the production of large amounts of DCD. Hence the venture profit for this alternative is low.

Flow sheet 8 represents the situation when large excesses of decane are utilized. Complete conversion of chlorine is achieved; hence no separation is required for its recycle. A small amount of DCD appears in the product. These advantages are offset by the costs associated with the separation and recycling of large amounts of decane.

In summary, three regions exist when comparing the eight flow sheets over wide ranges of conversion. At low conversions process flow sheet 6, which utilizes water as a means to separate HCl and chlorine, is the best. At intermediate conversion levels, process flow sheet 3 is best. Process 3 utilizes the feed and recycle decane as a solvent for absorbing chlorine from the waste hydrogen chloride stream. For high levels of conversion, process flow sheet 1, which discards unconverted chlorine, is optimal.

8.9 CONCLUSIONS

Using the basic principles of process discovery, we have been able to isolate the few more promising processing plans from an enormous number of possible alternatives. This is the same technique used by the experienced design engineer during the initial stages of process development.

It is necessary to go into much greater detail, summarized in Section 8.8, to distinguish among these dominant plans in order to isolate the process to be constructed. Such detailed analysis requires a much greater depth of training in engineering and economics than developed in this text. Our principles only give us an excellent beginning in process development, and much more must be learned to complete the job.

REFERENCES

Much of this chapter is taken from the paper by

G. J. POWERS, "Heuristic Synthesis in Process Development," *Chemical Engineering Progress*, **68**, no. 8, 1972.

Part of the economic analysis in Section 8.8 is from

R. R. HUGHES, "Mathematical Models for Process Design and Optimization," *American Institute of Chemical Engineers Today Series Lecture Notes*, New York, 1968.

PROBLEM

This class problem involves speculation on the technology that might arise from the following laboratory discovery reported in *Chemical Technology*, p. 198, April 1972.

> Vinyl chloride producers may now have the option of turning out a range of by-product, chlorinated solvents, for only a slight additional investment. G. DuCrest and associates at Pechiney-Saint-Gobain have found that significant amounts of trichloroethane and tetrachloroethane can be obtained in the oxychlorination step along with the usual dichloroethane (U.S. Patent 3,622,643). The trick is to pass mixed ethylene and vinyl chloride through the reactor, not the individual gases.
>
> The following data are obtained from an HCl–air system at 360°C over a $CuCl_2$ + KCl on Al_2O_3 catalyst.

	TOTAL CONVERSION %		SELECTIVITY %		
Feed	C_2H_4	C_2H_3Cl	$C_2H_4Cl_2$	$C_2H_3Cl_3$	$C_2H_2Cl_4$
Equimolar ethylene and vinyl chloride	96.9	96.4	42.2	47.5	3.2
Ethylene alone	79.7	—	72.2	3.1	0.6
Vinyl chloride alone	—	74.3	—	69.9	1.5

> U.S. production of dichloroethane, which exceeds 6 billion lbs annually, is used chiefly for vinyl chloride synthesis. Chlorination of trichloroethane under pyrolytic conditions yields trichloroethylene, for which yearly production is 550 million lbs. Its unit value is 7 ¢/lb according to the U.S. Tariff Commission, about double that of vinyl chloride.

Appendix A

CONVERSION FACTORS

TABLE A.1 Volume Equivalents*

	CU IN.	CU FT	U.S. GAL	LITER
Cu in.	1	5.787×10^{-4}	4.329×10^{-3}	1.639×10^{-2}
Cu ft	1.728×10^{3}	1	7.481	28.32
U.S. gal	2.31×10^{2}	0.1337	1	3.785
Liter	61.03	3.531×10^{-2}	0.2642	1

* *Source:* J. H. Perry, *Chemical Engineers' Handbook*, 3rd ed., McGraw-Hill Book Company, New York, 1950.

TABLE A.2 Mass Equivalents*

	AVOIR OZ	POUNDS	GRAINS	GRAMS
Avoir oz	1	6.25×10^{-2}	4.375×10^{2}	28.35
Pounds	16	1	7×10^{3}	4.536×10^{2}
Grains	2.286×10^{-3}	1.429×10^{-4}	1	6.48×10^{-2}
Grams	3.527×10^{-2}	2.20×10^{-3}	15.432	1

* *Source:* J. H. Perry, *Chemical Engineers' Handbook*, 3rd ed., McGraw-Hill Book Company, New York, 1950.

TABLE A.3 Linear Measure Equivalents*

	Meter	Inch	Foot	Mile
Meter	1	39.37	3.2808	6.214×10^{-4}
Inch	2.54×10^{-2}	1	8.333×10^{-2}	1.58×10^{-5}
Foot	0.3048	12	1	1.8939×10^{-4}
Mile	1.61×10^{3}	6.336×10^{4}	5280	1

* *Source:* J. H. Perry, *Chemical Engineers' Handbook*, 3rd ed., McGraw-Hill Book Company, New York, 1950.

TABLE A.4 Power Equivalents*

	Hp	Kw	Ft-lb/sec	Btu/sec
Hp	1	0.7457	550	0.7068
Kw	1.341	1	737.56	0.9478
Ft-lb/sec	1.818×10^{-3}	1.356×10^{-3}	1	1.285×10^{-3}
Btu/sec	1.415	1.055	778.16	1

* *Source:* J. H. Perry, *Chemical Engineers' Handbook*, 3rd ed., McGraw-Hill Book Company, New York, 1950.

TABLE A.5 Heat, Energy, or Work Equivalents*

Joules = 10^7 ergs	Kg-m	Ft-lb	Kw-hr	Hp-hr	Liter-atm	Kcal	Btu	Gram-calorie†
1	0.10197	0.7376	2.773×10^{-7}	3.725×10^{-7}	9.869×10^{-3}	2.390×10^{-4}	9.478×10^{-4}	0.2390
9.80665	1	7.233	2.724×10^{-6}	3.653×10^{-6}	9.678×10^{-2}	2.344×10^{-3}	9.296×10^{-3}	2.3438
1.356	0.1383	1	3.766×10^{-7}	5.0505×10^{-7}	1.338×10^{-2}	3.24×10^{-4}	1.285×10^{-3}	0.3241
3.6×10^6	3.671×10^5	2.655×10^6	1	1.341	3.5534×10^4	8.6057×10^2	3.4128×10^3	8.6057×10^5
2.6845×10^6	2.7375×10^5	1.98×10^6	0.7455	1	2.6494×10^4	6.4162×10^2	2.545×10^3	6.4162×10^5
1.0133×10^2	10.333	74.73	2.815×10^{-5}	3.774×10^{-5}	1	2.422×10^{-2}	9.604×10^{-2}	24.218
4.184×10^3	4.267×10^2	3.086×10^3	1.162×10^{-3}	1.558×10^{-3}	41.29	1	3.9657	1×10^3
1.055×10^3	1.0758×10^2	7.7816×10^2	2.930×10^{-4}	3.930×10^{-4}	10.41	0.252	1	2.52×10^2
4.184	0.4267	3.086	1.162×10^{-6}	1.558×10^{-6}	4.129×10^{-2}	1×10^{-3}	3.97×10^{-3}	1

* *Source:* J. H. Perry, *Chemical Engineers' Handbook*, 3rd ed., McGraw-Hill Book Company, New York, 1950.
1 therm = 100,000 Btu.
† Gram-calorie. Thermochemical calorie is defined as 4.1840 absolute joules.

Appendix B

ATOMIC WEIGHTS AND NUMBERS

TABLE B.1 Relative Atomic Weights, 1965, Based on the Atomic Mass of ^{12}C — 12.000

The values for atomic weights given in the Table apply to elements as they exist in nature, without artificial alteration of their isotopic composition, and, further, to natural mixtures that do not include isotopes of radiogenic origin.

NAME	SYMBOL	ATOMIC NUMBER	ATOMIC WEIGHT
Actinium	Ac	89	—
Aluminium	Al	13	26.9815
Americium	Am	95	—
Antimony	Sb	51	121.75
Argon	Ar	18	39.948
Arsenic	As	33	74.9216
Astatine	At	85	—
Barium	Ba	56	137.34
Berkelium	Bk	97	—
Beryllium	Be	4	9.0122
Bismuth	Bi	83	208.980
Boron	B	5	10.811
Bromine	Br	35	79.904
Cadmium	Cd	48	112.40
Caesium	Cs	55	132.905
Calcium	Ca	20	40.08
Californium	Cf	98	—
Carbon	C	6	12.01115
Cerium	Ce	58	140.12
Chlorine	Cl	17	35.453
Chromium	Cr	24	51.996
Cobalt	Co	27	58.9332
Copper	Cu	29	63.546
Curium	Cm	96	—
Dysprosium	Dy	66	162.50
Einsteinium	Es	99	—
Erbium	Er	68	167.26
Europium	Eu	63	151.96
Fermium	Fm	100	—
Fluorine	F	9	18.9984
Francium	Fr	87	—
Gadolinium	Gd	64	157.25
Gallium	Ga	31	69.72
Mercury	Hg	80	200.59
Molybdenum	Mo	42	95.94
Neodymium	Nd	60	144.24
Neon	Ne	10	20.183
Neptunium	Np	93	—
Nickel	Ni	28	58.71
Niobium	Nb	41	92.906
Nitrogen	N	7	14.0067
Nobelium	No	102	—
Osmium	Os	75	190.2
Oxygen	O	8	15.9994
Palladium	Pd	46	106.4
Phosphorus	P	15	30.9738
Platinum	Pt	78	195.09
Plutonium	Pu	94	—
Polonium	Po	84	—
Potassium	K	19	39.102
Praseodym	Pr	59	140.907
Promethium	Pm	61	—
Protactinium	Pa	91	—
Radium	Ra	88	—
Radon	Rn	86	—
Rhenium	Re	75	186.2
Rhodium	Rh	45	102.905
Rubidium	Rb	37	84.57
Ruthenium	Ru	44	101.07
Samarium	Sm	62	150.35
Scandium	Sc	21	44.956
Selenium	Se	34	78.96
Silicon	Si	14	28.086
Silver	Ag	47	107.868
Sodium	Na	11	22.9898
Strontium	Sr	38	87.62

ATOMIC WEIGHTS AND NUMBERS (*Cont.*)

TABLE B.1 Relative Atomic Weights, 1965, Based on the Atomic Mass of ^{12}C — 12

The values for atomic weights given in the Table apply to elements as they exist in nature, without artificial alteration of their isotopic composition, and, further, to natural mixtures that do not include isotopes of radiogenic origin.

Name	Symbol	Atomic Number	Atomic Weight	Name	Symbol	Atomic Number	Atomic Weight
Germanium	Ge	32	72.59	Sulfur	S	16	32.064
Gold	Au	79	196.967	Tantalum	Ta	73	180.948
Hafnium	Hf	72	178.49	Technetium	Tc	43	—
Helium	He	2	4.0026	Tellurium	Te	52	127.60
Holmium	Ho	67	164.930	Terbium	Tb	65	158.924
Hydrogen	H	1	1.00797	Thallium	Tl	81	204.37
Indium	In	49	114.82	Thorium	Th	90	232.038
Iodine	I	53	126.9044	Thulium	Tm	59	168.934
Iridium	Ir	77	192.2	Tin	Sn	50	118.69
Iron	Fe	26	55.847	Titanium	Ti	22	47.90
Krypton	Kr	36	83.80	Tungsten	W	74	183.85
Lanthanum	La	57	138.91	Uranium	U	92	238.03
Lawrencium	Lr	103	—	Vanadium	V	23	50.942
Lead	Pb	82	207.19	Xenon	Xe	54	131.30
Lithium	Li	3	6.939	Ytterbium	Yb	70	173.04
Lutetium	Lu	71	174.97	Yttrium	Y	39	88.905
Magnesium	Mg	12	24.312	Zinc	Zn	30	65.37
Manganese	Mn	25	54.9380	Zirconium	Zr	40	91.22
Mendelevium	Md	101	—				

Source: Comptes Rendus, 23rd IUPAC Conference, 1965, Butterworth's, London, 1965, pp. 177–78.

Appendix C

Glossary

Absorption A separation operation in which a soluble component of a gas mixture is dissolved in a liquid

Adiabatic process A process that involves no net heat absorption or removal

Adsorption A separation operation in which one component of a gas or liquid mixture is selectively retained in the micropores of a resinous or microcrystalline solid

Aerosol A dispersion of colloidal particles in a vapor

Agglomeration A particle-size-increasing operation in which powdered solids are brought together in a loose state of bonding to form larger particulates

Ambient The conditions of the surroundings, particularly of temperature and pressure

Analysis A deductive activity involving the decomposition of a problem into its parts to discover their function, relationship, or consequence

Azeotrope A liquid mixture of two or more components that volatilizes into a vapor with exactly the same composition as the liquid; that is, a mixture with unity relative volatility

Azeotropic distillation A separation operation in which an additional component is added to a difficult-to-separate mixture, forming a heterogeneous azeotrope with one of the original components that is more easily separated from the remainder of the original mixture

Boiling The rapid subsurface vaporization of a liquid

Boiling point The temperature and pressure at which the vapor pressure of a pure liquid equals the total pressure on its surface

Bottoms The lower volatility components of a mixture separated by a distillation operation

British thermal unit A unit of energy equal to the amount of heat required to raise 1 lb of water 1°F

Bubble point The temperature and pressure at which the first subsurface bubble of vapor is formed in a multicomponent liquid

Calciner A temperature-increasing operation in which solids are heated either directly or indirectly in a rotating horizontal cylinder

Cascade Any combination of processing stages connected in a manner to improve performance rather than increase capacity

Centigrade temperature scale A uniform measurement of temperature on which at 1 atm pressure water freezes at 0° and boils at 100°

Centrifugal compressor A vapor-pressure-increasing and volume-decreasing operation that produces kinetic energy by the action of centrifugal force and then converts this energy to pressure by reducing the vapor velocity

Centrifugal pump A liquid-pressure-increasing operation that produces kinetic energy by the action of centrifugal force and then converts this energy to pressure by reducing the liquid velocity

Centrifuge A separation operation that uses centrifugal force to exploit very small density differences between immiscible phases

Clarification A separation operation that exploits density differences to partially remove solids from a liquid–solid mixture

Classification A separation operation that exploits differences in the settling rate in a liquid of solids of different size or density

Cocurrent A contacting scheme in which the primary flows of two streams exchanging material or energy are in the same direction

Colloid A stable dispersion of small solid particles in a liquid

Compacting A particle-size-increasing operation involving direct pressure, as with presses, rolls, or molds

Condensation The transition of a vapor into the liquid phase

Conduction A mode of heat transfer characterized by heat flow through material or by direct physical contact

Convection A mode of heat transfer characterized by the mixing of fluids of different temperature

Cooler Any of a variety of temperature-decreasing operations

Cooling tower A temperature-decreasing operation in which the heat of vaporization required to evaporate part of a liquid is obtained from the sensible heat of the bulk of the liquid

Corrosion The thinning and loss of equipment construction material caused by combinations of erosion, stress, and electrochemical and biological reactions

Countercurrent A contacting scheme in which the primary flows of two streams exchanging material or energy are in the opposite direction

Crusher Any of a variety of particle-size-decreasing operations used for reducing very large solids

Crystal A highly organized state of matter characterized by arrays of atoms or ions in lattices

Cyclone A separation operation in which momentum differences are exploited to separate fine solid particles or liquid droplets from a vapor

Decanter A separation operation in which a dispersion of immiscible liquids settles and coalesces forming two distinct layers

Density The ratio of the mass of a substance to its volume

Dew point The temperature and pressure at which the first drop of condensate is formed from a multicomponent vapor

Dialysis A separation operation in which substances in solution having widely different molecular weights are separated by diffusion through a semipermeable membrane

Diffusion The tendency of a system of components at different concentrations to transfer mass within phases, across phase boundaries, or through membranes in such a way as to cause the concentrations to become uniform

Displacement compressor A vapor-pressure-increasing and volume-decreasing operation in which pressure energy is increased by displacing volume with a second fluid or by mechanical means

Displacement pump A liquid-pressure-increasing operation in which pressure energy is increased by displacing volume with a second fluid or by mechanical means

Distillate The higher volatility components of a mixture separated by a distillation operation

Distillation A separation operation that exploits differences in volatility to separate miscible liquid mixtures

Ejector A pressure-increasing operation in which the momentum and kinetic energy of a high-velocity fluid are employed to entrain and compress a second fluid

Electrodialysis A separation operation in which electromotive force is used to transport ionized materials through a semipermeable membrane

Electrostatic separator A separation operation in which solids capable of acquiring a surface charge are separated from a vapor

Emulsion A dispersion of small droplets of one immiscible liquid in another liquid

Endothermic Any process that absorbs heat

Energy The capacity for doing work

Energy balance A system for accounting for the total energy entering, leaving, and accumulating in a system based on the law of conservation of energy

Enthalpy A derived thermodynamic property equal to the sum of the internal energy and the product of the pressure and the specific volume

Exothermic Any process that produces heat

Expression A separation operation in which a liquid is separated from a solid by compression under conditions that permit the liquid to escape while the solid is retained between the compressing surfaces

Extract The effluent from an extraction enriched in the species more soluble in the solvent

Extraction A separation operation in which one component of a liquid mixture is dissolved into an added immiscible solvent

Extractive distillation A separation operation in which an additional low-volatility component is added to a difficult-to-separate mixture to increase the relative volatility of the original mixture by partially absorbing one of its components

Extrusion A particle-size-increasing operation involving both direct pressure and mixing, as with screws or augers

Fahrenheit temperature scale A uniform measurement of temperature on which at 1 atm of pressure water freezes at 32° and boils at 212°

Filtration A separation operation in which liquids are removed from a mixture with solids by passing through a permeable membrane

Flocculation A particle-size-increasing operation in which very fine solid particles dispersed in a fluid come together to form a larger precipitate

Flotation A separation operation in which air blown through a liquid suspension of a mixture of solids will separate air-avid solids

Fluid The liquid or vapor phase

Fluid mechanics The science dealing with the static and dynamic behavior of liquids and vapors

Fluidization Any process in which very fine solid particles are made to flow like a liquid

Foam A dispersion of a vapor in a liquid

Fog A dispersion of small liquid droplets in a vapor

Fractional crystallization A separation operation based on selective solidification of one species from a liquid mixture

Fractional distillation A distillation in which a vapor is continuously and countercurrently contacted with a condensed portion of the vapor

Gel A dispersion of a liquid throughout a solid matrix

Granulation Any of a variety of particle-shaping operations that form irregularly shaped dust-free clusters of particles

Heat A form of energy in transition, resulting from a temperature difference

Heat capacity The ratio of the enthalpy change to the temperature change of a substance

Heat exchanger Any of a variety of devices in which heat is transferred from one stream to another

Heat of combustion The heat of a complete reaction between a species and oxygen

Heat of formation The heat of a hypothetical reaction forming a species from its elements

Heat of fusion The difference in the liquid-phase enthalpy and the solid-phase enthalpy of a species at the melting point

Heat of reaction The change in enthalpy resulting from the procedure of a reaction under a pressure of 1 atm, starting and ending with all materials at a constant temperature of 25°C

Heat of vaporization The difference in the vapor-phase enthalpy and the liquid-phase enthalpy of a substance at the boiling point

Heat transfer The science of the transport of thermal energy

Heater Any of a variety of temperature-increasing operations

Heterogeneous azeotrope An azeotropic mixture that separates into two immiscible liquids of different composition

Heuristic An unproved rule of thumb based on experience sometimes useful in solving a problem

Homogeneous azeotrope An azeotropic mixture that forms a single phase

Immiscible liquids Two liquid phases that are not completely soluble in each other

Internal energy The intrinsic energy contained in a system as a result of molecular attractions and rotational, vibrational, translational, and other motions

Ion exchange A separation operation in which one component of a liquid mixture is selectively retained by ionic attraction to a resinous solid

Isothermal process A process that takes place at constant temperature

Kinetic energy The energy possessed by a system by virtue of its relative motion

Law of conservation of energy Except in nuclear reactions, energy can neither be created nor destroyed

Law of conservation of mass Except in nuclear reactions or at speeds approaching that of light, mass can neither be created nor destroyed

Leaching A separation operation in which a solute is removed from a solid by the use of a liquid solvent

Limiting reactant The reactant present in such proportions that its consumption will limit the extent to which the reaction can proceed

Magnetic separator A separation operation in which solids capable of being magnetized can be separated from a mixture of solids

Mass A measure of the inertial quality of matter

Mass transfer The science of the diffusion of material, especially across phase boundaries, to achieve uniform concentrations

Material balance A system for accounting for the total mass entering, leaving, and accumulating in a system based on the law of conservation of mass

Maximum-boiling azeotrope An azeotrope that boils at a temperature higher than the boiling point of any of its pure components

Melting point The temperature and pressure at which the solid–liquid phase transition occurs

Mill Any of a variety of particle-size-decreasing operations consisting of a rotating shell and a charge of balls, rods, pebbles, rollers, or other grinding media

Minimum-boiling azeotrope An azeotrope that boils at a temperature lower than the boiling point of any of its pure components

Molecular sieve A separation operation in which one very small component of a gaseous mixture is selectively retained on the lattice vacancies on a solid

Normal boiling point The temperature at which a liquid boils when under a total pressure of 1 atm

Optimization A process of selecting the conditions of temperature, pressure, flow rate, and other variables resulting in the most economic operation of a process

Packing Any of a variety of vapor–liquid contacting media used in continuous separating columns

Potential energy The energy possessed by a system by virtue of its relative position in a force field

Pound-mole The mass of a chemical species in pounds numerically equal to its molecular weight

Pressure The force per unit surface area exerted on or by a system

Radiation A mode of heat transfer characterized by the emission and absorption of electromagnetic energy

Raffinate The effluent of an extractor enriched in the species less soluble in the solvent

Reaction path The sequence of chemical transformations that lead from the available raw materials to the desired products

Recycle The practice of recovering and returning partially processed material for reprocessing

Relative volatility The ratio of vapor pressures of two components in a liquid mixture

Reverse osmosis A separation operation in which a low-molecular-weight liquid is removed from a solution through a semipermeable membrane under the influence of pressure

Screening A separation operation that exploits differences in particle size to separate mixtures of solids

Scrubbing A separation operation in which a soluble component of a gas mixture is dissolved in a liquid

Sensible heat The difference in enthalpy of a species at two different temperatures in the same phase

Slurry A dispersion of large solid particles in a liquid

Solubility The degree to which a solid, vapor, or other immiscible liquid dissolves in a liquid

Solute The dispersed solid or vapor that dissolves into a liquid solution

Solvent A liquid into which a solid, vapor, or other immiscible dissolves

Species allocation A system of routing the various chemical species in a process from the sources of raw materials, among the reactions, and to the desired products

Specific volume The ratio of the volume of a substance to its mass

Splitter A device that divides a stream into two portions, each with identical properties except possibly flow rate

Steady state The operation of a system such that there is no net change with time in material or energy inventories or flows

Stoichiometry The relationships expressing the changes in composition occurring in chemical reactions

Stripping A separation operation in which a volatile component of a liquid mixture is evaporated into a gas stream

Sublimation The transition of a solid directly into the vapor phase

Synthesis An inductive activity involving the discovery and assembly of component parts into a complete process

Temperature A measure of the internal energy of a substance

Thermodynamics The science that deals with the relationship of heat and work to the various forms of energy and their interconversion

Thickener A separation operation that exploits density differences to partially remove a liquid from a liquid–solid mixture

Tray Any of a variety of vapor–liquid contacting devices used in staged separating columns

Turboexpander A vapor-pressure-decreasing and volume-increasing operation in which the pressure energy is recovered as mechanical work

Valve Any of a variety of pressure-decreasing operations in which pressure energy is dissipated as friction of flow through an adjustable orifice

Vapor pressure The pressure exerted by the vapor phase of the species in contact with its liquid or solid phase

Vaporization The transition of a liquid into the vapor phase

Weight The attractive force of the earth on objects

Work The action of a force moving under restraint through a distance

Zone refining A method of continuous recrystallization by drawing the material to be purified slowly through alternate heating and cooling zones

Index

L

M

N

O

P

R

S

T